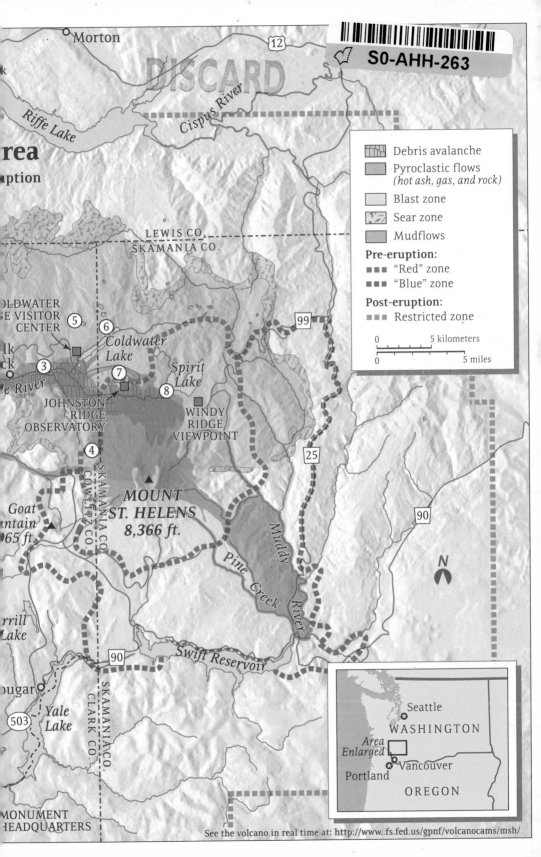

ECHOES OF FURY

ECHOES OF FURY

---

*THE 1980 ERUPTION
OF MOUNT ST. HELENS
AND THE LIVES IT
CHANGED FOREVER*

Frank Parchman

EPICENTER PRESS

Our mission statement: Epicenter Press is a regional press publishing nonfiction books about the arts, history, environment, and diverse cultures and lifestyles of Alaska and the Pacific Northwest.

Publisher: Kent Sturgis
Acquisitions Editor: Lael Morgan
Editor: Don Graydon
Jacket design: Kate Thompson, DillonThompson LLC
Text design: Victoria Sturgis
Mapmaker: Robert Cronan, Lucidity Information Design
Proofreader/Indexer: Sherrill Carlson
Printer: Transcontinental Printing

Library of Congress Control Number: 2004118235
ISBN 0-9745014-3-3

PRINTED IN CANADA
First Edition, March 2005
10 9 8 7 6 5 4 3 2 1

Distributed by Graphic Arts Center Publishing Co.

To order single copies of ECHOES OF FURY, mail $24.95 plus $7.00 for shipping and handling (WA residents add $2.80 state sales tax) to: Epicenter Press, PO Box 82368, Kenmore, WA 98028. Visit our online bookstore at www.EpicenterPress.com or call our 24-hour, toll-free order line at 800-950-6663.

green
press
INITIATIVE

The text pages for the first printing of 8,250 copies of ECHOES OF FURY were printed on 14,688 pounds of recycled, acid-free paper with 100% post-consumer content. According to the Green Press Initiative, of which Epicenter Press is a supporter, use of this recycled paper made it possible to avoid consumption of 176 mature trees (averaging 40 feet tall and 6-8" inches in diameter), 8,205 gallons of water, and 29,930 kilowatt hours of electricity. For more information about the GPI, visit www.GreenPressInitiative.org.

To my father, Frank Parchman Sr.; my mother, Rittie;
my friend Gary Eisler, who was there;
and my wife, Sheila, who made it possible.

*And after the earthquake a fire . . . and after
the fire a still small voice.*

—I Kings 19, verse 12

"Vancouver, Vancouver, this is it!"

—Last radio transmission of David Johnston from USGS
observation post near Mount St. Helens

CONTENTS

THE CHARACTERS

Where they were on Saturday, May 17, 1980

Volcanologist DON SWANSON has persuaded a friend and colleague to take his shift at the forward observation post five miles from the volcano, which has spewed steam and ash since March. But today all is quiet.

Tree thinner JIM SCYMANKY and his crew work a hillside thirteen miles from the mountain. He looks forward to finishing the job Sunday and heading home to his family.

ROALD REITAN AND VENUS DERGAN, carefree and in love, are camped on the Toutle River, thirty miles from the mountain. The volcano is the last thing on their minds.

A single working mother, DONNA PARKER looks forward to spending time with her son, visiting her mother, and catching up with her brother Billy by phone.

A rookie reporter on the Longview, Washington, *Daily News*, ANDRE STEPANKOWSKY plans to hike along the Columbia River with another reporter.

A graduating student at the University of Washington in Seattle, PETER FRENZEN spends the day drinking beer and having fun at the forestry department's annual picnic. He plans to spend Sunday recuperating.

A young unemployed radio technician attracted to danger, ROBERT ROGERS is camped seven miles from the volcano. He wonders what new adventure Sunday will bring.

PROLOGUE

1980

NO ONE NOTICED when the mountain first came alive on March 20. The earthquake that hit at 3:47 p.m. measured 4.1 on the Richter scale. This region of Western Washington averaged three quakes a day. Seven days later, 123 years since the volcano's last significant activity, Mount St. Helens exploded with a blast that ripped off a dome of ice and hardened lava and sent plumes of ash two miles into the sky. A summit crater formed, one hundred and fifty feet deep and two hundred feet across. Officials huddled together and made evacuation plans for the towns nearest the volcano. Sheriffs' deputies sealed off approaches to the mountain. Residents within a Red Zone established by the governor were ordered out.

Geologists said it was not a major eruption. They disagreed on what would happen next. Some believed the mountain would have a series of these minor eruptions and then return to its slumber. Others predicted a day of doom when the volcano would blow fire, smoke, ash, and rock into the air and send floods of mud and melted snow raging through the valleys below.

In the weeks that followed, the mountain vented steam and ash in several minor bursts. The airspace around the volcano became congested with planes and helicopters carrying newsmen, geologists, and sightseers. Novelty items went on sale: a mountain-shaped ashtray that emitted smoke; a bumper sticker that said "Get Your Ash Together"; a painting of George Washington wearing a surgical mask. The most popular item was a T-shirt that said "I'm a Mount St. Helens Survivor."

Thousands of tourists clogged the highways around the mountain every weekend. Families picnicked in the meadows outside the restricted zone as they watched the volcano send its plumes of smoke harmlessly into the sky. One of the world's foremost volcanologists, Frenchman Haroun Tazieff, arrived. "It's spectacular, particularly for the people who are not used to it," he said of the volcano action, "but absolutely not dangerous." He called the evacuations from the Red Zone "utterly unnecessary."

A geologist for the U.S. Geological Survey team, thirty-year-old David Johnston, called the volcano "a dynamite keg with a fuse lit." However, a team of Dartmouth scientists measured the sulfur dioxide in the gas from the volcano, found it low, and indicated a major eruption was not imminent. As April waned, so did public interest. It was no longer front-page news. Some USGS officials talked of going home. On May 17, authorities permitted homeowners to briefly enter the Red Zone to pick up belongings. A second trip was planned for the following day.

At 8:32 a.m. on May 18, the mountain erupted with a force over one thousand times greater than the atomic bomb dropped on Hiroshima. From a USGS observation post, David Johnston called by radio to his Vancouver, Washington, headquarters: "Vancouver, Vancouver, this is it."

PART 1
2000, 1980

A Terrible Beauty

So this is what it's like to die.

—Jim Scymanky

CHAPTER 1

Death and the Mountain

DON SWANSON TURNED off Interstate 5 onto the road to the mountain and tried to subdue the thought that kept recurring since he left the airport. If it hadn't been for what most people would call a twist of fate, he would be dead and his friend David Johnston could be driving the rented red Toyota toward Mount St. Helens to an observatory named after Swanson, instead of the other way around.

Swanson sped through the tiny tourist town of Castle Rock, Washington, with its fast-food outlets, Mount St. Helens Motel with its "Hope You Had a Blast" sign, and Cinedome Theater featuring shaking seats and a deafening sound system where the volcano had erupted continuously for years in the same documentary, shown on a thirty-five-foot-high screen.

Since the eruption twenty years before, he had told himself that Johnston's death, and his own role in it, had been one of those things beyond anyone's control. It was like missing a flight that later crashed with a good friend aboard. But it was not so much that he believed in fate as in the quirkiness of life, where a snap decision could have unforeseen impacts. As a scientist he had seen nature's randomness at work in occurrences for which there was no explanation.

Past Silver Lake the highway wound through lush valleys, with ridge tops hiding the mountain beyond. Near the outskirts of Toutle he passed a former mobile home park where ancient house trailers sat forlorn, listing to one side. On their walls, near the caved-in roofs, mud rinds marked how high the floodwaters and mudflows had risen after the eruption. Now a school, gas station, private volcano museum, and a few other buildings made up the heart of the town on one side of the road.

On the other side, in front of a few rundown houses and mobile homes, rows of tables were piled with trinkets, junk, and leftover souvenirs—faded T-shirts, volcano ashtrays, and tiny packages of "Authentic St. Helens Ash." The souvenirs were from stores, long since out of business, whose owners had banked on millions of tourists who never showed up.

Usually this time of year the town would be getting ready for Volcano Daze and its parade of clowns, costumed children, marching bands, and floats of smoke-spouting, papier-mâché volcanoes. But the year before, for marketing purposes, the town had renamed the event Mountain Mania and moved it from the May anniversary of the eruption to June when it could attract more tourists.

Outside of town a wooden shack sat rotting in the weeds, the abandoned first home to the radio station that provided early warning of eruptions. The woods thickened on both sides of the road as early morning sunlight scattered through the trees and dappled the pavement in front of Swanson's car. Around a bend, sitting in a hollow, he came upon a cluster of businesses whose main attraction was an A-frame cabin still half-buried in dried mud from the eruption. A twenty-five-foot-tall statue of Bigfoot stood to one side, a weathered relic from an era before the eruption when the mythical ape-man and scenic beauty had been the biggest tourist draws for this area of southern Washington state.

Now the Toyota climbed a slight incline toward Hoffstadt Bluffs, where the county had opened a visitor center with eruption photos from the local Longview *Daily News*, a restaurant, and amusement rides on helicopters, llamas, and even dogsleds. He passed the Weyerhaeuser tree farm with its acres of uniform firs sweeping up hillsides, seamless, like a green blanket. An old-growth forest once stood here, until the eruption left those trees stripped and fallen like so many toothpicks in orderly, swirling rows that followed the direction of the exploding mountain's wind-wave of pumice and rock.

Beyond the tree farm the car climbed steep, gray hills as it traveled around hairpin turns, with desertlike valleys below in the volcano's blast zone. Swanson caught his first full view of the mountain in the distance. A surprise snowstorm had blanketed its sides in white. It looked like a mound of vanilla ice cream with its center scooped out as it rose from a pumice plain whose greenness startled him. Where once there had been only the grayness of ash for miles, grass had sprouted.

Waves of emotion he could not quite understand swept over him as he drew closer to the volcano. On past anniversaries he never felt anything like this. He had managed to get through each anniversary without dwelling on his feelings about the eruption that drastically changed his life and robbed him of one of his best friends and colleagues. He had kept busy, treating it like just another workday.

Maybe he was feeling his own mortality, the winding down of a brilliant career the mountain had forged. But at age sixty-two, he didn't feel that old. With a ruddy face, impish blue eyes, fiery red hair, and beard peppered with gray, there was still something of the kid in Don Swanson—the kid you knew in science class who was always tinkering, taking things apart to see how they worked, or mixing chemicals that unexpectedly exploded.

He had grown up the son of a pharmacist in the shadow of Washington's Cascade Mountains. In the small town of Eatonville, he lived at the end of a street with nothing but woods beyond, and 14,411-foot Mount Rainier towering over them. His father encouraged him to explore the woods, where he often walked on muddy trails collecting rocks, leaves, and insects. When he was ten the family moved to Centralia, a little over thirty miles from the turnoff to Mount St. Helens. They hiked the mountain's trails and on sunny days picnicked near its most popular recreation attraction, Spirit Lake.

The rocks that he collected along Spirit Lake Highway during a Christmas break from college got him interested in geology. A professor at graduate school sparked his curiosity about volcanoes. In 1963 he hiked to the summit of Mount St. Helens for the first time, to see what a dormant volcano looked like. During research on the rocks of the Cascade Range for his doctoral thesis, he became increasingly intrigued with the mountain. A year after school he joined the U.S. Geological Survey and his work took him to places like Italy, Japan, and Hawaii, where he studied volcanoes. But St. Helens always beckoned him back. It was his pioneering work after the eruption that propelled him to international recognition.

Swanson pulled into the parking lot below Johnston Ridge Observatory. He wasn't sure what to expect. He had been invited to speak at events leading up to the twentieth anniversary. During past anniversaries he usually spoke as a scientist, but this time the organizers included him with a group of survivors of the blast who were to speak the next day. It felt strange that people considered him a survivor, he thought, as he walked up to the observatory to hear a group of scientists presenting a public symposium.

Water from melting ice cascaded down the side of the steep, slick sidewalk. At the top a torrent of wind blew cold off the mountain and through the breezeway leading to the entrance. His parka blew against him as he leaned into the wind. In a courtyard, teenagers on a tour threw snowballs from the melting snow in the hot morning sun.

He had a free pass, but he paid as he always did because he wanted to support Forest Service efforts to educate the public about the volcano. While other scientists after the eruption had wanted Mount St. Helens to remain in a pristine state for study, with only limited public access, he had come to advocate reaching out to everyone. He believed the site could be preserved for scientific inquiry while ensuring that the memory of what happened there would never be lost.

The symposium was under way inside a large movie-style theater complete with padded seats and a sound system to rival that of any cineplex.

He took a seat near an aisle with a few empty seats between him and a group of senior citizens with cameras and binoculars dangling from their necks and glossy Mount St. Helens brochures in their hands. A scientist was speaking about how life had returned to the mountain. Red drapes thirty feet high and fifty feet wide hung behind the speaker. This was the auditorium where the Forest Service showed its special-effects-laden film about the eruption. The fifteen-minute film was automatically cycled by computer.

The audience of tourists, scientists, and Forest Service personnel peppered the speaker with questions. Several times Swanson raised his hand, but tentatively, as though he didn't want to intrude, and the speaker never called on him. Suddenly the speaker was caught in midsentence by dramatic music booming through the sound system. He rushed to get out of the way of a huge screen unfurling from the ceiling. The room darkened. Clouds swirled on the screen. The sound of wind roared from the speakers, fading with the rise of an actor's voice speaking the last words of David Johnston. Then with an explosive boom that seemed to rock the seats, the mountain erupted on the screen.

After the carnage, a much younger Swanson appeared on the screen with other scientists. It had been years since he had seen this movie. He studied himself with interest, as if he were trying to find clues to that man he had been twenty years before. On the screen he was helping stretch a measuring tape across a large, smoke-belching crack on the crater floor, and it was obvious he relished his work. On film he could hardly contain his smile while the narrator explained how St. Helens had brought about the greatest advancements yet in understanding volcanoes.

When the film ended, the screen rose and the red drapes parted to reveal a huge window. Like the scene in *King Kong* where a curtain rises and the chained beast is shown for the first time, people in the audience now gasped at the view. There was the mountain, majestic with its mantle of new snow, towering in the crystal air over everything.

As the symposium broke for lunch, a reporter approached Swanson to keep an interview appointment. Swanson offered to take him to lunch at a visitor center a few miles away. This time the employee selling tickets recognized Swanson. "I just saw you on the *Today Show*," he told Swanson, who had been featured on the show broadcast live from Hawaii's active Kilauea volcano, where Swanson was chief scientist in charge of the Hawaiian Volcano Observatory. The Forest Service employee would not let Swanson pay. The man leaned over and whispered something to a fellow worker. Within moments other employees streamed forward to shake the hand of the scientist. It was like a homecoming.

Swanson led the reporter to the cafeteria past displays where Disney-like automatons of forest rangers and animals told the story of the volcano before entranced tourists. Swanson hardly had time to eat his sandwich. The questions were more probing than he had expected, especially those about his relationship to Johnston, though he didn't feel uncomfortable in answering them.

After the interview he drove the reporter back to the observatory. On the way he pointed out the stumps on the hillside and explained the difference between the stumps that were smooth on top, remnants of a logging clear-cut, and those that were splayed, showing they had been blown down by the erupting volcano. A few hundred feet from the entrance of the observatory parking lot, he stopped the car. He led the way up a path to the top of the ridge.

"There used to be a plaque up here," Swanson said. "I'm sure it must still be here." But he couldn't find it. He took the reporter to the edge of the ridge and pointed into a gully. "That's where we think he died, but we're not sure." And he thought back twenty years and how he should have been the one killed, not the man who reluctantly took his place and died a few hundred feet from where they now stood.

CHAPTER 2
Sunday, May 18, 1980

THAT MORNING, Don Swanson should already have been on his way to the mountain, but the supplies he needed were late in arriving. He had just attended the daily briefing on the second floor of the Forest Service building in Vancouver, Washington, where the U.S. Geological Survey set up a makeshift headquarters after the initial volcanic activity in March. Nothing unusual was reported at the meeting. David Johnston had radioed in from Coldwater II, an observation post five miles northwest of the crater. He said it was a clear day with no clouds. He had a perfect view. Johnston reported little change in the growth of a four-hundred-foot bulge on the side of the mountain that was being monitored with an electronic distance-measuring device.

Two days before, the forty-two-year-old Swanson had bumped into the younger Johnston in a hallway at headquarters. Swanson was looking for someone to take his place at Coldwater II for a single night. A graduate geology student visiting from Germany was scheduled to leave on Sunday and Swanson had some things he wanted to show him on Saturday, the day Swanson's stint at the observation post was to begin.

Johnston hesitated. While neither man had yet served a turn at the post, both had expressed their reluctance for the duty. Johnston didn't mind taking chances in gathering data. He had been into the crater to collect gas samples several times. But he had a healthy respect for what he considered a dangerous volcano. He saw no reason to take risks just sitting and watching the mountain.

While Swanson too respected the power of the volcano, it did not begin to approach Johnston's fear. In fact, after the initial blast in March, when scientists needed fresh ash from the crater, Swanson took on the task. Holding a long-handled ice ax with a soup ladle attached, he had a helicopter hover a few feet over the lip of the crater while he leaned out into buffeting winds and scooped up the ash in the moments between eruptions. His biggest complaint about duty at the observation post was that he thought it would be boring. Clouds usually obscured the mountain.

Johnston, though, understood how seriously Swanson took his mentoring of young geologists. "Well, it's not the safest place," Johnston said, "but I guess I can spend a night up there, if you will replace me the next day."

After the Sunday briefing Swanson went downstairs, killing time until his supplies arrived. He joined another scientist in the almost deserted room where the seismographs were kept. Suddenly the machines came alive with the scratching noise of their pens. He had never seen the pens record such wide arcs. A major earthquake was under way and it was centered on the northwest side of the mountain near Coldwater II. It was precisely 8:32.4 a.m.

He rushed upstairs to the radio room and grabbed the mike. "Coldwater II, this is Vancouver, this is Vancouver. Do you hear me, Dave. What's going on?" There was no reply. "Dave come in. This is Vancouver." He tried several more times and then gave up.

Swanson ran back to the seismograph room, now crowded with other scientists. They quickly decided this was something big. A Forest Service spotter plane and pilot were available at nearby Pearson Air Field. Swanson, who had the most experience using the plane in his work, was chosen to go see what Mount St. Helens was up to. He grabbed a notebook and ran from the room.

Even with the chain saw buzzing in his ears, Jim Scymanky would never understand why he didn't hear the volcano go off. He and three other men had been working near Camp Baker for the past four weeks as tree thinners subcontracted to Weyerhaeuser, the giant timber company. With his chain saw gliding through the thin Douglas firs, his thoughts turned to his wife and four children in Woodburn, Oregon. This would be the crew's last day. He had hoped to finish the job the day before. He usually phoned his wife when he was delayed, but he had worked late and failed to call. Now he was anxious to get home.

Scymanky, thirty-six, was a tall, rangy man with kind, hazel eyes who still had his good-old-boy manner and the hint of a soft-spoken accent he acquired while growing up in Georgia. He had worked as a school maintenance man, a mechanic at the Boeing plant in Seattle, and a baker. But nothing compared with the satisfaction he felt working in the woods. "It's the only way to go," he told his friends.

The man working only a few feet away had taught him everything that Scymanky knew about working in the woods. Leonty Skorohodoff was a short, stocky Russian with a beard. His parents emigrated early in the twentieth century from Russia to China, where he was born. With other members of the Russian Orthodox denomination known as the Old Believers, the family then fled to South America before taking up permanent residence in Woodburn and the nearby town of Mount Angel.

Their foreman was working in the trees below Scymanky and Skorohodoff. Evlanty Sharipoff, a lifelong friend of Skorohodoff, was also born in China. A short, thin man who took pride in his strength, Sharipoff often showed up men half his age. The three men had worked together for the past year. Nightly they would build a fire, talk, and eat before sharing the tent where they slept. They had grown close.

Looking up from the tree he was cutting, Scymanky spotted Jose Dias, the fourth member of the crew. Dias stood in the road below, watching them. He refused to work on Sundays. "Why don't you work Sunday and get the job done faster?" one of the men had asked him in a kidding way earlier that morning. "What's the matter? You afraid the volcano is going to go off?"

"No, no. I'm not going to work today. I've got to have a day off. One day a week."

Dias, short but powerfully built with a goatee and black, curly hair, had been with the crew for only a few weeks. A friend of Leonty from South America, he was still learning how to thin trees. With his carefree nature, the rest of the men couldn't help but like him.

For weeks they had teased each other about the volcano erupting. They were convinced it would never touch them. As the men took a break, Dias climbed up the hill to their pickup and went to sleep in the back seat. When the men returned to work, the trees toppled one after another with a snap, their branches breaking their fall as the crew worked its way up the hill. The coolness of the morning still hung in the air now beginning to warm in the bright sunshine. Scymanky took a deep breath of the air scented with cut fir and thought how beautiful a day it was.

Screaming broke the serenity. At first Scymanky thought someone had cut himself. He turned to see Dias running barefoot down the hill, screaming in Spanish "¡El volcan esta explotando! ¡El volcan esta explotando!" The volcano is exploding!

Scymanky was dumbfounded. "How the hell can the volcano be exploding and we're here working?" he asked himself. "I never heard anything." But when he saw Dias's eyes bulging with fear as the panicked man ran past him, Scymanky knew something was seriously wrong.

—✳—

Robert Rogers was on the mountain May 18 almost by happenstance. The day before on a whim he decided to drive to St. Helens. He didn't bring a sleeping bag or gear. It was just a nice day for another adventure in trespassing.

Ever since the volcano came alive in March, Rogers had been evading law enforcement officers and forest rangers to get to the mountain. On the afternoon of that first volcanic activity he drove to the top of Portland's West Hills. He was amazed to see the top of Mount St. Helens poking through a cloud layer and ash pouring from it. Rogers made up his mind to go see it up close.

He had no specific reason to go. He wasn't a photographer, and he had no scientific training. His only experience with the mountain had been a stay with his parents at a Girl Scout camp when he was a youngster and some subsequent climbing. There was just his overdeveloped curiosity about how things worked and a compelling urge to go to the mountain— the same urge that drove him to climb radio towers, bridges, railroad trestles, and high-rise buildings. He often climbed to the top of one of Portland's municipal bridges in the middle of the night. There he was on top of the world as he looked at the city lights reflecting off the Willamette River, the silhouettes of the other bridges he had conquered, and the mountains beyond. The fear of capture by police only added to his adrenalized high. He had never been caught.

Rogers had long, moppish, brown hair, an aquiline nose, and an owl-like face with blue eyes that could grow large and bat at you behind his glasses. His friends considered him an eccentric genius who never quite fit in. But he could be fun to be around if you didn't get too close to the flame that burned brightly within Rogers but never seemed to consume him.

A few days after that first eruption, he lost his job as engineer for a local radio station. It didn't bother him much. At twenty-nine he was still unmarried and saw himself as an unfettered spirit. Now he was free to pursue his interest in St. Helens. The fact that authorities had established a restricted Red Zone around the volcano added to its allure.

Spirit Lake Highway on the north side of the mountain gave the quickest access, but it was heavily guarded. So he started going in through the small town of Cougar on the southwest side, driving his battered little four-cylinder French Simca. With its rear engine, it was ideal on rough logging roads. Usually at night he would drive through Cougar, past Merrill Lake, and around the base of Goat Mountain. From Goat Mountain he would don cross-country skis and ski to timberline on his way to the volcano.

After his first trip he concentrated on mapping the logging roads he could use for rapid escape if he was discovered. He found it easy to travel on unused logging roads at night with his headlights off. Even with no moon, at this altitude a little bit of skylight filtered down. There was only one problem. The system of logging roads was like a tree with branches, ending without any connections. More than once he got lost.

When the volcanic activity slowed, the media began giving more coverage to trespassers being arrested. Many of these people boasted of their exploits, but Rogers avoided the media. He struck up a relationship with graduate students and professors in the geology department at Portland State University. By this time he was taking clandestine hikes into the summit crater itself. No one in the department had access to the mountain, so they eagerly awaited Rogers' ash samples, rocks, photographs, and descriptions of the crater. In exchange they put up with his constant stream of questions about volcanoes.

When NBC News hired a professor from the university as a consultant to accompany a crew to the crater in a helicopter, the teacher took Rogers along. Members of the department received a contract to set up tiltmeters for the Trojan nuclear plant, which was about forty miles from the volcano. They used Rogers to guide them.

On May 15 he hiked to the crater and took a series of photographs. He printed them back in Portland at a U-Develop Darkrooms shop and was pleased with his seven-shot panorama. That Saturday, May 17, he decided on the spur of the moment to drive to the mountain. It would just be a day trip for more mapping.

The snow level was very high now and he found almost all the roads open. He drove his Simca farther in toward the volcano than ever before. On the side of Goat Mountain he was surprised to run into a middle-aged couple. They let him use their binoculars. The binoculars had obviously been dropped many times and the lenses were not aligned, giving a split, almost surrealistic view of the mountain. Rogers liked the effect. "Those are pretty nice binoculars," he told them. He grabbed a set of the crater photographs from his car and spread them out for the couple.

"Those are pretty nice pictures," the man said.

"You want to trade?" Rogers asked. The couple traded the binoculars and even threw in a Weyerhaeuser map of the area. Rogers had made his first photo sale.

Rogers drove on. He saw a Ford Pinto parked along the road and pulled over to check it out. An ice ax lay in the back seat. Around here, he thought, that could only mean someone was going to try to climb St. Helens illegally. When Rogers found the owner off the road looking up at the mountain, he was at first startled. The man was in a Forest Service uniform. He turned toward Rogers.

Rogers spoke first: "You know, you really shouldn't let anybody see that ice ax in the back of your car."

The man smiled at him. "That shouldn't matter."

"Don't be dumb," Rogers told him, thinking of the authorities. "This is the last place you want to take any chances."

The man, Francisco Valenzuela, had just transferred from New Mexico to work for the Gifford Pinchot National Forest as a recreation coordinator. As they talked about mountains and climbing, it became obvious to Rogers that Valenzuela had but one intention.

"If you're going to try climbing that sucker, I can help," Rogers said as he pointed to St. Helens, which had shown no significant activity for three days.

The men decided on a reconnaissance trip in Rogers' Simca. They drove down to the South Fork of the Toutle River, then up along the river toward the mountain as far as they could go. They stashed the car in the only patch of trees left in a clear-cut. Rogers looked across the river. There, three-quarters of a mile away, stood a tiny tent with no one around. "Mountain climbers use those small tents," he told Valenzuela. "They must be up in the snow."

The men located a trailhead in the Sheep Creek drainage. "We should come back here at 3 a.m. when it's dark and go right up to timberline," Rogers said. This would allow Rogers to finish mapping the last two road systems while giving Valenzuela the lay of the land. They drove back to Goat Mountain and found a view spot on a high ridge, where a man and his wife had set up camp with a Dodge van. Call letters for a ham radio operator were stenciled on the back door. The site was seven and a half miles from the summit.

Ty Kearney and his wife, Marianna, were volunteers. They had been camped there for a week as part of a network of ham operators working with the Washington Department of Emergency Services to provide early warning of any eruption. Rogers and Valenzuela kept quiet about their own intentions. Valenzuela loaned Rogers a sleeping bag. The sky was clear. The men slept on the ground under a blanket of stars.

Well before dawn on Sunday they left without waking the couple. Rogers drove the Simca back up the South Fork of the Toutle and hid his car again in the patch of trees. The men hiked up the Sheep Creek drainage trail until they reached some clear-cuts that Rogers had found in his earlier mappings.

The early morning was peaceful, with no wind and only a slight nip in the air. The birds were quiet. Light shone from stars packed tightly in the sky. Valenzuela lay back on a fallen log in a cluster of trees. Rogers wanted to climb up to timberline with his camera. Valenzuela, relaxed and half asleep, decided not to go. At timberline the sun rose over the mountain with a suddenness and majesty that caught Rogers by surprise. He took a photograph. It was 5:32 a.m.

Valenzuela was awake and agitated when Rogers returned. "We got to get the hell out of here," he yelled. "There's too much light. They'll throw you in jail and I'll lose my job."

On the way back to the car, Rogers noticed a pickup truck parked next to the tent they had seen across the river. Down the way from the tent was an old green station wagon with a man standing next to it. The car was facing downhill. Rogers saw no tent and no gear, only a tripod for a camera. It was a setup for a fast escape, he thought.

"I think I'll hike over there and tell those people they need to go to higher ground," Rogers said. The geologists at Portland State had told him that one of the last places you want to be if a volcano explodes is near a river bottom, because of flooding and heavier-than-air gases. "No way," Valenzuela said. "We don't have time. We got to get out of here."

They drove to the bottom of the South Fork and back up a switchback to the ridge where they had met the Kearneys. Marianna Kearney was working at an easel, painting the mountain. As they got out of the car, Rogers heard a voice in his head that said, "Robert, get your camera." But he ignored it.

"Hey, we thought you guys had gone back to Portland," Ty Kearney said. "Where you been?"

Rogers made up an elaborate story about fishing. Kearney squinted his eyes toward the Simca, looking for fishing poles, Rogers thought. Then Kearney began to interrogate them like a cop, or so it seemed to Rogers. He was relieved when Kearney's ham radio started squawking. It was another ham operator, a retired Navy radioman named Gerald Martin who was situated near David Johnston at Coldwater II.

Martin believed he was seeing more steam activity than before. "Now there's a new one that's just opened up there," Martin radioed in a matter-of-fact military manner as he watched a steam vent erupt.

"Uh huh. Well, I reported it yesterday, but that's okay. You're seeing the same thing I'm seeing," Kearney radioed back as he sat in the van with the door open, Rogers and Valenzuela standing next to it.

"It's coming out of the crater, going straight up, going straight up that south wall of the crater and coming over the top," Martin said. "Oh, oh, I just felt an earthquake, a good one shaking . . . uh, there's . . ."

Rogers spun around. There was no noise, but ash was starting to pour from the volcano.

"Now we've got an eruption down here," Martin was saying as Rogers ran to get his camera. "Now we got a big slide coming off. The slide is coming off the west slope . . ."

Rogers tried to focus the camera. He was no longer sure what he was seeing. He just kept clicking. Valenzuela screamed out, "Look, there it goes. The whole north side is sliding away."

Martin radioed, "Uh, now we've got a whole great big eruption out of the crater . . ." Rogers pressed his shutter release button. The camera jammed.

The two reporters sat in the Pancake House in Longview that Sunday, more than forty miles from the mountain. They had just ordered breakfast and were looking forward to a hiking trip down the Columbia River Gorge.

Jay McIntosh and Andre Stepankowsky were young, ambitious journalists, just starting their careers. But Stepankowsky felt his career was going to end almost before it began. He was foundering. He had come to the Longview *Daily News* straight from the prestigious University of Missouri School of Journalism. The problem was he couldn't write in the concise, punchy manner required of news stories. He harbored ambitions to write like Wordsworth and other literary luminaries and it showed in his writing. Editors tore apart his flowery, overwritten prose. What was worse, he was assigned stories that didn't particularly excite him.

Stepankowsky was an outsider. Raised in northern New Jersey, he carried the region's distinctive accent with him. He had that aggressive, blunt, almost bullying manner that Westerners identified with the other side of the country. Short and stocky, with bushy, black hair billowing out from around a serious, but open, dark-complexioned face, he had the disheveled look of an intellectual unconcerned with his appearance. Before taking up journalism, he had trained to be a concert pianist.

Now he was working in Longview, Washington, a city of thirty thousand people that had been founded as a company town in the 1920s and had blossomed into something more with the decades of growth in the timber business. An expansive park along a lake graced the civic heart of the town with its Georgian-style brick government buildings and Carnegie Library. The family-owned *Daily News*, which got its start along with the timber industry, also grew and prospered. Now with a twelve-reporter news staff, it was considered one of the best newspapers in the region.

The veneer of respectability couldn't hide the smell of pulp from belching mills and the fact that Longview was, at its core, a blue-collar, no-nonsense working town. This was fine with Stepankowsky. He had worked summers as a carpenter and liked earthy, straight-talking people who did not put on airs. It was as good a place as any to get his start in achieving his aim of making what he called "a contribution to society."

Only now he doubted he was going to be doing it much longer as a journalist.

As he and McIntosh sat talking, Stepankowsky was startled as he glanced out the window. The sky, crystal clear a moment before, was turning black. They watched the smoke gather into a massive cloud that grew larger with each second. This was different from anything they had seen before. Something terrible had happened but they could not yet conceive that the volcano, so far away, had caused it.

They called off the hike and hurried to the newspaper office. Stepankowsky could not shake feelings of dread as they watched the black cloud snuff out the day. The newsroom was deserted. The next edition, due to come out Monday afternoon, would not be put together until Monday morning. Stepankowsky called managing editor Bob Gaston at home. Gaston knew he didn't have time to waste trying to track down any of his star reporters. Instead he told Stepankowsky to flip a coin to see which of the two young reporters would get the assignment and meet the paper's photographer at the airport. A plane would be waiting for them. Stepankowsky won the toss.

—✳—

It took Peter Frenzen a few moments to realize what he was hearing on the radio. Something about Mount St. Helens. He was nursing a hangover from the annual picnic of the University of Washington School of Forestry the day before. It had been a day of ax throwing, log rolling, and drinking beer for the soon-to-graduate senior.

Frenzen, at twenty-three, was one of the bright lights of the school's forestry program. He had been raised in the suburbs of Chicago, where his father worked as a meteorologist. Growing up in a science-oriented family, there was lots of exploration of the local forest preserve. But the oak woodlands and hillsides were nothing compared with the majestic beauty he found during a visit to the Pacific Northwest. Here he discovered old-growth forests with trees rising hundreds of feet into the air. When he asked the name of a mountain he was looking at, he was told it was no mountain, it was a hill. And there were lots of hills and mountains teeming with lakes, streams, forests, and wildlife. It was a paradise where he could see spending his life.

He transferred his credits from the University of Chicago to the University of Washington. At first he pictured himself as a forest ranger. With his rugged good looks, mop of curly black hair, and athletic build he was almost a poster boy for the stereotypical ranger standing next to Smokey the Bear. But one summer he got a job working with forestry researchers involved in ecosystem science. Frenzen became intrigued with all the components it took to make a forest work. Ecosystem science brought into play his knowledge of biology, botany, and geology. He couldn't see himself getting bored with this work, and it would still place him in the outdoors.

One summer he climbed Mount St. Helens. From a distance he was awed by the mountain whose perfect symmetry and beauty rivaled that of Japan's revered Fujiyama. That night on the summit he slept in the open. Stars filled the sky. A meteor shower, like spurting fireworks, streaked overhead. Then, to the north, the sky came alive in curtains of shimmering white light. It was the northern lights, the aurora borealis. Entranced, he lost sense of time. Only when sunlight peeked over Mount Adams in the distance and the sky turned golden and red did he realize he had not slept. It had been a deeply moving spiritual experience he never forgot.

By the time he was a senior, he had his choice of two of the nation's top forestry schools, Yale and Oregon State University, for graduate study. He chose Yale. But first he needed to complete a study that summer of an old mudflow on the slopes of Mount Rainier, one of the region's great volcanic mountains.

He looked at the clock. It wasn't yet 9 o'clock.

"Reports are still sketchy," the radio broadcaster reported in a shaky voice. "But apparently there has been a huge eruption on Mount St. Helens. No deaths have yet been reported but we will keep you posted throughout the day."

Frenzen didn't own a television set. He ran from his apartment to a nearby apartment house, where he pounded on the door of a young woman he knew. "St. Helens has just erupted and I've got to borrow your TV," he told her. She nodded her head, a quizzical look on her face. He pulled the plug and cradled the set as he turned to go out the door. "Thanks," he said to his speechless friend.

The television coverage frustrated him. There was little new information. Thinking of his summer work ahead on Mount Rainier, he realized how much more interesting it would be to study mudflows made by a live volcano. He wondered if he should go down and see what was happening.

The young couple was not aware of Mount St. Helens that morning. The mountain was out of sight more than thirty miles away.

When Venus Dergan went on a trip with Roald Reitan she could never tell where she might end up—like the previous day when the two lovers, only a year out of high school, had been going to the Green River and Roald suddenly remembered a great fishing hole outside the town of Toutle.

Venus adored the slim, muscular man who loved to fish, hunt, and camp. Roald, at nineteen a year younger than Venus, was happiest when he was with the pert, shapely brunette. They shared a love of spontaneity and the thrill of adventure of a young couple throwing off parental shackles and discovering the world. These past months had been some of the happiest of her life.

After a six-month stint in New York City following graduation, she had returned and began dating Roald again. She found a job and her own place in Tacoma, Washington, where she and Roald had attended school. She prized her newly found independence almost as much as she did her relationship with Roald.

From the moment Roald found the special place on the Toutle River, Venus loved it. It was on a bend of the river where fish gathered in a deep spot called the Jericho Hole. On Saturday they had pitched their pup tent on a grassy knoll surrounded by trees, a few feet above the sandy riverbank. It was only two miles out of town. Above the tinkle of the low-flowing South Toutle, they could hear traffic on the nearby road.

They drove up some logging roads and fished the North Fork of the Toutle River, with no luck. Back at the campsite the day was growing hot and the fish weren't biting, but neither of them cared. It was so beautiful here, the air was crystal clear, and they were in love. They drank beer, talked and laughed, and soaked in the sun.

That night they built a fire, ate and drank, and smoked cigarettes as the sky faded into a blackness pierced by a million stars. Above the gurgle of the river and the crackle of the fire they talked about family, friends, and dreams for the future. Then they climbed into Roald's 1968 Oldsmobile Cutlass and turned on the radio. The car was so huge that she teased him about it being a boat. Late into the night they snuggled and listened to rock and roll. They went to bed at midnight.

Roald woke about six on Sunday morning and left the tent. It was another beautiful day. He stuck his head back in the tent and tried to talk to his groggy partner, who was not going to have anything to do with fish or getting up. He told her what a great day it was. She didn't stir. But he wasn't too disappointed as he climbed back into the sleeping bag.

Later that morning Roald shook Venus awake and she heard a siren. "Do you think it's a fire?" she asked. Then she became aware of a more disturbing sound—rushing water just outside the tent. Venus was first to crawl out of the tent. What she saw shocked her. The river was rising rapidly, like water in a bathtub. The water was almost to her feet.

"Roald, you better get out here."

The news flash came on the television at Donna Parker's home in Molalla, Oregon. Mount St. Helens had erupted with a massive explosion. She got her son up and prepared breakfast. Then she and her boy got ready to drive to her mother's house in the hills high above Molalla. It had a picture window with a view of the mountain.

When Parker first read about activity on the mountain, she didn't think about its destructive power so much as she thought about her early childhood. She carried faint memories of St. Helens, of the nearby lake her family lived on, and of her father, an engineer for the train that hauled logs off the mountain. But now she lived well over a hundred miles away, where the volcano couldn't touch her or her family.

Except for the memories, she gave the volcano only passing notice. Parker, a busy single mother in her early forties, was an assembly worker at the Tektronix plant in nearby Wilsonville. A petite woman with a thin face and brown hair she often wore in a bun, she laughed easily but had a steely nature that she tried to hold in check when angered. Her life revolved around her son, her mother, a brother in Portland she was especially close to, and her work.

Now Parker stood in the living room of her mother's house, unable to comprehend what she was seeing out the window. The volcano spit out a continuous roll of billowing clouds that rose thousands of feet and blackened the sky.

Over the years her brother Billy had climbed the mountain several times. "Billy's gone clam digging this weekend," her mother said. "Wasn't that a nice Mother's Day card he brought me?" Billy had visited the previous weekend.

Looking at the roiling clouds off the volcano, Parker felt a knot in her stomach. "Anybody there is dead," she told herself. "The poor families," she said to her mother. "I can't imagine how awful this is going to be for them."

The thought depressed Parker. Her family had experienced more than its share of deaths and it wasn't hard to imagine the rituals of grief that would be performed in the days ahead. She was just thankful that Billy had gone to the coast.

Chapter 3

May 18: Holocaust

THE SOUND WAS A HISS at first, then grew louder as if thousands of pieces of gravel were being shot through the trees, then roared as it overtook the four loggers.

"Let's get out of here, let's get out of here," the hysterical Jose Dias screamed as he stumbled through the brush. Jim Scymanky dropped his chain saw and hollered to Evlanty Sharipoff, who stood frozen in place. "The volcano is exploding. Let's go." Leonty Skorohodoff joined them. The rows of trees they had cut now formed barricades to their escape. They stumbled over the trunks as branches cut their faces, moving only eight feet before the air around them turned hot and black.

Scymanky didn't know what knocked him down, only that he was now buried in ash and fallen limbs. The hiss returned as heat swept over him, gradually growing hotter through his clothes. The lava, he imagined, was moving down the hill. He waited for it to take his life. It grew even darker. It seemed like he was at the bottom of a deep pool, the water growing hotter as he strained to hold his breath. He heard the muffled screams of his friends. Images of his family flashed before his eyes. "Is this really possible?" he asked himself. "It can't be. A minute before, it was beautiful here, just perfect. This can't be happening."

A spot of light spread across the darkness. He turned his head to look at the sky now turning a shade lighter. Down the hill the other men began to stand, silhouetted. As Scymanky stood, he choked on the ash that blew in his face, almost blinding him. He struggled to breathe. The heat seared his lungs and penetrated his clothing. He stumbled back to the ground, then picked himself up and walked and crawled to the men.

Scymanky convinced them that they needed to get their burned bodies into water as soon as possible. He led them to a stream about twenty-five yards below. He felt constant pain as he crawled through the brush and over logs. Finally they all lay down in the water. The mountain stream that he remembered ran cool and clean, but now this water was a gray ooze. Though it ran warm over his burning body, it was cooler than the air.

They huddled in the stream for half an hour. Then they decided to try to reach the pickup truck almost two hundred yards away. Hot ash reached their ankles as they walked. More ash fell from the sky. A fine gray, almost white powder blew up from the ground with each step. They breathed it into their lungs.

Both sides of the road were filled with logs lying black, dusted with ash. The air had the smell of burnt embers cooling. The men were silent as they moved up the road. They found the pickup blown into a ditch, blanketed in ash. They tried to get the tailgate down so they could crawl in under the camper canopy, but the tailgate was mangled and wouldn't open. The sky began to blacken once again. Hot, fine pieces of rock fell. Scymanky managed to jar open the driver's door. The men crowded into the cab, their raw flesh burning with pain as they pushed up against each other. The darkness grew and soon they could not see each other.

Scymanky felt his pain intensify. He recoiled in agony against the steering wheel and then the cab door. Scymanky now knew they would never leave the pickup. "So this is what it's like to die," he thought. "Why is it taking so long to die? If the volcano would just make one big blast and get it over with." The men began to pray, Scymanky in English, Dias in Spanish, and Skorohodoff and Sharipoff in Russian. "Oh, God, help us in this hour of need . . . "

Don Swanson and the pilot came over a hill in the twin-engine plane and suddenly saw it. A stream of black ash boiled out of the crater and broadened into masses of dirty cauliflower clouds that balled upward for miles. Lightning streaked through the column of ash. Swanson was transfixed. He felt like an explorer entering a strange, beautiful, but dangerous land for the first time. Few other modern volcanologists on record had ever observed a stratovolcano in all its terrifying glory so soon after a major eruption. Unlike the relatively flat shield volcanoes he had known in Hawaii with their eruptions of slow-moving but graphic streams of lava, the steep-sided stratovolcano exploded violently, blasting debris over miles with waves of hot, lethal gases and heat. It was much more dangerous.

"The whole situation is overpowering," he radioed back to USGS headquarters. For several minutes he just let the mike sit in his lap as he tried to comprehend the scene. "It looks like a bomb hit the area."

There was something very different about the mountain aside from the eruption cloud. As they came closer, it hit him. More than 1,300 vertical feet of earth had been blown off the original 9,677-foot-high peak of Mount St. Helens. It now had a flat top with a gouged-out center.

Wind from the west was sweeping the smoke away enough so he could see that the southern flank of the mountain was mostly intact. Small flows of superheated rock, ash, and gas occasionally spewed down the south side. But on the north side the wind was blowing the ash to the northeast in what seemed an impenetrable wall. The pilot flew into the wall. Particles of rock ricocheted off the skin of the plane. Ash blinded their sight. The pilot swung around. He tried it once more, without luck.

Swanson was frustrated. This was perhaps the greatest opportunity any scientist ever had to observe an eruption and now he was locked out of the heart of the action. He tried to find where David Johnston had been stationed but could see nothing through the smoke. It was then he realized that Johnston was probably dead.

Feeling the awe, frustration, and now sadness, he tried to keep his emotions in check. He had a job to do. He was supposed to radio observations back to headquarters, but he kept finding the mike in his hand on his lap. Overwhelmed, he tried to find words to describe what he was seeing. He wasn't keeping good records, such as descriptions of photographs he was taking. He worried that he was going to miss something important.

Lightning bolts branched down from the billowing ash cloud to ignite fires on the east side of the mountain. Hot rocks hurled to the east also contributed to the flames engulfing whole forests, sending more smoke and ash to the already darkened sky. He recorded as much of it as he could with the old Bell and Howell 16 mm windup movie camera they had given him. He found himself cursing more than once as he fumbled to load another hundred-foot roll of film. The radio crackled with reports of other witnesses.

The plane started traversing back and forth from the South Fork of the Toutle River on the west to the Muddy River on the east. It suddenly dawned on him that he had heard nothing but the drone of the plane, the crackle of the radio, and transmitted voices. Swanson strained to hear if the volcano was making any detectable noise. He couldn't hear anything. It was like a surreal silent movie. Back and forth they went, stuck in the only pattern the volcano would allow. As he saw the same powerful images over and over, he tried to imagine what it must be like on the north side, the side they couldn't fly into.

At 12:17 p.m. there was something new. "The color of the Plinian cloud has started to change from medium gray to dirty white," Swanson radioed Vancouver. The change in color of the high column of ash and gases, named after an ancient founder of volcanology, he thought indicated that the volcano had started to release continuous, large masses of molten rock, or lava.

—✳—

When Robert Rogers' camera jammed, Ty Kearney was signing off his radio, preparing to flee. As Kearney jumped from the van to grab a stove and folding table, his wife gathered up her easel and artwork. He hopped back in to the van and brought out his camera. He took several shots capturing the full development of the eruption's lateral plume. Then everyone rushed for their vehicles.

Down the mountain, Rogers' tiny Simca led the way, followed by Francisco Valenzuela's Pinto, with Kearney's Dodge van a distant third. For a mile the road headed straight for the volcano. Its boiling black cloud expanded to within a quarter mile of them. Rogers could see that the area where the radio operator Gerald Martin had been, near Coldwater Peak, was engulfed in black smoke. Tall white alders along the road were bending in the eruption wind. The road branched into a Y and Rogers turned right, followed closely by Valenzuela. But something was wrong. He glanced back to see Kearney turn off onto the other road. Rogers slammed on his brakes. The car spun and ended up in the middle of a wide mud hole. He gunned the accelerator and the rear-engine car he called the Mountain Goat flew out of the mud.

Valenzuela was not so lucky. His Pinto was stuck. He jumped into Rogers' car. Rogers floored the Simca and continued down the same road. They couldn't see the volcano but they could hear its rumble.

The two men came to a saddle off the road. They jumped out of the car, trying to get their bearings. Across the river was the ridge where they had seen the pickup and station wagon that morning. The pickup was still there but the station wagon was gone. The whole side of the ridge was smoking. Entire stands of trees had blown down and were on fire. Lightning shot from the darkened sky into the ridge. Rogers now realized he had taken the wrong turn—that Kearney had taken the right road to get to Merrill Lake and beyond to Cougar. But he was too caught up in what he was seeing to get back into the car.

Rogers thought about the people he had supposed that morning were mountain climbers. Their lives must have been snuffed out. Then he looked on his side of the river to where he and Valenzuela had stood earlier that day. The area was untouched. The blast zone must have ended across the river.

An immense plume moved toward them. The sound of the volcano had grown to a deafening roar. He reloaded his camera and got it to work. The plume grew larger as it reached higher into the sky. It took him ten shots with a 50-millimeter lens to get as much of it as he could. He took another series of photos while the cloud expanded. He screamed into Valenzuela's ear, "There's no rocks in that. They would be coming down. It's just ash."

He couldn't believe how high the vertical column had grown. Minutes before, back up on the hill with Kearney, it had been a lateral blast north toward Seattle. Now the two men leaned back and tilted their heads up, trying to see it all. They couldn't.

"That's up in the stratosphere," he yelled at Valenzuela. "It's got to be twelve miles or more up there." Then he looked down and realized what was different. "The whole damn top of the volcano is gone," he said in disbelief.

A tremendous wind rushed in to feed an updraft into the column of the ash cloud above them. It pulled on their clothes and sucked papers out of the backseat of the Simca. He thought about Valenzuela's Pinto. "We can't leave your car back up there," Rogers said. "The authorities would make sure we never got back in and you'll probably get fired."

"Everything I own is in that car," Valenzuela screamed back.

They drove back around the side of the hill to the stranded car. They got their ice axes out and dug around the tires. Valenzuela tried the car. Its wheels spun in the mud. Rogers used some nylon rope to tie the front bumper of the Pinto to his rear bumper and eventually pulled the Pinto from the mud.

Rogers wanted to reload his camera but couldn't find his last roll of unused film. He decided to drive back up the mountain to look for the film where they had camped the night before. Valenzuela now thought the man he was with was crazy, but he didn't have much choice but to follow. Rogers knew the roads.

Reporter Andre Stepankowsky had lost his view of the volcano. He could see nothing but haze from the plane he rode in with staff photographer Roger Werth. It looked like a black drape had been hung over the mountain. Then they saw it after they came over a bank of hills. The volcano belched smoke like a locomotive. A column of black ash towered eighty thousand feet directly in front of them. The column billowed through a cloud layer overhead. But he could not see the summit.

Ash plumes drifted east. One plume slid down the side of the east flank while another floated lazily above it. Flashes of lightning sparkled and zigzagged in the ash columns. Stepankowsky could see the electrical charges spark fires in the forests below. He filled page after page in his notebook. This was a story he wasn't going to blow.

The summit emerged from the clouds. The pilot moved in for a closer look. Ash churned up from the crater in a mushrooming cloud. The plane tilted and Werth held down the shutter release on his camera as the automatic advance swirled forward in a series of clicks.

Stepankowsky sat transfixed as he tried without success to spot Spirit Lake through the ash clouds. An eighty-four-year-old named Harry Truman lived at his resort on the lake. The cantankerous old man had caught the nation's attention with his refusal to leave. Now he must be dead, the reporter thought.

Ash blocked the sun on the east and northeast of the mountain. When the plane attempted to reach the north side, it felt like trying to fly into a sandstorm deep in the night. The plane was rebuffed and the pilot tried again. Stepankowsky reminded himself that he was only twenty-four, that he was about to get married. He didn't want to die. The plane turned back once more, but now Stepankowsky could taste his breakfast. He asked for an airsickness bag and threw up part of a Spanish omelet.

Stepankowsky imagined the ash cloud as a ghost slinking toward them, arms outstretched, grasping at their small plane. It amazed him to look south and east and see Mount Adams and Mount Hood rising in splendor in a clear blue sky, just as Mount St. Helens had only hours before.

The pilot gave Stepankowsky an assignment. There were now so many planes in the sky that he was worried about a collision. He needed the reporter to become a spotter. Stepankowsky took his spotter duties seriously. Planes buzzed all around.

Thick ash clung to glaciers on the mountain. Fingers of mudslides now slipped down its flanks. The radio crackled with a transmission: a wall of water was rushing down the Toutle River. As they headed west toward the Toutle River Valley, he noticed that the volcano had blanketed mile upon mile of Douglas fir forest with a mantle of gray ash.

They followed the river down to Weyerhaeuser's 12-Road Logging Camp on the South Fork of the Toutle and found that mud covered half the site. Logging trucks, cranes, pickups, and buses were strewn like toys, on their sides, upside down, and some at odd angles sticking out of the mud. Huge logs rested helter-skelter across the camp. A log train with thirty-nine cars had been demolished.

The wall of water had taken a railroad bridge with it, followed closely behind by much of the camp's stockpile of logs, some equipment, and debris that included uprooted, old-growth trees more than eight feet in diameter.

"What the hell is going on?" Roald Reitan asked as he followed Venus Dergan out of their tent. The sky was no longer blue. It was a dirty gray. He looked at the rising river. The water had turned a milky white. Tree branches, broken boards, and other debris swirled in it. From upstream, Roald could hear what sounded like a huge monster tramping through the trees and brush, breaking branches as it made its way straight for him and Venus. The water was starting to slosh at the sides of the tent. "We're getting the hell out of here," he yelled. Venus wanted to grab her purse from the tent. Roald did not even think to return to the tent to put on his shoes and socks.

The sound grew louder. The river curved in an S-shape up from their camp, with high, sheer walls of bare rock hiding their view upstream. But he could see the tops of trees that had to be over a hundred feet tall beyond the rock walls. The treetops shook, and he heard one crashing sound after another above a growing roar as the trees fell from view. "Get into the car," he shouted. The emergency siren continued with its unrelenting *wranaah, wranaah, wranaah*. The rising water was turning into small rapids. He grabbed the tent and with one pull ripped it from the ground. He dragged the tent with one hand and picked up the two lawn chairs with the other hand.

They ran for the car. Venus climbed in as Roald tossed the tent and chairs into the trunk. He got behind the steering wheel, pulled the choke, pumped the accelerator pedal a few times, and turned the ignition key. Nothing happened. Venus thought they may have left the radio on the night before and drained the battery. Roald tried once more. Again nothing happened. Water was rising around the car.

"Aw shit," he said as he looked upriver. A railroad bridge was floating around the bend a hundred yards above them. It swung like a gate closing to become lodged between the two banks. The bridge held back a huge pile of uprooted trees and debris, including ragged boards ripped from buildings, stripped logs from the Weyerhaeuser camp, brush, parts of a roof, and logging equipment, all jostling in the torrid river as they pushed against the creaking bridge. Behind the debris dam, a wall of water rose.

The bridge held for a few moments, then broke. Logs popped into the air and rolled end over end down the river. The grayish white river water became a torrent of chocolate-colored mud. They heard a boom as a forty-foot log crashed down and lodged against the front of the car. Water now surrounded the Oldsmobile. "Jump up on the roof," Roald said. The water was just above their ankles, but in seconds it rose to their waists. They clawed to the top of the car and crouched down. Roald's nostrils filled with the dank smell of mud mixed with something that smelled like freshly cut trees.

Another log rammed the car, then another, and another. The car began to tilt. With each jolt they fought for balance. Venus didn't scream. She just followed Roald's orders and sometimes cursed. Water started to lift the car as it was pushed into the torrent by the logs. A huge log headed straight at them. "Jump," Roald screamed. As the log upended the Oldsmobile, they leaped.

Roald landed on a log. Venus hit the water between the log he was on and another one. Then she was gone. He straddled the bobbing log like he was riding a wild bronco. The river had changed to the consistency of mortar. A log pressed in from the side and started to crush his knee. At almost the same time, a tree beneath the water snagged his other leg. The pain was unbearable. He yelled for Venus over and over, telling her to hang on. There was no response. He thought about his parents. They didn't know where he was. He was going to die out here and they would never know where or how he died.

Venus looked up for something to grab onto. On her left side the car swept past. She thought how odd it was that she had always called it a boat, and now here it was floating down the river. She knew she was going to die. Huge logs were coming at her. The current picked her up just as she realized Roald was yelling at her, "Hang on, hang on." She still couldn't see him.

The hot, brown muck churned like a huge washing machine, logs crashing down around her, rubbing against her, scraping off skin as she was swept downriver. She didn't understand what was happening, or why. All she could do was try to hang on, keep her head above water, and pray the pain would not grow any worse as she died.

Something hit Roald hard on his back and wrapped him against the log he was on. A smaller tree had rolled across his log. He sat up again. He heard a *boom, boom, boom*. Another tree, only larger, had skipped over the log and hit him again in the back. He bent over with more pain and grew dizzy. "I give up. I give up. God, do it now. Get it over with." Just then the pressure came off one leg. He pulled the leg up and the other one also came free. "Son of a bitch," he said in surprise.

He rose to his knees. He didn't know why, but something made him crawl to the end of the log. He saw her floating. Venus looked up at him, her face and hair caked in mud. He grabbed her shirt. But the log rolled and he lost his grasp. She was gone again.

He clambered to a larger log. He steadied himself on the bouncing log and tried to run. He fell several times onto his forearms. He no longer felt any pain, operating now on adrenaline, fear, and instinct. He had to find her.

Out of the corner of his eye he saw her shoe floating between two logs that formed a V. He jumped onto one of the logs. Roald peered into the water and spotted what he thought was her nose. Then he could see the outline of her face. He tried to grab her shoulder but got her hair instead. She never made a sound as he pulled. He got her halfway out before she slipped from his muddy arms and went back under. "God, I'm sorry," he yelled. "I'm sorry."

Venus was being carried away. Debris seemed to be holding her up. Her head was between two logs, and he knew they were going to smash her. The logs moved up to her face, ripped skin off, and then parted.

She was sucked under once more. A log trapped her beneath the water. She tried to raise her hand. Another log pinned her wrist against the log she was under. The pain was excruciating as the logs jostled, repeatedly smashing her wrist but never parting enough for her to free herself. She could no longer hold her breath.

By now Roald was at the end of one of the logs. The logs parted and there was her hand sticking up from the muck. He grabbed her wrist and pulled. She came out all the way as he crouched as close as he could to grab her under the arm and hold on without falling over. He knew he wouldn't have another chance.

Robert Rogers spotted the yellow film canister on the ridgetop. It was standing on end in the ash dust. As he drove toward the canister, he opened the car door, slowed down, bent over like a polo player as he held the steering wheel with one hand, and scooped up the film.

Valenzuela was right behind him in the Pinto. As they turned to go back down from the ridge, the ash cloud dropped upon them. They turned on their windshield wipers and lights as they crept forward. Rogers could barely see the road ahead. He came to another road and saw a huge Caterpillar earthmover that he thought he remembered was next to a road leading down to Merrill Lake. He turned right.

Visibility grew worse as the ash rained down. For what seemed like forever he pushed ahead, not really seeing where he was going. Up ahead he saw a yellow outline, as though in a mist. It was the Caterpillar. They had driven in a circle.

Rogers was lost. He got out of the car. The stifling hot air had the musty smell of sulfur. "If we go out into this again," he told Valenzuela, "we are only going to get lost and maybe into something more serious."

So they sat and waited, each in his own car for a fast getaway. About a quarter of an hour later, Rogers saw a tiny spot of light sky within the blackness, to the west. The light grew rapidly. "I know what's going on now," Rogers told himself. "Prevailing winds are from the west. This stuff is going to blow by."

When the ash cleared, they saw they were in the middle of a clear-cut. Nearby, across the Toutle River, trees still stood, only covered with ash. But in front of them, row after row of trees had fallen. And there were pockets of destruction behind them. Whatever had knocked down the trees had spared the two men. Rogers looked down below and saw the road system was intact all the way to Merrill Lake and Cougar.

The sky grew brighter and clearer. Haze lifted from the volcano. They could now see it perfectly, belching smoke. They decided to stay. It was like having a front-row seat at one of the greatest events of all time. They were just over eight miles from what was left of the summit of Mount St. Helens.

They heard the buzz of a helicopter. The pilot saw them and started down. "Oh, god, he's going to try to rescue us," Rogers said. "Don't act excited. Don't make him think we need to be rescued." The helicopter neared the ground just as a tornado of ash engulfed it. The chopper rose high into the sky and flew off.

The governor's helicopter had just landed at the airport in Kelso, a city of ten thousand adjacent to Longview. The plane carrying Andre Stepankowsky made its runway approach. On the ground he found Washington's governor, Dixy Lee Ray, but she didn't have much to say. She was overseeing what was shaping up to be one of the worst natural disasters in the nation's history. Yet he found her strangely reticent.

Stepankowsky went back to the *Daily News* office in Longview and sat down to write what he knew could be the biggest story of his career. An editor told him his fiancée, Paula LaBeck, had called wanting to know if he was okay. "I told her you were out flying."

He thumbed through his notes and started typing. He got partway down the page on his computer screen, read what he had written, and stopped. It just wasn't good enough. He started over. Every few minutes he would have to stop, grasping for words. What he had seen was so overwhelming that words describing it seemed like exaggerations. "People are not going to believe this," he told himself. Finally he decided to just let the story flow. He wrote page after page. Adjectives piled up. He reached for metaphors that could somehow capture what he had witnessed. He liked the image of the volcanic plume as a ghost. It contributed to the eeriness of the experience he was trying to convey.

Two and a half hours later he was finished. He looked over his copy and transmitted it to Bob Gaston's computer. More staff members arrived. Society reporters found themselves checking hospital admissions. Sportswriters were recruited to cover press briefings by emergency workers.

He called his fiancée, whom he had met at the school of journalism in Missouri, and they spoke briefly. She was now editor of the *Inland Register*, a Catholic newspaper in Spokane, Washington. They were to be married in a month.

As he worked the phone and took notes, he glanced at Gaston. The editor, known as a perfectionist, furiously typed away. But there was no way for Stepankowsky to know if the editor was working on his copy or somebody else's.

Donna duBeth, the paper's lead reporter on the volcano, was the latest to fly over the disaster and ears perked up to hear what she had to say. She strode over to Gaston's desk. "We're all going to be flooded out," she said loud enough for everyone to hear. "It's going to come right down to where we are. The mudflows are going to wipe out the town."

Gaston caught Stepankowsky's eye. The editor shrugged. Stepankowsky shrugged back. Everybody returned to work.

The loggers sat in the pickup for almost an hour, shifting their bodies constantly, trying to find relief from the pain. As it began to grow light again, Scymanky could make out the silhouettes of fallen trees scattered like so many children's pickup sticks among the boulders. He tried to start the pickup. Nothing. For a long while he sat and then grew angry with himself for having thoughts of giving up. This cab was not going to be his coffin. When he pushed open the cab door and yelled "Let's go," no one argued.

The logging road was covered in ash, rocks, and trees. The men had difficulty following its route until Scymanky remembered that it traveled next to a small stream. They walked along the stream bank, trying to stay close to the road. Every once in a while, one of them would fall into the stream and have to crawl out. They choked on the ash that still hung heavy in a haze around them. The ash blotted out the sun, leaving only a dim glow high in the sky.

Scymanky tried to keep Dias from getting too far ahead. At one point he lost sight of him and feared the worst. "Dias, Dias," he yelled out. The men picked up the pace. When they found Dias, Scymanky noticed the man's bulging eyes and hyperactive behavior. Dias seemed like a wild, wounded animal trying to outrun a predator. Scymanky told him to stay close to the group, and for a while Dias did so.

Rows of fallen, smoldering trees blocked their path. As they tried to crawl over the trees, ash on the bark burst into flame from their touch and rose as sparks. Scymanky could no longer feel his blistered hands inside the heavy work gloves singed black. He began to grow cold and his teeth chattered. He looked at the other men. Their bodies shook, too. The air was not cold. He hoped they weren't going into shock.

Ash was thick on their hair, faces, and bodies. Looking like ghosts grown gray in the night, they walked faster to keep warm. For several miles they trudged on, with ash still falling from the sky. They found a spring spurting clean water from a crack in a rock wall. They drank the cold water until their bellies were full, then tried to vomit up the ash that clogged their throats and lungs. But they found themselves only gasping for breath through hacking coughs that failed to clear their airways.

They felt revived by the water and hated to leave the spring, but Scymanky convinced them they needed to keep walking. The road was bordered on one side by a mountain that rose thousands of feet and on the other by a cliff that dropped into a valley. The valley looked like a battlefield. Trees stood smoldering, black against the sky. Thousands of other trees, uprooted, lay on their sides in tidy rows that followed the direction of the wind blast from the volcano. Smoke rose in wisps from the blackened ground. A haze of steam evaporated from the river, swollen with debris and ash as it ran through the valley. Lightning shot through the sky.

The fallen trees grew fewer as they made their way down the road. Scymanky thought they had a chance. A major highway was not far away. "Hurry," he said. "We're going to get out of here." But their burned and stiffening bodies could not go much faster. Then around a bend they dropped to their knees. Blocking the way was an immense pile of rock that had avalanched across the road. They were trapped.

Scymanky felt as if he had been punched in the stomach. He fell flat to the ground and the others joined him. While he lay in the warm ash, his body shaking from chills, he studied the other men and heard their moaning. They looked as if black shoe polish had been smeared on their faces and then dusted with gray talcum powder. Their clothes were in tatters. They shook as badly as he did.

They gathered some green branches to lie down on, back to back, trying to stay warm. But the pain grew so bad that they had to get up and walk around. It wasn't long before they had to lie down again to rest and get warm. They went through this routine a dozen times before Scymanky worried that they needed to keep moving to survive. "Why don't we go back up and get some water. It's higher up there, anyway," he said to the men. "We might have a better chance to get rescued."

When they walked back up the road, they were surprised to see something they had missed on the way down. The road split. Down below was a vast valley.

"Where are you going?" Scymanky yelled at Sharipoff, who started down the road. The Russian stopped for a moment and looked at Scymanky. There was something about the man's eyes that Scymanky had not noticed before. They had a crazed look. "There's more water down there," Sharipoff said as he turned to go.

"That water is dirty down there," he screamed at Sharipoff.

"No, there is more water down there," the Russian said.

Scymanky caught up with him and grabbed him by the arm. Sharipoff fell to the ground. "You are just going to get killed down there. Now get up," Scymanky said as he tried to pull Sharipoff up, but the Russian pushed him away.

Sharipoff lay with his face partially in the ash. "I'm going down here for water . . . I'm going down here for water," he kept muttering. Skorohodoff, who now joined them, also collapsed. Out of the corner of his eye Scymanky saw Dias continue up the road toward the water.

"You're both crazy," Scymanky yelled at the Russians. "You're just going to die down here. At least up above, we have water."

"The water up there is bad for you," Skorohodoff said. "You are going to get pneumonia."

Scymanky realized that Sharipoff was no longer listening to him. One more time he tried to pull the man up, only to be fought off again. Scymanky realized he had no more energy to give. He turned and started back up the road.

Scymanky heard footsteps behind him. It was Skorohodoff. When the two men got back to the water, Dias was lying next to the spring. They all drank hungrily. Then they began again the ritual of getting up and lying down, getting up and lying down, trying to find some way to relieve the pain. They talked about dying.

Dias decided to go back to where they had encountered the avalanche debris. He was going to try to find a way over it. "You are just crazy," Scymanky told him. "If there was a way, I would go out. But there is no way." Dias ignored him.

Though nightfall was hours away, Scymanky became obsessed with a fear of dying in the darkness with no heat. If he was going to die, he wanted to die warm. There were matches in his shirt pocket, but his burned arms had grown so stiff he didn't know if he could get the matches out. He strained against the pain, barely holding on to the matches. They were soggy. But he soon realized it wouldn't have mattered, anyway, even if they had been dry. He started laughing. His gloves were burned to his hands. He would never have been able to hold a match to strike it, with the rigid work gloves stuck to his burnt, stiffening fingers.

—✳—

When Don Swanson returned to USGS headquarters in Vancouver, he answered questions and tried to fill in gaps. This new information was taken across the hall to where the emergency services departments for

Oregon and Washington were trying to cope with the disaster. The antiquated phone system for the Washington operation had failed, hampering rescue operations and leaving the news media to dispense what information they could dig up for the state's anxious citizens.

Early in the evening Swanson was asked to talk with the man who had been on duty at the Coldwater II observation post before being relieved by David Johnston. Harry Glicken was a graduate student who had just finished a stint of two and a half weeks at the post when Johnston, filling in for Swanson, had arrived. Johnston was Glicken's friend and mentor.

Glicken was overwhelmed with guilt and grief. Swanson did not particularly want to talk to the inconsolable man. But some people thought Swanson might be able to help Glicken, and they could use whatever information the graduate student could provide. The two men moved out onto a fire escape for privacy.

Glicken told Swanson that at about noon that day, he had arrived at Toutle High School, the staging area for rescue teams and helicopters. With no official authority, he persuaded a chopper crew to take him up to look for Johnston. Glicken and the crew were soon lost. Maps no longer corresponded to anything they saw on the ground. They spotted a car with people in it near a logging camp. The chopper landed. Glicken and some paramedics rushed to the car. But as they touched the bodies of the people they hoped might still be alive, skin fell off onto their hands. The victims had been mummified by the heat of the initial blast

Glicken pushed the crew on to look for Johnston. Clouds of ash blew into the helicopter and the pilot finally gave up. Back at the staging area, the distraught Glicken talked another crew into making an attempt. But all he discovered was a land of gray when he attempted to find any point of reference to help in the search for the observation post. Again they turned back.

Still, Glicken would not give up. He corralled yet another helicopter crew. The result was the same. He pleaded with crews to take him in once more. They all declined. Conditions were dangerous, he was told, and there were now many reports about possible survivors needing rescue.

For Swanson, facing Glicken was difficult. Glicken was miserable with the knowledge that he had not been able to find the observation post. The thought that Johnston was still there and might need help haunted him.

The young geologist, recently graduated from Stanford, had an appointment to be in California on May 18 to discuss graduate work at the University of California. Johnston, filling in for Swanson, had replaced Glicken the night before so the student could fly out that morning. In his grief-stricken mind, Glicken thought that if he hadn't been scheduled to go to California, Swanson would have asked him to stay on another night instead of asking Johnston to take the duty. At the very least, Glicken thought, he might have stayed on with Johnston, his role model, and somehow that might have made a difference.

"It's not your fault," Swanson said. It wasn't anybody's fault, he told himself. It was just a roll of the dice.

Then his empirical scientific mind kicked in. There had been no confirmation of the number of dead or who they were. At this point, all they know for sure was that this had been an enormous eruption and there had been fatalities.

CHAPTER 4

May 18: Deliverance

"HOLD ON, HOLD ON," Roald kept saying to Venus. He held her tightly, his body over hers, as they lay prone on the huge, bobbing log. He wasn't going to lose her again.

Venus Dergan looked up at the chaos around her. She tried to grasp the fact that she had just been in this swirling mass of debris, full of dead animals, bridge parts, bashed-in trucks, logs, uprooted trees, and crushed houses. She could hardly believe she had survived. There was no pain. Her body was numb. She felt nothing but the shock of realizing she was still alive.

Venus had no idea what caused the destruction. She thought the world was coming to an end. She saw a rowboat sitting untouched on a mound of debris. "Look at that boat," she said to Roald. "How can that be? Everything is ruined . . . and that damn little boat doesn't even have a scratch." The absurdity of it all hit them. This couldn't be real.

Roald Reitan was surprised that Venus could talk. She looked like a mud monster, he thought, like something out of *Creature from the Black Lagoon*. From her hair to the soles of her feet she was encased in mud. Only her piercing brown eyes showed through.

With logs still jostling them, they floated into an area that Roald recognized as a park. The mudflow slowed as the channel widened. He saw their chance. He thought they might be able to jump into the muck near the edge of the flow and make it to shore. But first they had to get to the edge. They were still near the middle of the river.

"We're going to jump from log to log and as soon as I'm certain it's shallow enough, we're going to jump into this shit and wade to shore," he said. She didn't say anything, but her eyes showed her concern through the mask of mud. "We'll be okay," he told her. "Just remember to jump when I tell you."

He was ready to pull her along if she didn't move. But every time he yelled "jump" she sprang off the log in unison with him, his arm around her waist, like trained athletes. They jumped several logs in succession.

They came to the edge of the flow. He hesitated. He wasn't quite sure that it was shallow enough here, that it might not sweep them back into the flow or drown them. Venus looked up at him. He grabbed her under the arm. "Jump!"

They leaped into the mud. It came to his chest but it almost covered her neck. He tried to prop her up, to keep her face out of the mud, as they waded to shore.

The hillside beyond the bank was steep. "Come on, we've got to climb, we've got to get to higher ground," Roald said. They would have to use their hands and feet, like four-legged animals, to get through the brush and trees.

"Oh, god," she cried out. She had discovered that her wrist was broken. He tried pulling her up the hill by her other arm. Then he pushed her up the hillside from behind. They reached level ground. He thought they were high enough now.

They looked at each other in amazement, trying to catch their breath. "What in the hell just happened?" Roald asked. They never suspected the volcano.

Roald checked her injuries. The logs had ripped at her face. Blood oozed through the thick layer of mud, dirt, and debris. An open gash ran from her wrist to her elbow. It looked like raw meat. Beneath the split flesh she could see bone. She started to slip into shock. "Oh, my god, I'm in trouble," she said. She slumped to the ground. He sat with her.

Venus looked at her bare feet, covered with blood and mud. She stared down the steep hillside. She couldn't comprehend how they had climbed it with neither of them wearing shoes.

Roald started pacing. "We've got to get out of here. We've got to get out of here," he said over and over. The flood had carried them under a bridge, and he hoped to find a way back to the span.

Venus stared at the ground and told herself, "I am not going to survive if I have to spend the night out here."

"Maybe we can go back upstream to the bridge we floated under and get on that road and walk to town," Roald said.

He pulled Venus to her feet. They started along an elk trail. A plane buzzed overheard and they started yelling. Then there was another buzz and a *whop, whop, whop* as another plane and then a helicopter came into view. But nothing happened. They realized no one could see them through the brush and trees. "We need to find a clearing where somebody can see us," he said.

Roald now almost ran. They heard more planes and their hopes rose. "Come on. Keep up," he kept yelling at Venus. The mud that encased her body felt like concrete. Her arm was swelling and her leg throbbed. "Roald, I can't go any faster. I can't walk. I can't keep up."

They found an overgrown logging road that led down to the river. Roald ran ahead. She suddenly heard his voice, screaming for help.

Roald had found the bridge, with a sheriff's car in the middle of it and an officer standing by with two other men. He yelled and waved his arms until he was hoarse, and finally one of the men pointed at him. The officer reached into the car, pulled out the handset for his radio, and started speaking into it.

The men came to Roald and followed him back to Venus. They picked her up and carried her down to the bank. They wiped some of the mud from her eyes and examined her arm. The gash had widened. One of the men wanted to lance her swollen arm. "You're not going to touch it, you're not going to touch it," she screamed. "Leave it alone."

"Just calm down," one of them said. "We've got a helicopter coming and we're going to airlift you out first."

"Do you know what happened to you?" one of the men asked. "Mount St. Helens erupted."

Venus and Roald looked at each other in surprise. "What does that have to do with us?" Venus asked. The men explained that the eruption had sent mudflows down the river. For the first time, their ordeal made sense.

Venus couldn't stop shaking. Her teeth were chattering. She started to convulse.

"Where's the goddamn helicopter?" Roald screamed.

Soon a green Huey descended, the hurricane-like wind from its rotor beating against them. There wasn't enough room to land. One of the men waved it off.

Two minutes later a small Ranger helicopter arrived, a Weyerhaeuser insignia on its side. The pilot lowered the chopper and held the front of the skids against the hillside while keeping the rotor going. The door slid open. A man jumped out, picked Venus up, and placed her in the helicopter. He assured her that someone would come soon for Roald.

"You okay, buddy?" one of the men on the ground asked Roald as the chopper rose into the gray sky. "How badly you hurt?"

"I'm okay. I'm just worried about Venus."

"Buddy, you're hurting, whether you know it or not," the man said. He was looking at Roald's feet. Mud, sticks, and leaves were stuck to his bare, bloody feet. Blood flowed from the mud-encrusted wound on his knee. Roald could see the ligament through the wound, but he didn't feel any pain.

"We got word," the man said, "that there's something coming down after this. They say it's far worse than the flood we just went through."

Jim Scymanky's body grew stiffer as he lay on the ridge next to Skorohodoff, who was mumbling in pain. Then they heard a slight rumble in the distance. As the sound grew closer, Skorohodoff turned to Scymanky, fright in his expression. Neither man spoke. The sound grew into a deafening roar. Skorohodoff looked into the valley below and cried out, "Oh, god. It's over. It's going to get us. It's going to come right up here and get us."

Scymanky looked down the cliff. A mass of dark ooze, hundreds of yards wide, moved through the valley, coughing up and carrying huge boulders with it and uprooting entire stands of stripped trees. Wisps of steam rose from its surface. Skorohodoff said he could feel the heat. But Scymanky didn't know if the Russian was imagining things or if his own burnt skin was now so numb that it was no longer sensitive to heat. Scymanky didn't feel any rise in temperature as the wave neared. Both men thought it was lava from the erupting volcano.

For a long time Scymanky stood, mesmerized by the wave. Then he watched it sweep over the road where they had last seen Sharipoff. "He's dead," Scymanky thought. "If it comes up this way, we'll just go higher." Then he decided the flow was not going to reach them. Scymanky lay back down with Skorohodoff, both of them squirming in pain.

Scymanky heard a buzz in the distance. It was a helicopter. Another chopper came into view, then a third. Both men ran up and down the ridge, waving their arms. The helicopters began flying patterns over the valley, coming close to the men but apparently not seeing them. Scymanky grew angry as he watched the helicopters fly away. "Fools," he thought. "Why won't you come over here?"

The talk of dying began again. They both knew this night would be their last. With each hour, their bodies stiffened even more as they lay in the ash. They closed their eyes but could not sleep. Then, through the stupor of pain that had blocked his senses, Scymanky recognized the sound of helicopters. He opened his eyes to see two choppers fast approaching. They buzzed overhead. Scymanky raised up on one arm and began waving. The Russian next to him only moaned, with his face in the ash. One of the pilots waved back and Scymanky jumped to his feet.

"Get up, get up, Leonty. They're coming for us," he screamed with a hoarse voice. But his friend didn't move.

One helicopter hovered over the black ooze flowing just below them. A man jumped from the helicopter into the mud. Scymanky and Skorohodoff realized that what they had seen earlier was not lava, but a mudflow.

Finally the helicopter landed and the pilot jumped out. One of the crewmen shouted to Scymanky to come down to the chopper.

"We can't move," Scymanky said. "How are we going to get down? Man, we're burned."

The crewmen finally comprehended the situation and started climbing to the burned men. The second helicopter landed uphill. When the first crewman from below got to them, he picked up Skorohodoff like a child in his arms and carried him up the road to the second helicopter. Scymanky followed, with a man helping him under each arm, forty yards back up the road he had come down seven hours earlier. He told the men about Dias and Sharipoff and they transmitted this information by radio. Then they strapped the two loggers in the back seat of the chopper. The two men leaned against each other. Skorohodoff was unconscious.

The crewmen asked Scymanky how he felt. He could hear them but could not see them. Their faces would flash before him and then dissolve into whiteness as though someone was turning a TV set on and off. During periods of consciousness on the flight to St. John Hospital in Longview, he caught glimpses of the devastation below. Everything was ruined, covered in gray. He no longer cared.

The pilot, Captain Jess Hagerman, flew for both Weyerhaeuser and the National Guard. Only hours earlier he had fought to continue rescue missions into the blast zone. An Air Force Reserve pilot had returned from the devastation and told officials there was no hope of finding survivors. Hagerman argued with his superiors and the officials to keep looking, and he won.

Robert Rogers and Francisco Valenzuela watched the volcano endlessly spew its plumes of black and gray to the sky. Finally Valenzuela said, "You know I work for the Forest Service, and they are really going to want me to report for duty."

"Yeah. You got to show up," Rogers said. They hopped into their cars. Rogers led. He stopped after several miles, just before the place known as Four Corners. It was a good spot to take what he was now calling volcano-hero photos. He told Valenzuela to stand on a stump. Rogers lay down on the road and shot up at Valenzuela, with the ash plume towering over him. Then he had Valenzuela take photos of him from the same angle.

At Four Corners he slowed down. A road from there led up to within half a mile of timberline and the volcano from the south side. A locked gate blocked this road. He considered getting a wrench from his car to take the hinges apart. He even thought up the story he would tell if he was caught beyond the gate. "Hey, our road got washed out. We're looking for a way out," he would tell authorities.

But he kept going down, regretting his decision to leave the mountain with each mile he drove. The roadblock at Cougar was now heavily manned by law enforcement officers. He knew he would not be getting back in this way anytime soon. As they traveled through the river valleys, their vision of the volcano and its billowing cloud became obscured.

Valenzuela left to report for work. Rogers took I-5 south toward Portland. After several miles he saw a line of cars parked along the freeway. He turned to see what everyone was looking at. There, rising like the cloud from an atom bomb to the edge of space, was the volcano plume. It was the first time he had been at a vantage point to see it in its entirety. He said to himself, "So that's what it looks like."

The small helicopter with its bubble cockpit rose straight into the air with Roald Reitan and the pilot in front. Below him Roald saw for the first time the immensity of the destruction he had survived.

"Do you want to see what almost killed you?" the pilot asked.

Before Roald had a chance to answer, the pilot spun the chopper around. Roald recoiled and let out a long "ahhh." His jaw dropped. The volcano was putting on a light show as it belched a continuous cloud of gray and black smoke that turned blue and purple as lightning shot through it, leaving an incandescent glow. "Look at that son of a bitch," the rescuer behind him said.

The pilot flew toward the plume. Rocks in the air hurtled past them. Roald wondered again if he was going to make it out alive. "This ain't no safe place to be," the pilot said, turning back. Roald looked down to where he had been five minutes before. A wall of water and mud, even higher than the one that had swept Venus and him away, cascaded across the land. The flood lifted up the bridge and carried it away.

They flew to the high school playfield in Toutle, site of an emergency center. Huey helicopters were perfectly aligned in a row. State police and sheriff's cars idled with exhaust pouring into the air. Ambulances were lined up, red lights rotating on their roofs.

The field was full of activity. White-coated medical personnel bent over blanket-covered bodies. Military personnel in camouflage uniforms and polished boots were erecting tents. Reporters seemed to be everywhere along with television cameramen who jostled for position.

Venus Dergan lay on a mat. People milled around her, talking. She couldn't catch what they were saying but then she distinctly heard a man say, "Ah, look at her." She could barely open her mud-encrusted eyelids. A woman tried to wipe some mud off, but it had dried like concrete. The woman then tried swabbing her eyes with a Q-tip. Silicon in the ash only bit into her cornea. The lady gave up.

Venus saw a helicopter land. Roald jumped out and a man ran for him. "Say, you wouldn't happen to have a cigarette?" Roald asked. Medical people converged on him. "Hey, what are you doing to me?" he yelled out.

"You need to lay down," one of them said.

"Hey, I'm all right. How's Venus?" He fought them off. He started pacing in circles. He was still hyped with adrenaline from his ordeal. Venus could see he was muttering to himself but she couldn't hear what he was saying. The medical people brought a gurney and forced him to lie down. He sprang back up and tried to walk. All at once he started to collapse as exhaustion overwhelmed him.

The numbness was leaving Venus's body. She was feeling more pain. They strapped her and Roald onto gurneys and loaded them into an ambulance. An attendant hooked them up to monitors. The driver and attendant tried to figure out how to get to St. John Hospital in Longview. Bridges were out. Interstate 5 was shut down. It was going to be a long ride.

As the siren blared, Venus slipped in and out of awareness. Dried mud pulled her skin tight. She thought her teeth were going to break from the chattering she couldn't control. Her body shook and she went into convulsions again.

"What's happening to her? What's happening to her?" Roald kept screaming.

"She's going into shock," the attendant told him. The attendant quickly hooked her up to an IV. It took more than an hour to reach the hospital.

The hospital emergency area swarmed with patients. They waited for what seemed an endless time to see a doctor. Finally they were taken to separate treatment rooms. Roald kept asking about Venus. "Where did they take her? Where is she?"

"She's being taken care of," a nurse told him. "Let's worry about getting you taken care of." A physician determined that Roald's injuries were not life-threatening. They cleaned the mud off, irrigated his knee and feet, and told him he could go home.

"Do you have somebody who can pick you up?" the doctor asked.

Roald called his father. "Dad, I have something to tell you. Venus and I are in a hospital in Longview. We were in the eruption from St. Helens. I'm okay but they don't know about Venus."

"You're what? You're where?" his father asked. Roald explained again. "We'll be right down, son."

He waited in a wheelchair in the emergency area for word of Venus. A reporter started asking him questions. "Get the hell away from me," he screamed. He didn't want to talk to anybody. He was afraid Venus was going to die.

Nurses tried to cut the clothes off of Venus with knives and scissors. They had difficulty cutting through the mud-encrusted fabric. They pulled and tugged at her clothes but she felt little pain. She had been given morphine. After the nurses finally got her clothes off, Venus was shocked to see the extent of her injuries. Hours went by as she underwent tests. The hospital staff didn't know what to do with her. They had never treated injuries involving volcanic ash and silica.

Roald watched as more patients were wheeled through the double-swinging doors of the overwhelmed emergency department. Ambulance sirens were almost continuous. Finally a doctor came to talk to him about Venus. She had abrasions and ripped skin on more than half of her body. A large chunk of flesh had been gouged from her calf. Her face had been badly scratched and cut, with pieces of skin torn off. Layers of skin were missing from both sides of her forearms and the top of her hands. Her wrist was broken. But the doctor said that miraculously she had suffered no internal injuries. She was going to live. Roald could see her after they washed the mud from her wounds.

She screamed with pain as the nurses wiped her wounds. Nothing had ever hurt so much. The morphine no longer helped. The silica in the mud rubbed against her raw skin with each swipe and dab of the sponges. The nurses scrubbed for a long time trying to get the glasslike silica grains from her hair and scalp. But they couldn't get them all. Their biggest fear was that the ash and silica encrusting her eyelids would scratch her cornea.

Finally Roald was allowed to see her. They were left alone. She looked up at him. "Roald, how did we ever live through that?"

A man with a camera strapped around his neck suddenly approached them. "I hear you two survived the eruption of Mount St. Helens." He was from the Vancouver *Columbian* and wanted to take their photograph and ask a few questions. Before they could protest, the man aimed his camera and clicked the shutter. Roald and Venus tried to answer the man's questions.

"I don't think he was supposed to be in here," Venus said as the man left.

They assigned her to a permanent room and she was given more morphine. Roald stroked her forehead. "You need anything else, babe?" he asked.

Her eyes flashed up at him. "Thank God we're alive, Roald, we're alive," she said. "And you get to go home."

Roald worked the remote of the television while Venus rested. Between news flashes, his parents walked into the room. He saw the shocked looks on their faces as they peered closely at him and Venus. "Are you all right? What on earth are you doing down here?" his mother asked. "What in the hell is going on?" his father asked. His parents were wearing their best clothes. They had been dressed to go out when they were called. "All I want to do is go home," he told them.

A doctor asked Roald and his parents to leave while Venus was examined again. The doctor found traces of mud and ash still embedded in her wounds. "If you don't scrub these out cleaner, she is going to be tattooed," he told the nurses. "It's going to cause terrible scarring."

This took a few moments to sink in. "Oh, God, no," Venus said with a terrified face. They wheeled her back into the emergency room. "Please, please, don't do this again," she yelled at them. The nurses had tears in their eyes. They began scrubbing her wounds. She screamed and sobbed uncontrollably as they held her down and tried to get the last bits of debris out of her raw skin. A nurse gave her twice the dose of morphine she had received earlier, but the pain persisted. She prayed she would pass out.

It was dark before Venus's mother and father arrived. She looked up to see her mother burst into tears at the doorway. The look on her father's face was excruciating. She had never seen her parents in such pain. "Thank God you're alive," her mother said, sobbing as she stood over her daughter, not knowing where it was safe to touch her.

It was time for Roald to leave. He bent over and kissed Venus. "They'll take good care of you here, babe. Let me know what's happening."

She smiled up at him. "I'll keep in touch, don't worry about that."

As Roald and his parents made their way down the hall, his mother kept turning to look at the gurneys that carried the bodies of eruption victims, covered with sheets. Roald's father and a medical assistant used a fireman's carry to lift Roald from his wheelchair into the car. His knee was swollen and his feet throbbed from wounds. "You can't believe how hard it was," his mother said, "to look at those bodies with the sheets over them and realize that could have been your son."

On the long drive toward Tacoma, bypassing the closed freeway, all Roald could think about was Venus and about how good it felt to be alive. Finally they stopped at a little motel. Roald was now in agonizing pain. His father carried him into the room. When they got his clothes off, his parents were startled to see his injuries. He had fallen several times jumping from log to log. Abrasions covered his forearms. His head and back were bruised and ached intensely where he had been hit by flying logs. Wounds that looked like red raspberries spotted his body. The wounds were still filled with mud and grit. His knee and feet looked like swollen chopped meat.

His father emptied a wastebasket, washed it out, and filled it with warm water and soap. He scrubbed his son's wounds while Roald tried to keep from crying, then passed out.

At the hospital, Venus lay in bed, trying to fight off the pain. It was difficult to even talk. Another doctor examined her. She and her parents were shocked when the doctor said she would be released in the morning. But the hospital staff had to make room for victims they knew would soon arrive, many fighting for their lives.

After her parents left, she switched from channel to channel with the TV remote. They all featured Mount St. Helens. With pain medication, she often dozed, only to awake to a blaring television with its images of an erupting mountain. The glass IV bottle rattled against its stand throughout the night.

Scymanky kept asking for something to deaden the pain. He lay next to Skorohodoff in the Longview hospital. He could hear phones ringing over the sound of policemen's walkie-talkies. Physicians, nurses, respiratory technicians, and others stood around him and Skorohodoff. Nobody seemed to know what to do with them. Their bodies were covered with soot and ash. They had extensive burns on the back, arms, part of the face, and the top part of the torso. They had been exposed to ash inhalation for seven hours.

The group of people suddenly parted. Emergency medical technicians and a flight nurse from Emanuel Hospital Medical Center in Portland had arrived. They were part of the hospital's Life Flight helicopter team. They quickly took control of the situation. First they had to decide if the men could handle the flight to Portland. What worried the flight nurse more than the burns was the ash. "With that amount of ash on their bodies, I wonder how much they inhaled?" he said. "I'm afraid that once they start on IVs that their windpipes and trachea might swell shut."

The flight nurse tried to talk to Skorohodoff but could not understand the Russian's halting English spoken between moans. The nurse kept repeating the words "pain" and "airway" as he gestured to Skorohodoff. The decision on whether the two men were in a condition to fly could be a life-or-death one.

The flight nurse checked both men's blood pressure, respiratory status, and level of consciousness. Skorohodoff was in the more serious condition, he said. He was not going to place a tracheal tube in the Russian's throat. "I think they are stable enough to fly," he said. "Let's see what we can do about getting these guys out of here." The EMTs placed the two men on stretchers.

It looked like a parade on the sidewalk leading to the helipad. Nurses, physicians, technicians, and Scymanky and Skorohodoff were strung out as they made their way between the solemn onlookers who lined both sides of the sidewalk. Cameras flashed at them. The flight nurse walked with his hand on Skorohodoff's chest to check his breathing. The nurse kept turning around to make sure that the aide who carried the IV bag connected to Skorohodoff stayed up with them.

On the trip to Portland the flight nurse kept monitoring IV lines, catheters, and urine output, reporting his findings over the radio to a physician back at Emanuel. Skorohodoff kept repeating how cold he was. Scymanky kept asking for something for the pain. "I know you're hurting," the flight nurse told him. "But it will be a lot better if you can wait until we get to the burn center." He tried to explain that the hospital's burn center would have a better chance of prescribing treatment if Scymanky was not drugged. But Scymanky in his pain was not sure he understood.

They unloaded Skorohodoff first and then Scymanky. With IV bags attached, they began the descent down the winding ramp from the helipad. A mob of journalists was on the scene. Everywhere there were flashes and the strong lights of TV cameras.

A doctor accompanied the injured men on an elevator. He was Philip Parshley, director of the Oregon Burn Center at Emanuel, the only such center in Oregon and one of only a handful in the United States outfitted to treat severely burned patients. His training as first a physician, then a surgeon and burn specialist, had equipped him well to head such a center. He was a driven man who led the campaign to establish the center in 1974 and hovered over it like a father. The center had special beds, treatment rooms, tubs, heat shields, and other devices for treating the burned.

Parshley expected a great deal from the center's nurses but he also could be protective. On the national average, burn nurses lasted no more than two years in their highly stressful jobs. His lasted much longer. They were like a family and they worked well together. Now the center with its team of physicians, burn nurses, social workers, dietitians, psychologists, and occupational, physical, and respiratory therapists was recognized as one of the best burn centers in the United States. If Scymanky and Skorohodoff were to have a chance at life, they could have been brought to no better place.

Parshley saw that the burns were bad. Most burn victims are given some type of treatment within an hour after their injury. Skorohodoff and

Scymanky had received none for over seven hours. But the doctor was more concerned about the damage from ash inhalation.

The men were taken to private rooms. A nurse undressed Scymanky. She expressed surprise there were no burn holes in his tattered clothing. His body had been burned as if someone had taken a hot iron and pressed it against his clothing, burning the skin beneath but leaving no mark on the cloth. The ash was different from what she expected. It was like dark beach sand and hard to get off. His burns were similar to ones she had seen on patients burned by water.

The volcanic ash that fell from his clothing made the linoleum floor slippery. She warned other nurses coming into the room to be careful.

A nurse who introduced herself as Melissa Metcaff came to his bed and began asking questions. The National Guard was trying to find the other members of his crew and needed more information. He answered her in detail. Then she checked the extent of his burns. He was burned on 47 percent of his body. She put a dressing on his burns, made sure his IVs were running properly, and checked his vital signs. She gave him medication for the pain.

Throughout her shift Metcaff monitored Scymanky's breathing and vital signs. She continued to talk to him, orienting him to where he was and what time it was.

The National Guard pilot shook his head and looked again. A man was riding an uprooted fir tree floating lengthwise in the massive mudflow below. Captain Darald Stebner flew his helicopter lower. He brought one skid of his aircraft to the floating tree, at the same time watching to avoid logs, house parts, and other debris. Stebner tried to keep pace with the moving log without losing control of the chopper. He could see that the man on the log was badly burned.

A crewman leaned out with a hand extended. Jose Dias rose up, but when he grabbed the hand a jolt of static electricity threw him back onto the tree and the rescuer back into the cockpit. Stebner pulled the chopper up. He turned on his FM radio in the hope it would defuse the static. He made another pass and the crewman leaned out again and this time managed to pull Dias aboard. Dias had survived the violent trip that took him miles from where he was swept up into the mudflow while trying to escape the mountain.

Dias was taken to Emanuel Hospital, where doctors immediately placed a tracheal tube down his throat to aid breathing. Nurses found a wallet in his back pocket that contained several pieces of identification with the name Raymond Casillias. He was admitted as Raymond Casillias.

Don Swanson finally had time to call his wife, who was at their home in Cupertino, California. "Barbara, I'm okay, I'm okay," he blurted out. There was a long silence. She had not yet heard about the eruption.

At his apartment in Seattle's University District, the forestry student Peter Frenzen gradually learned more from television news reports. It became apparent it was a good thing he had not traveled to St. Helens after hearing of the eruption. People of the area were dealing with unprecedented destruction.

He learned that nearly ninety thousand acres of trees, some five hundred years old, had been mowed down. More than thirty thousand acres of forest were glowing with fires.

A series of mudflows had rolled over hills 250 feet high and sloshed 360 feet up the sides of valleys along the upper Toutle River. Eight of the ten bridges on the river had been yanked from their foundations to be swept along with all manner of debris. A fully loaded logging truck was reported moving down the river upright as if it were being driven on a liquid highway. Spirit Lake was thought to have been completely buried under a mass of bubbling gas and mud over forty feet deep. One television reporter interviewed a pilot who said, "It's just gone."

No one knew how many people were dead. More than two thousand people had been evacuated from homes near the mountain. Emergency rooms were jammed. Hundreds of emergency vehicles were disabled by the ash. Thousands of motorists were stranded.

The fifteen-mile-high plume of the volcano drifted toward Eastern Washington, dropping ash like a gray snowstorm, plunging cities into midday darkness. Lack of visibility caused a sixteen-car accident near Spokane. Every major highway in the eastern part of the state was shut down. More than two hundred people turned up at Yakima's hospital with breathing difficulties.

Frenzen still wanted to go to Mount St. Helens to study the mudflows. It was just a matter of when.

Donna Parker did not stay up to see the last newscast that night. After gazing in astonishment at the ash plume through the picture window at her mother's house, she returned home. She had to get up early the next day to go to her job at Tektronix in Wilsonville, Oregon. By late evening, reports on television confirmed that nine people had been killed. The thought of death, grieving families, and funerals haunted her restless sleep.

He couldn't believe how calm the newsroom seemed that evening. Even with journalists from the around the nation wanting to use their facilities, everyone on the Longview *Daily News* staff just did their jobs in a quiet, deliberate manner. The catastrophic flood that reporter Donna duBeth had predicted had not materialized. Even so, the adrenaline flowed and Andre Stepankowsky was wired.

After dinner he was given another assignment. The eruption had sent mud, logs, ice, and debris down the Toutle River and into the Cowlitz River. Water was rising rapidly to the twenty-one-foot flood stage. Officials were afraid dikes would break, sending cascading waters through an area known as Lexington, a community of housing developments and mobile home parks.

Stepankowsky drove to the area two miles north of Kelso, a city adjoining Longview. The main highway to the area had been closed. Red emergency lights spun in the darkness. Residents lined the banks of the Cowlitz. They seemed not so much in a state of fear as stunned disbelief as they watched homes float past.

About midnight the river was within three feet of the top of some dikes and rising. Officials ordered an evacuation. Sheriff's deputies roused people from their beds and from behind television sets. Fire engines wailed, going up and down the streets.

Not everyone was going to leave on his own. The owner of an apartment complex on the river joked, "I own these apartments, so I guess the captain has to go down with his ship." A tenant standing with her belongings chimed in, "But the crew is ready to jump ship." Nearly five hundred families were evacuated.

Stepankowsky wasn't tired. He was still hyped-up by the day's events. He watched a barge with a grapple try to pull logs out of the river. It was eerie to see the barge's huge searchlight sweep across the darkened river with all its devastation. Several miles below Lexington, the Cowlitz poured into the Columbia River, where a logjam nearly half a mile wide and twenty miles long now drifted toward the Pacific Ocean.

Helen Scymanky made a big dinner for Jim. She expected him to return home at any moment on that Saturday night. Whenever he was late or expected to spend another day working in the forest, he called. But this time the call never came.

After she and the children went to Mass the next morning, she turned on the radio. Mount St. Helens had exploded. At first her body went numb except for a tightness in her stomach. But as she kept listening and heard there were few casualties and all the Weyerhaeuser workers had been evacuated safely, she felt relieved. "No big deal. He'll be home this afternoon," she thought.

As the day wore on, she began to worry. She postponed a shopping trip. In her gut she felt something was wrong. Waiting by the phone, though, she tried to convince herself that all this worry was foolishness. Jim would soon call to say he had been delayed in traffic.

By dinnertime she decided she had to get some things from the store. She asked her eleven-year-old son Danny to listen for the phone. When she returned she saw Danny standing in the driveway with tears streaming down his cheeks. His father had been burned by the volcano and was being taken to a hospital, said the lady who had called. He handed his mother a piece of paper with the name Emanuel Hospital written on it and a phone number.

It was only when she ran into the house that she realized how tightly she had been clutching her keys. She flung them to the floor. She started crying and told herself over and over, "It can't be true. My Jim can't be hurt." She called Jim's father. With a calm voice he settled her down and told her he would contact the hospital and call back.

As she waited she got ready to go to the hospital. The phone rang. Jim's father said the doctors recommended they not go to the hospital that night. Jim's sister came over and helped her comfort the four children before Helen put them to bed. She tried to get some rest herself but kept seeing a mountain blow up, over and over again.

CHAPTER 5

May 19–23

THE SCIENTISTS AND TECHNICIANS were stunned by data they saw on Monday, the day after the eruption. The information came from a St. Helens flyover on Friday of a Convair 580 plane with instruments that could detect hot spots indicating molten rock, or magma, near the surface.

The flyover was part of the federal government's planning for storage of nuclear reactor waste in rock vaults deep beneath the mountain. Normally the data would have been analyzed the next day, but this was put off because it was a Saturday and would involve overtime pay for workers of the private contractor doing the work.

Now the science professionals pored over the data that gave startling new details about magma that had been deforming the north flank of the mountain, creating a four-hundred-foot bulge. The magma was at such shallow depth that it was heating surface rocks and lubricating the bulge with melted water, the data showed. The new information also showed that the bulge was dangerously loose and that a slight tremor would send it cascading down the mountain. If this information had been in hand that Saturday, there would have been time to evacuate people on or near St. Helens.

From a helicopter the day after the eruption, Don Swanson saw a sea of gray ash sweeping to the horizon in every direction. Puffs of white steam surged from vents on the mountain. Clouds of mist hung in the air. New waterfalls cascaded over bare rock ridges. Rivulets of steaming water branched down the side of the mountain to join into wide ribbons of brown, wiggling through the gray in new riverbeds carved from the earth. Huge chunks of ice buried beneath hot ash exploded, sending geysers four hundred feet into the air and leaving the land pocked with craters more than a hundred feet wide. A mass of mud bubbled beneath thousands of stripped, fallen trees where Spirit Lake had been. This could have been what the land looked like at the dawn of time, Swanson thought as they continued their flight.

For a long time Swanson and two other scientists in the helicopter looked for David Johnston's camp at Coldwater II. They found nothing but devastation. One of the scientists was Peter Lipman, who had taught Johnston and looked on him as a foster son.

Plumes of smoke rose from the mountain's horseshoe-shaped crater. The opening at the mouth of the two-mile-long crater was nearly a mile wide. The scientists could tell by the size of the hole that there had been a huge landslide or series of landslides. Swanson had been wrong about there being lava flows. There had been too much water in the blasting magma.

Clouds soon enveloped the helicopter. The pilot decided to descend. They crested out on top of a ridge and broke through the clouds. What they saw astonished them. The immense debris avalanche had ridden up and over the ridge and down into the next valley. Millions of tons of debris had flowed a thousand feet *uphill*.

Everybody was talking about the eruption when Donna Parker went to work on Monday. That morning she got a call from her mother. Her brother Billy had not shown up for work at the Pacific Northwest Bell office in Lake Oswego, outside of Portland. One of Billy's coworkers thought Billy and his wife might have been on the mountain. Donna couldn't imagine where they were. She knew they couldn't be on Mount St. Helens because authorities had the area blocked off. But deep down, she could no longer be sure.

The burn center was almost overwhelmed when Leonty Skorohodoff's family arrived with uncles, aunts, cousins, children, wife, and the mother who led them all. The women were dressed in bandannas, aprons, and full, wide skirts. The men with their full beards wore pantaloons, tunics, boots, and peasant caps. A palette of bright yellows, reds, violets, and purples swirled around the nurses as the relatives all seemed to speak at once. The strange language baffled the nurses, who got through to the family that they needed to have a single spokesperson. They selected Skorohodoff's wife, Tanya, because she spoke English. But the nurses soon discovered it was the mother who did most of the talking through her daughter-in-law.

The tiny Russian woman with the weathered and wrinkled face had bright blue eyes that could turn very hard when one of her demands was challenged. The rest of the family followed her orders without question. She didn't hesitate to order the nurses around.

Helen Scymanky and Jim's relatives arrived a short time later. She went up to Tanya first. Often they had celebrated together with their husbands and families when a job was completed. "I can't believe it," Tanya said. But Helen found she could only stare at the woman wearing the colorful native dress and bandanna. Words would not come. She turned and walked away.

Scymanky's father, mother, and two sisters joined Helen to listen to the doctor's prognosis. "I have a feeling that Jim is going to make it," Parshley told them. But it would be a long and difficult road. There could be much suffering for the patient and those closest to him. Scymanky's father and Helen prepared to go in to see Jim. After they walked through the burn center's yellow double doors and into the restricted area, they washed their hands and put on gowns. But nothing could have prepared Helen for the shock. It was not so much the blackened skin swelling on Jim's body as it was the excruciating pain he was in. He looked up at her as she tried to control her emotions. "Why did God do this to me, Helen?" he mumbled over and over through a morphine haze.

Andre Stepankowsky couldn't sleep. He could not get visions of the day's destruction out of his mind. He got up early the next morning and arrived at the newspaper office before 7 a.m. Once more he went to Lexington. The dikes had held. But other areas were not so lucky.

Green Acres was a subdivision on the Cowlitz River in the tiny town of Castle Rock, a dozen miles north of Longview. Mud and water had pushed several houses off their foundations. He found a row of four houses sitting in the muck at the edge of the mudflow.

He balanced himself as he walked across sheets of plywood that led to the first house. The plywood led to a ladder going up the side of the house. He climbed the ladder and crawled over the roof to another ladder, which led down to more plywood. At the end of these sheets was a ladder going up the side of the next house. Proceeding in this way, he heard people when he reached the fourth house. He walked in through a sliding glass door. "Hi. I'm Andre Stepankowsky from the *Daily News*, and I'd like to interview you," he said to the people, who were startled as they stopped to look up at him.

The house had been knocked twenty feet off its foundation and tilted at a thirty-degree angle. The fireplace hearth had sunk into the floor. Mud covered the kitchen floor. The left wing of the home had collapsed and the garage had separated from the house. But the family felt fortunate to be alive. "We were lucky the eruption didn't happen in the middle of the night," Jerry Cripe told Stepankowsky.

Cripe and his wife, Karen, and two children had watched the mud and water cresting above the riverbank the night before and decided to leave.

They couldn't believe the damage they found when they returned the next morning. "I love this house," Karen Cripe said as she packed her children's belongings. There was little chance they would return. Insurance would cover only a third of the damage.

Robert Rogers called the Cowlitz County Sheriff's office. He gave township, range, section, and quadrant of where he had seen the station wagon, the photographer, and the pickup across the river the morning of the eruption. Rogers provided times and descriptions, and said he believed that people associated with the vehicles were probably injured or dead. He didn't give his name.

When Roald Reitan woke up, he had no problem remembering where he had been the day before. He felt like he had been beaten. His parents took him to their doctor in Tacoma.

Roald couldn't walk. His knee was crushed. The doctor couldn't tell from x-rays if the ligaments had been torn or just stretched. Because of his abnormally large wounds and how close they were to the bone, surgery was ruled out. There was the potential of staph infection. Bruises and cuts covered his body. The x-rays revealed that miraculously Roald had broken no other bones. His father had gotten out most of the ash when he washed the wounds.

The doctor felt the best course for Roald would be physical therapy, but first he was going to send him to an orthopedist. The doctor was not pleased with the limited care the hospital in Longview had provided. He didn't know when Roald would walk again.

Venus Dergan, too, was struggling. After being discharged from the hospital, she was taken to the family physician in Centralia. He was outraged by what he found. She should never have been released in her condition. She didn't want to be admitted to another hospital. But daily outpatient physical therapy would be required and possibly surgery on her wrist. And her body needed to be cleaned again. The hospital had not removed all the ash from her wounds.

Her mother made a bed for her on the couch where it was thought she would be most comfortable. Her father carried her wherever she needed to go in the house. The only place he could hold her was beneath the armpits. Anyplace else caused pain. It was like she was a big infant, Venus thought. The loss of her independence was beginning to sink in. She was happy to be alive but she couldn't understand why all of this had happened to her. Late in the day she called Roald.

—✳—

As they flew in the helicopter, examining the carcass of the mountain, Don Swanson and the other scientists also looked for survivors. Near Spud Mountain facing St. Helens, they spotted a blue Datsun. Tracks in the ash led a hundred yards away. They could tell that someone had walked away from the car but then run back, judging by the length of the returning strides.

The helicopter tried to land but couldn't because of the whirlwinds of ash it kicked up. The helicopter backed off as a larger military chopper tried to land. It had the same problem. The military helicopter crew dangled a rope to the ground, and a man rappelled down. The man found a body in the car and marked the vehicle with tape indicating a fatality. Swanson wondered whose body it was.

Swanson's helicopter crew then spotted a pickup with a camper shell. The helicopter was able to land because the ash was wet. There was no one in the pickup.

The scientists roughly outlined the blast zone. It was huge, unprecedented. In a fan-shaped area extending eight to twenty miles out and fifteen miles wide, trees stripped of bark lay in orderly rows, sometimes swirling where whorls set up by the blast wind had left them. Splotches of red, rust, orange, off-white, and purple adorned the flow of superheated rock, ash, and gas where various materials had mixed together.

The helicopter swept over Spirit Lake for a closer look. It was not just filled with mud and logs, they determined. There was still a great deal of water beneath the logs. A huge jam of mud and debris had dammed up the outlet. If the dam broke, it would release water and debris that could wipe out the cities below.

At dusk they returned to Vancouver. They had been in the air for twelve hours, ignoring requests from other scientists who wanted their crack at St. Helens in the USGS's only helicopter. The scientists reviewed what they knew. Debris trails and the blast zone indicated that the avalanche with the initial eruption may have been one of the largest in recorded history.

The blast, however, had gone way beyond the avalanche. Intense wind from the eruption had burned and blown down millions of trees. The huge plain of pumice—lightweight, porous volcanic rock—indicated there had been pyroclastic flows known for their superheated rock, gas, and ash. The blast had been accompanied by phreatic eruptions, explosions caused by the pressure of steam from heated underground water. The scientists also found indications of strong magmatic activity—movement of magma, or molten rock. While they now had a general idea of all that had taken place, Swanson knew that the next day it would be time to start working on the details.

With the sun setting as they flew home, they came upon some elk. The animals hobbled among the dead bodies of the rest of the herd, some still standing, mummified into black, carbonized statues.

When he returned to the office, Andre Stepankowsky got into an argument with Donna duBeth. She had entered the newsroom after flying that morning, raised the palms of her hands dramatically, and proclaimed, "Lexington is gone."

"What do you mean gone? I was just out there," he told her. The discussion escalated to a heated argument. He was a novice reporter who had made his share of rookie mistakes. She was a seasoned journalist and one of the paper's lead reporters. But he was not going to let anyone intimidate him when he knew he was right.

Stepankowsky worked on his feature about the couple abandoning the wrecked house at Green Acres. Then he made phone calls for another story. He looked up to see copies of that day's edition being distributed in the newsroom. He grabbed one. The editors had confirmed that Lexington still existed, and his story had run. But in the first pages of the newspaper he could not find the story on his previous day's flight over St. Helens during the eruption. He found it buried on page seven. The story had been edited down to six inches of type appearing in a two-column box headlined, "Plane followed eruption's path."

He had been the first newsman to observe the eruption from the air. But it had not been enough. Stepankowsky did not have time to brood. He was given another assignment.

Longview's water treatment plant was clogged with volcanic silt. Drinking water was being trucked into the city. With water supplies limited, the biggest fear was fire. Neighboring fire departments had put all personnel on duty to help if needed, Stepankowsky was told.

An even worse threat, the reporter learned, was flooding. Scientists had discovered that Spirit Lake still existed. A huge dam of mud and debris was all that held back billions of gallons of water. Officials did not know if the dam was going to hold.

Orders had been issued to prepare for evacuation of the region's largest cities, Longview and Kelso. If the dam broke, floodwaters might destroy both cities and everything between them and Spirit Lake, about forty miles away. Officials would have a three-hour warning for evacuation, Longview police told Stepankowsky.

Even with three shifts and thirty highly trained nurses, the burn center was swamped. Many of the nurses would work a full shift, go home for a short rest, and then return to help provide the high-intensity care required by burn patients.

Jim Scymanky often asked himself what he had done to deserve this. But sometimes he would let his head sink back into the pillow and tell himself how lucky he was to be alive. He knew he wasn't going to die, but there was no escaping the pain. When a nurse changed his dressings for the first time and carefully unwrapped the bandages around his chest, he clinched his teeth and tried not to scream. As she applied cream to his skin, it felt like he was being burned again.

Scymanky from the description the nurses gave, knew the man they called Raymond was Jose Dias. But he had known the man only a short time. He didn't know which name was correct and did not say anything. On television that night, he watched the news attentively but there was nothing about the fourth member of his crew, Evlanty Sharipoff. He asked the nurse. Nothing.

Donna Parker was now fairly certain that her brother Billy and his wife, Jean, were on the mountain. Billy's best friend and coworker, George Gianopoulos, had told her that Billy left a note saying they were going to camp near Mount St. Helens. But even so, Parker tried to reassure herself that they were all right.

Billy was a knowledgeable outdoorsman. He had a camper truck equipped like a second home with stocks of food, pots, pans, guns, fishing poles, climbing equipment, extra clothing, camping gear, first-aid kit, cameras, a radio, water, hidden extra cash—even a metal container for his toothbrush. She figured that Billy and Jean were trapped somewhere. But she wasn't worried about them starving or needing any of the necessities of life.

Monday night she talked on the phone to friends who hunted in the area. They had detailed maps. She sent her son to get the maps. The next morning Parker and her family tried to reach the Forest Service and the Cowlitz County Sheriff's office by phone. All they got were busy signals. "I know how to get through," her brother's boss told them. When he called them back, he was furious. Cowlitz County representatives had been rude to him and even refused to take information about Billy.

Parker skipped work and drove to Portland to meet with Gianopoulos. The note her brother had left told where Billy and Jean would be camping. Parker sat with Gianopoulos and tried to figure out from the maps where her brother might have gone. They looked at the maps for a long time but could not find any place that corresponded with the note.

On the Tuesday after the eruption, ash clouds, steam, and smoke from hundreds of fires kept scientists from getting a view of the crater. But some geologists explored the blast zone. The smell of sulfur and ashes permeated the area. There was no sound of birds, of rushing water, of insects, or even the wind. There was only the whacking sound of helicopter blades and the buzz of search planes. Fifteen people were rescued from the blast zone that day. Billy and Jean Parker were not among them.

The next day Don Swanson and two scientists flew in a helicopter to look at the crater, now visible. They found that the crater had a flat floor at an elevation of approximately 6,000 feet.

They flew on to Spirit Lake. The lake, covered with suspended sediment, was now two hundred feet higher than before the eruption and had shifted position. But it was still a lake. They landed and took measurements. Material near the lake surface recorded a temperature of 92 degrees Fahrenheit, while the ash surrounding it was 120 degrees at a depth of two feet. They heard small explosions and saw clouds of debris floating toward them. They believed the explosions were caused by water coming in contact with hot rock.

The dam of mud and debris rose 150 feet above the lake. They thought the water level had dropped somewhat from heights reported the previous day. The possibility remained that the dam could leak or break, sending flash floods downstream.

The Red Cross office in Longview was overwhelmed with more than thirty-four hundred callers in forty-eight hours. Call-takers often wrote down incomplete information about names, phone numbers, addresses, and descriptions of the missing and their last known whereabouts.

The nurse in the burn center hung up the phone, trying to figure out what was going on. She had just returned from vacation. The center had been receiving mysterious calls. The callers would not identify themselves or leave a message or phone number. They wanted to know if the center had a patient by the name of Jose Dias, and the answer was always no.

Later that day she looked up from reviewing charts to see a man standing before her. "Do you have a patient by the name of Jose Dias," he asked.

The nurse looked at him suspiciously. "No, we don't."

"How about a Spanish-looking man?"

"Not that I know of, but I'll check."

The nurse found a Spanish-looking man in one of the rooms and returned to the station. "Yes, we do have a Spanish-looking man here," she told the visitor. "But I'm sorry, his name is Raymond Casillias." She had no way of knowing the patient was actually Dias from Scymanky's tree-cutting crew. A driver's license and other documents carried by Dias showed the name Raymond Casillias and he had been admitted under that name. Because doctors had placed the tracheal tube down his throat, he had no chance to say anything about the name.

"I'm Raymond's cousin," the visitor said. "I would like to come in and see if that's him."

"That's Jose," the man said when he saw Dias.

The nurse was growing more confused. "Well, what do we call him?"

The man started walking away. "Just call him Raymond," the man said over his shoulder.

Within a day a short, stocky woman, plainly dressed, showed up. The distraught woman spoke only Spanish. The nurse on duty called for an interpreter. The woman said she had traveled by bus from Stockton, California. Dias was her husband, she said. The nurse looked at the woman's left hand. There was no ring, but obviously there was a strong emotional tie between Dias and this tearful woman, so the nurse decided not to press the issue.

The woman, named Maria, had little money. Arrangements were made for her to stay at the hospital's apartment complex. From then on she sat at Dias's side as much as possible.

When Dwight "Rocky" Crandell finished his news conference at the Forest Service building, Robert Rogers approached him. Crandell was one of the chief USGS scientists with the Mount St. Helens team. Rogers showed the scientist a sequence of six slides that captured the initial phases of the May 18 eruption. He hoped the slides would be useful in the study of the eruption and might even help him secure a pass into the Red Zone.

Crandell pulled a magnifying loop from his shirt pocket. He took one of the slides from the holder. "You need to handle these more carefully," the scientist said. "They have already been damaged."

Crandell handed the slides back as though their lack of perfection made them useless. He turned to take a reporter's question. Rogers was miffed. He had taken a chance showing the scientist the slides. He was beginning to believe that the USGS, the Forest Service, and law enforcement officials were in cahoots to keep him from the volcano.

But he wasn't going to be near the volcano anytime soon. The next day he would be on his way to help build a radio station in Park City, Utah. He had met a man in Seattle earlier that year who had done the impossible in Rogers' eyes. He had managed to get a construction permit for a radio station and a license to broadcast without using the services of a consulting engineer or an attorney specializing in broadcast law.

The man had little money. Rogers so admired him that he volunteered himself and a friend to do the work for nothing. He made the offer before Mount St. Helens came alive in March. Tearing himself away from the volcano was going to be one of the hardest things he had ever done, but he had given his word.

At first Venus Dergan was intimidated by her physical therapist. He was the biggest man she had ever seen, and all muscle. She wondered what pain he would put her through. But the man, who looked to be in his thirties, with blond hair, blue eyes, and a handsome face, had a friendly smile. "We're going to be together every day in therapy and everything is going to be okay, Venus," he said in a soothing voice.

He carried her into the whirlpool room in the clinic. "We need to put you in the whirlpool for thirty to forty-five minutes a day," he said. "This is going to be a very slow process. We don't want your wounds to heal up too fast with thick and heavy scabs."

She stared at the whirlpool. "It's going to hurt a little bit at first," he said, "while your wounds are fresh and open. But we need to do it to make sure the wounds stay clean and heal correctly." Venus knew she couldn't undress herself. The man sensed her distress. "It's all right," he kept telling her as he slowly took off her clothes, being careful she was not in pain. Then he picked her up. She thought she had never known anyone so gentle. He placed her into the tub. At first it hurt, but once she became used to it, the water was almost soothing.

Roald, when he called, told her about his own therapy. They had packed ice on his knee, then given it electroshock treatment. He still was unable to walk.

Children with dirty, smudged faces played around the makeshift camp that looked like something out of John Steinbeck's novel of Depression-era migrants, *Grapes of Wrath*. Tents and lean-tos were scattered among aged cars, camper pickups, and larger RVs in Hudson Park on the Oregon side of the border, across from Longview. Baby cribs sat in the afternoon wind beneath the trees. The latest volcano news blared from portable radios as campers read, talked, and joked. The sharp odor of Coleman stoves and smoldering firewood caught in Stepankowsky's nostrils.

The people were all refugees. Some had been flooded out. Others feared being caught in another massive eruption or in the floods should the dam at Spirit Lake give way.

"It's better to be a safe chicken than a dead duck," Darlene Greer of Longview told him. "At least up here we have water," she said.

"We sleep up here better than in town," her husband, Terry, said. "The situation is pretty unpredictable now. They tell you one thing on the radio and then tell you something else." The couple was like many who had flocked together in dozens of camps that had sprouted up in Oregon's Columbia County parks.

But some had chosen a camp on the hills off Oregon Highway 30 overlooking Longview and Kelso. The camp had no running water or toilets. Here he found Theresa Fortenbury of Kelso, holding her four-year-old daughter. It had rained the night before and winds off the Columbia River whipped against them. The mother was worried about the traces of pneumonia her daughter had. "I'm just trying to keep her warm as possible," she told him. "Since the rain quit, it's been better. The hard thing is keeping the kids from getting sick."

She told Stepankowsky that she and her husband would stay at the camp until the uncertainty about Spirit Lake was over. "I never thought I would have to move out of my home because of a volcano," she said.

The young married couple hesitated before Swanson in the USGS office. They were friends of a University of Idaho graduate student, Jim Fitzgerald. Swanson had become a friend and colleague to the young geologist when they both worked on projects in Idaho.

Fitzgerald was now dead, the couple told Swanson. The three of them had come over together from Idaho to view the volcanic activity. On the evening before the May 18 eruption the couple decided to go into town to spend the night while Fitzgerald camped at a place outside the Red Zone. He was caught in the blast the next morning.

Swanson was shocked. He had not known that Fitzgerald was anywhere near the area. He asked the couple to describe Fitzgerald's car. He learned that it was the blue Datsun he spotted from the helicopter the day after the eruption—the car they were unable to land near because of all the ash. He had even taken a photo of it.

Swanson pieced together what had happened. The windshield and grill were blown out by the blast, but Fitzgerald survived. Footprints in the ash indicated Fitzgerald had walked away from the car and the mountain, but then ran back to the vehicle, perhaps fearing the ash cloud moving toward him. An autopsy indicated he died of ash inhalation.

Fitzgerald's ordeal, and the imagined image of the student sitting dead in the car, haunted Swanson. He wondered what his reaction would have been if he had been the one to discover the body of his young friend. Then Swanson thought about how he and the couple from Idaho were alike. They had escaped death by a quirk, a chance decision. He was coming to believe that no one could really plan one's own life.

Dr. Parshley carefully grafted skin onto Scymanky's body where it would serve as a temporary skin closure. But his patient, under anesthesia, was having a bad dream. Scymanky was riding backward on a roller coaster, going down into the bowels of the volcano where a brilliant white glow shone. Metal robots with square heads and bodies and long arms lined the track. The robots swung their arms, hitting his elbows, arms, and sides as he passed. The roller coaster started going faster. The blows from the robots came harder and faster. Scymanky put his hands out against the walls of the volcano to slow the roller coaster. But the white light at the bottom of the volcano grew bigger and brighter.

By Thursday 118 people were listed as missing. Some families had given up on the search-and-rescue teams and were illegally entering the restricted zone to search for relatives. That morning, Donna Parker's mother called. From the tone of her voice, Parker was afraid of what was coming next. "Donna, they got a picture of the truck in the *Oregonian*," her mother said. "It says two people were found dead in it."

"Are you sure it's Billy," Parker asked.

"I don't know."

The picture was a close-up of a truck that looked like Billy's, sitting with ash up to the bottom of its frame, tires half-buried, and the top of the camper shell smashed in. The windows had been blasted out. The grainy photograph showed two figures sitting upright in the front seat. No facial features were distinguishable. She asked her mother to contact the *Oregonian* and find out if the newspaper had any photos that showed the license plate.

A little while later her mother called again. Phone calls had been coming in from friends in Washington, Oregon, and California, anywhere Billy had been with the old truck he loved and kept immaculate. They had seen the photograph in their local newspapers. The *Oregonian* had called back. The newspaper had other photographs of the truck. Parker's half-brother and stepfather went to the newspaper office and looked at the photos. They made a positive identification of Billy's truck from photos that showed the license plate. The *Oregonian* called in the number to the motor vehicle office. The truck belonged to Billy.

That night newscasters told of plans by local, state, and national officials to turn Mount St. Helens and the eruption area into a huge public attraction. "I think we'll get millions of tourists this summer," said Dave Seesholtz, who had been sent by the Forest Service to scout potential visitor center sites and begin negotiations for use of private land. Some people envisioned a Disneyland-type attraction with aerial trams, tourist centers with talking, animated characters and interactive displays, and reality rides to bring home the experience of being caught in a volcanic eruption. President Jimmy Carter, who visited the disaster area, thought Mount St. Helens could outdraw the Grand Canyon as the nation's number one natural attraction.

Washington governor Dixy Lee Ray took the opportunity to charge that the volcano's victims were basically thrill seekers who had violated her Red Zone restrictions and brought on their own deaths. Carter took his cue from Ray. "One of the reasons for the loss of life that has occurred is that tourists and other interested people, curious people, refused to comply with the directives issued by the governor, by the local sheriff, the State Patrol, and others," the president told reporters. "They slipped around highway barricades and into the dangerous area when it was well known to be very dangerous. There has been a substantial loss of life."

Donna Parker couldn't understand why these people were saying such things. Her brother never trespassed or ignored warnings. His hunting buddies teased him because he refused even to cross private land to get to game. It just didn't make any sense.

Every day Peter Frenzen bought all the newspapers. With fellow forestry students, he studied the photographs of Mount St. Helens. "There, I think that's one," he said one day, pointing to what looked like a ribbon coming down the side of the mountain. "That looks like a mudflow."

He realized it would be some time before any scientists other than USGS geologists were authorized to study the volcano and its aftereffects.

However, there had been unauthorized scientific forays. One scientist had parachuted into mud-covered Spirit Lake for water samples.

As Frenzen prepared for his summer project on Mount Rainier, he could not get St. Helens out of his mind. He decided to send an amended application to the National Science Foundation to see if he could get support for a trip into the eruption area. Sanction by the NSF could help him gain needed approvals once the area was opened to more scientists.

Stepankowsky found the family exhausted, and he could relate to that. When the reporter reached a hundred hours of work that week, he quit counting. But now his weariness seemed a small thing compared with what Robert Manthe and his family had experienced.

Manthe was a popular local dentist. His dedication to his patients had cost him many holidays with his family. He treated patients without consideration of their finances. He was not a rich man. With a lot of his own sweat and toil, he had seen his 5,500-square-foot dream home and landscaped grounds become a reality over a ten-year period. It had taken him, his wife, and two children years to clear the land. He had built the barn with his own hands. The family had furnished the house one piece at a time.

Now everything was a total loss. The barn was no more. The manicured grounds were buried under fifteen feet of pumice, trees, and mud. The house stood with crushed walls and damage beyond repair. Federal flood insurance would not come close to covering their loss. The only way to retrieve belongings was by helicopter.

At sunrise that morning, members of the family had helicoptered into their former home. Ten pickups driven by friends waited on the closest solid ground to take retrieved possessions. The family dug through mud to find their treasures. The helicopter ferried streams of goods to the waiting trucks—silverware, dressers, desks, game trophies, guns, family photographs. Manthe's wife monitored the radio in case the mud dam at Spirit Lake broke.

All day they dug through the mud, becoming elated as they unearthed precious belongings. "I might not sell this land. I had boys and they will have boys. No, I'm not dead. I'll come back," Manthe said as he pulled family photos from the mud and fought to hold back the tears.

Stepankowsky interviewed the family at their rented home as they unpacked boxes and tried to deal with their loss. The family was receiving many calls of assistance. "Those calls mean a great deal to me," Manthe said. "Everything happens for a purpose. It shows what wonderful friends we have." Manthe's son Chris added, "It makes the family tighter." Hearing this, Manthe cried again.

Stepankowsky found himself deeply touched by the family's resolve and humanity. As he went back to the office, he hoped he could find the words to do justice to what he had seen and experienced.

"What was splendor was now immersed in a sea of mud," he wrote. "The mudslide that gobbled up the dream home of Castle Rock dentist Robert Manthe wiped out ten years of hard work Sunday. But again, like the sea, the mud coughed up a fortune more valuable than the one it destroyed."

The story and the work that week had left Stepankowsky emotionally and physically drained. But he wasn't going to get much rest that Sunday, his day off. He was going to be moving from his tiny apartment to a larger one. He was to be married within a month.

The inside of her brother's house looked like the home of people who have left for just the weekend. Donna Parker noticed a few coffee cups on the kitchen counter, a couple of to-do notes, clothes draped across a chair in the bedroom.

Parker was looking for Billy's written instructions for his funeral. There had been much death in the family, including Billy's only child, a son who died in an auto accident four years before. After that her brother sat down and wrote extensive directions for his own funeral and burial.

She found the document. Billy wanted a large funeral in the family tradition with song, prayer, and remembrances. That Sunday the owner of the mortuary was scheduled to retrieve the remains of Billy and his wife, Jean. The only problem was, nobody was quite sure yet where the bodies had been taken.

He looked up from his hospital bed to see a priest standing over him. Jim Scymanky recoiled. He hadn't asked for a priest. Didn't they know he was going to be all right now? He could feel it. But then he wondered if they knew something he didn't.

CHAPTER 6

May 24 – June 1

EARTHQUAKES SHOOK the land around the mountain. The helicopter sat nearby, its blades whirling, ready to whisk the scientists away at the first hint of trouble. No one knew when the mountain would erupt again. The geodetic network Don Swanson had established to measure ground deformation with a laser and targets had been destroyed. While it was still too dangerous to place any targets in the crater, Swanson had permission to make new baseline readings on the few targets that were left at what was called Butte Camp. The location was only two miles southwest of the steaming crater, close enough that an eruption in this direction could kill him and the scientists with him.

It was hard for him to believe it was not quite a week since the eruption and David Johnston's death. Working eighteen to twenty hours a day had not left him much time to grieve. As he worked at Butte Camp, he knew he could not allow fear to grip him or cloud his judgment. At the same time, he did not take the threat lightly. He knew well St. Helens' past.

Though he lived in the shadow of the mountain while growing up, Swanson often told people he didn't even realize it was a volcano until he was in high school. After that his fascination with everything to do with Mount St. Helens never left him.

The volcano, he learned, had been named by British explorer George Vancouver in 1792 after he and his crew spotted the mountain from the deck of his ship the *Discovery*. The politically correct Vancouver christened the peak Mount St. Helens after Britain's ambassador to Spain, Alleyne Fitzherbert, also known as Baron St. Helens. At that time, Spain still claimed this part of the world.

Indians who had lived in the area for at least ten thousand years had other names for the occasionally smoke-belching behemoth. To the Salish north and west of the mountain, the Klickitats to the south and east, and the Cowlitz to the southwest, the mountain was Lawelata (One from Whom Smoke Comes), Loo-wit (Keeper of the Fire), and Tah-one-lat-clah (Fire Mountain).

A young brave's initiation into manhood often required a hike to timberline where he would walk back and forth in an attempt at communion with the Great Spirit who lived on the mountain. Hiking to the peak itself was taboo and would raise the wrath of spirits guarding their domain. To the Cowlitz, the mountain was dreaded for its race of cannibals and for the pristine blue-green lake at its base where evil spirits and souls of the most wicked were banished into a native purgatory as they donned shrouds of clouds to drift for eternity.

St. Helens also played a more heroic and benign role in native legend. St. Helens was considered the most sacred of the area's volcanoes, which also include Mount Hood, Mount Adams, and Mount Rainier. In a legend central in one way or another to all the tribes' mythology, brothers Hood and Adams fight over a lovely maiden represented as St. Helens. As they fight, thunderbolts crash down, fire and smoke fill the sky, rivers turn black, and all becomes dark as animals flee in terror. The maiden St. Helens tries to stop the battling brothers but fails until all three mountains fall in exhaustion to the earth. The Great Spirit arrives and surveys the destruction with anger that does not extend to St. Helens. Her reward for trying to preserve the tranquillity of the land is an eternal life as a beautiful and shapely mountain, a young maiden with an ancient and all-knowing soul who has the ability to continually renew herself after the periodic eruptions that only temporarily mar her stunning beauty.

In a way the legend was accurate about the mountain's history and cycles of destruction and renewal. In the 1960s scientists found evidence that Mount Hood and Mount Adams had, indeed, erupted at about the same time along with St. Helens in ancient history, raining massive destruction upon the land. St. Helens had experienced at least three long cycles of devastation where its famous cone was destroyed and then built up again over the ages.

The peak blown away on May 18 was twenty-five hundred years old, a short span in geological time. The mountain itself is the youngest of the Cascade volcanoes, only forty thousand to fifty thousand years old. Tracing back over four thousand years, scientists discovered that the Cascade Range was one of the most active in the world, averaging one to two eruptions per century with St. Helens leading the pack. Recurrence cycles for St. Helens were somewhat irregular, with eruptions occurring every 40 to 140 years.

The last span of eruptive activity started in 1800 and lasted for fifty-seven years, with intermittent periods of calm. But the initial destruction of the peak must not have been too great because William Clark of the 1805 Lewis and Clark expedition wrote in his leather-bound journal that he had just spotted "the most noble looking object in nature," Mount St. Helens.

Just before the end of the nineteenth-century eruptive phase, Thomas J. Dryer, editor of the Portland *Oregonian*, in 1853 became the first person documented to have climbed to the top of the peak. For a man whose occupation was steeped in words, Dryer found himself almost speechless in detailing the experience.

He wrote that the mountain was "sublimely grand and impossible to describe. . . . It would be futile to attempt to give our readers a correct idea of the appearance of the vast extent of the country visible from the top of the mountain." But he tried anyway: "The ocean, distant over one hundred miles, was plainly seen. The whole Coast and Cascade ranges of mountains could be plainly traced with the naked eye. The snow-covered peaks of Mts. Hood, Rainier, and two others seemed close by."

If Dryer's powers of observation were somewhat dimmed, it might have been because of his nose bleeding, his ears ringing, and his lungs gasping for breath as a result of the climb and high altitude, which he explained in his story for the newspaper.

Dryer's climb blazed a trail for the exploitation that followed. Commercial timber cutting began in the early 1880s in the mountain's lowlands. By 1895 a railroad hauled logs from eight logging camps on or near the mountain. In 1901 a wagon road to Spirit Lake was smoothed and widened for use in nearby copper mining on St. Helens. At the same time climbing parties of fifty or more ascended the mountain from the nearby urban areas of Portland and Seattle.

A YMCA camp rose on the shores of Spirit Lake in 1909 and the laughter of children fishing, hiking, and paddling canoes became a harbinger of the recreation mecca the mountain was to become. In one summer season alone in 1938, some twenty-six thousand people were counted visiting St. Helens.

The mountain became an emblem for nature and the outdoor life in the Northwest with its breathtaking beauty emblazoned in image after image on calendars, postcards, picture books, and newspaper recreation sections. It became an icon for those who lived in the region, becoming steeped in modern myths to rival those of the Indians. Bigfoot, also known as Sasquatch, was supposedly born here and still roamed its environs, according to a legend that grew to sweep the nation. It was the alleged resting place of hijacker D. B. Cooper, last seen bailing out of the belly of a jet over or near the mountain along with bags stuffed with currency.

Commercial exploitation kept pace with recreation. By 1949 the mountain had over six hundred miles of logging roads. By 1980 St. Helens had thirty-seven hundred miles of road and little old-growth timber left. From the air its picture-perfect beauty was now lost in checkerboard squares of clear-cuts. The clear-cuts reflected the complicated holdings of public and private lands established by Congress in the late nineteenth century. Congress encouraged the advancement of railroads in the West by providing title to specified parcels of land every few miles in exchange for laying track. Much of the railroad land had been sold to Weyerhaeuser Timber Company, which first came to log the mountain in the 1920s. By 1970 almost all the forests on state and private lands in the area had been logged. The vast majority of old-growth timber still standing was in public areas managed by the Forest Service. That agency launched an aggressive program of allowing road building and logging that even reached high upon the sides of the mountain where regeneration would be impossible.

In the years just before the May 18 eruption, environmentalists launched an effort to save the small amount of old-growth forest left on both public and private lands. Weyerhaeuser reacted by increasing the rate and amount of its cuts beyond a sustainable yield and initiating its own lobbying and public relations campaign, branding the environmentalists as killers of the local economy and jobs.

Scientists also were attracted to Mount St. Helens. In the mid 1950s two scientists whose lives were still intertwined with the mountain, Dwight "Rocky" Crandell and Donal Mullineaux, began a series of explorations that greatly advanced the science of volcanology. They borrowed techniques from other fields to conduct, like detectives, one of the most exhaustive inquiries ever made into a volcano's history.

Using the discovery that ash from every eruption throughout history left a unique fingerprint of its composition, the two scientists began studying ash and soil layers that they found most visible in road cuts and river canyons on the mountain. They refined a technique for differentiating the debris from different eruptions to the point it could be done in the field with equipment packed in a knapsack. They also carbon-dated chips of burned wood found in the layers to date eruptions.

Reading pyrotechnics became the specialty of Mullineaux while Crandell concentrated on divining the history that lahars, or mudflows, provided in putting together the story of the volcano's past. The geological background of the area became apparent as it was discovered that landslides from previous eruptions had dammed rivers that created lakes, that plains were really the tops of two-hundred-foot-deep lahars, and that canyons were the result of rushing waters from melting glaciers on the mountain's sides. The two scientists found evidence of previous eruptions spewing ash that landed in substantial amounts as far away as Eastern Canada. They were able to approximate the dates of past eruptions and their destruction over the millennia.

In 1978 these pioneering scientists published what became known as the Blue Book on Mount St. Helens. Considering past eruption cycles, they believed an eruption on Mount St. Helens was overdue. An eruption could occur "perhaps even before the end of this century," the book warned. A "Notice of Potential Hazard" that was included with the Blue Book and a cover letter was sent to government officials in the state and also ended up in newsrooms.

Officials who didn't bother to read the entire report assumed the worst and thought Mount St. Helens could erupt any day. News stories with incomplete information stirred the fears of local citizens. A special meeting was called in the state's capital, Olympia, just days after release of the report to prepare for the coming doom and destruction. The USGS rushed to alert state government that an eruption was not imminent but that it would be wise to make preparations. The result was that some in state government became skeptical of scientific warnings, and the various departments involved with emergency preparedness did nothing.

To Swanson and other scientists, the incident demonstrated that long-range, open-ended predictions were of little practical value in preparing a population for an eruption. That was one of the realizations that drove him now.

The next day, exactly one week after the first eruption, the mountain erupted again, sending billowing clouds of ash west and south. Wet ash rained down just after dawn that Sunday, throwing the land back into darkness. The wet ash was a perfect conductor of electricity. Sparks of electricity arced across power lines and short-circuited power stations. Tens of thousands of people lost power in the biggest outage in Cowlitz County history. The cities of Longview and Kelso went dark. Train traffic halted as rail switches circuited out. Semitrucks on the nearby I-5 freeway left rooster tails of ash a half-mile long. Vehicles piled up. Police cruisers and ambulances on the way to the accidents lost power as ash clogged engines.

It was a terrible day to be moving to his new apartment, but Andre Stepankowsky had little choice. He had finished packing in the darkened apartment he was leaving. The ash had gotten into everything, including the meager possessions he had thrown into a couple of boxes. The ash had even settled into the few armloads of clothes he owned.

As he drove to his new apartment, he could not escape an eerie feeling. The streets, covered with a slick ash paste, were deserted. Residents had shuttered themselves in their homes or became trapped when they couldn't get their cars out of garages with electric doors. Block after block of downtown buildings remained dark and desolate, shrouded in gray ash. It was as if he had been plunked down in the middle of a *Twilight Zone* episode.

The lethal eruption of the previous Sunday had, for the most part, sent ash north and east. Longview had been little affected by falling ash. But Spokane, where his fiancée, Paula, and her family lived, had been hard hit. Paula's mother had been at an auction May 18 when the huge ash cloud descended. People kept leaving the auction as she nabbed one great buy after another, until she was the only one left. She wandered outside. It was dark in the middle of the afternoon. She asked a passerby what was going on. "Haven't you heard? Mount St. Helens erupted," he said. "No, I hadn't heard. But what's it doing over here?" Stepankowsky loved this story. With all the death and destruction, he had to keep some sense of humor.

Yet now with ash in his hair, on his face, and coating the clothes he wore, he could find nothing funny about the situation. Later, rainfall started mixing with the falling ash, and drops of mud poured from the sky. It took only sixteen-hundredths of an inch of rain to send the mud-engorged Cowlitz River to a twenty-four-foot height, where it lapped at the top of dikes in nearby Castle Rock. By the afternoon in Longview and Kelso, a few gas stations opened up. Word got out. Hundreds of vehicles joined lines at the pumps. Their occupants were leaving town.

Everhart Funeral Home in Molalla had handled the arrangements after most of the deaths in Donna Parker's family. The funeral director—first the father, then later his son Michael—had comforted the bereaved while meticulously attending to every detail of the family's large and tradition-laden funerals. Parker trusted the funeral director implicitly.

Michael Everhart called her that Sunday to let her know he would not yet be able to pick up the bodies of Billy and Jean. He still didn't know where they were. He had thought they were in Toledo, Washington, where a temporary morgue was set up at the airport. But he learned they were no longer there.

Then he told her what he had found out about the condition of the bodies. The coroner had determined that her brother died from a large rock that smashed through the camper of his pickup and hit him in the back of the head. Ash was found clogged in his wife's air passageway, but none in her lungs, which meant she must have already been unconscious or dead when the ash wave hit. Both bodies had been cooked. Their internal fluids had evaporated, leaving mummified carbon corpses that, while perfectly formed, flaked away with a touch.

To preserve the bodies, they had been frozen. "Your brother has been cooked and frozen, Donna. If we have a funeral service, I have no way to keep him from thawing. There would probably be an odor," he said. He advised her to have just a graveside service.

"But Billy's will specified he wanted a big funeral," she said. "He even had his pallbearers picked out."

Scientists determined that the mushroom cloud from the second eruption reached forty thousand feet before dispersing around the globe to join air streams already carrying ash residue from the explosion the Sunday before. Geologists had just taken the first steps in trying to piece together what happened during the May 18 eruption. Plans were being made to analyze scores of substances, from rock to melted plastic taillights, to help determine the heat, velocity, and direction of the first eruption.

The second eruption only reinforced what Don Swanson already knew. He sat listening at the morning meeting of USGS scientists. Gas tests showed the volcano emitting the highest levels of sulfur dioxide it had ever recorded, and the hydrogen sulfide levels were just as high. St. Helens was now a virtual blast furnace. It was going to be some time before he could get into the crater to take measurements. Some of his colleagues argued that work inside the crater should be abandoned for the foreseeable future because it was just too dangerous. The death of another scientist could severely cripple the work of the USGS.

Yet Swanson was not about to lose one of the first opportunities scientists had to study a living stratovolcano, the kind that most threatened human life. In the past, volcanoes of this type usually erupted in remote areas of the world, with scientists often not arriving until weeks later and without the technology for a thorough study. The volcanoes he studied in Hawaii were shield volcanoes, flatter than the steep-sided stratovolcanoes. The shield volcanoes emitted spectacular lava flows and fountains but not violent, supersonic blasts like St. Helens.

Whenever St. Helens died down enough that he was not chancing certain death, he was going into the crater, Swanson promised himself. He needed to see if he could find some way to help predict future eruptions. He owed it to David Johnston and the others who had died.

The crane lowered Jim Scymanky into the steel tank filled with red foam. He tried not to tense in anticipation of the pain. This was a new experience. He was going to be tubbed. Slowly the red liquid rose around his body. It was part of his Betadine world where he was scrubbed with the iodine-based disinfectant, sprayed with it, and now bathed in it. He relaxed his body. This was not going to be as bad as he thought. But when they brought him out of the Betadine, he was freezing. They quickly dried him off and placed his body in a sling where he was weighed as he froze again.

Scymanky looked up as they wheeled his gurney back to his room. He thought he caught a glimpse of Skorohodoff. But it happened so fast, he couldn't be sure. The man hanging by his arms and legs had been a blur.

Day after day the gray ash rained down. Six inches of ash lay on the ground. The mountain, with its eruptions, continued to haunt her. Venus Dergan leaned up over the back of the couch and looked out the window to watch the ash flutter from the sky like gray-white snow. The television soap opera she was becoming addicted to, *General Hospital*, droned on in the background.

Her initial euphoria at still being alive had worn off. "Why me?" she kept asking herself. She and Roald had not taken chances. They had just gone camping in an area they thought perfectly safe, and they had not heard any warning from the authorities. Now she had lost one of her most prized possessions, her independence. What was going to happen to her job and the house she rented? She had just gotten started on her own in life, and now this.

"Goddamn ash," her father said as he came in from outside. "It's screwing up my engine."

Venus looked over at him. She loved him and what he was doing for her. But it was hard when you couldn't even go to the bathroom by yourself. She looked back at the darkened sky and fought off the tears.

Often the three women sat together in the waiting room. Besides her native Russian, Tanya spoke English and Spanish. Maria only spoke Spanish and Helen only knew English. When Tanya was there, which was most of the time, she could translate. But when she wasn't, Helen had to communicate with Maria through gestures and sign language. The three women often discussed how their men were doing and the doctor's latest prognosis. On the good days they shared hope. On the bad days they tried to comfort one another. But mostly they shared the silence of waiting.

The nurses were surprised one day to see a tall Anglo priest who asked for Maria and wanted permission to see Jose Dias. When the two entered Dias's room, the priest spoke to him in Spanish. "I am a priest. Do you understand, Jose?"

With the tube for breathing stuck down his throat, Dias answered by nodding his head. "I am going to say a prayer for you," the priest said. Dias nodded again. After the priest finished praying, the marriage ceremony began. "Jose, do you take Maria to be your lawful wedded wife?" The man in the bed lay still. Twice more the priest asked the question without getting a response. The priest asked one more time and Dias shrugged his shoulders. "Si," the priest said. He blessed the marriage. Maria bent over and patted Dias on the shoulder as she held his hand.

He was on his way to cover the police beat when he noticed a cop in front of the station looking under the hood of the police car. "What's going on?" Stepankowsky asked. He peered into the engine and saw a piece of panty hose tied from the air intake manifold to the frame. "What in the hell is that all about?" he asked the cop, who looked at him with a sheepish grin.

"Oh, that's how you keep the ash from blocking it up because it acts as a prefilter and then the vibration shakes it out."

Stepankowsky went back to the office to write up the story. He also wrote how the ashfall had local government officials baffled. "How do you deal with something like this?" Longview mayor Jack McCullough asked. The ash was too thin to scoop and too pasty to sweep. He also wrote a story about protecting cattle from eating and breathing the ash. On the positive side, agricultural officials said the ash, with its aluminum and sand, eventually would improve the local clay soil.

The afternoon *Daily News* was delivered to his desk. In it was a photograph of Roger Werth holding up a *Time* magazine. On the magazine cover was a photograph of the May 18 eruption plume billowing up to the sky like an atomic blast. Werth had taken the photograph when he was in the plane with Stepankowsky. An accompanying story played up the rising fame Werth was gaining with his photos. Werth, who was almost the same age as Stepankowsky, was even scheduled to testify before Congress about the eruption photos.

The next day Stepankowsky's story about the innovation of the Longview cop, which was picked up by a wire service, sparked a raid on panty hose across the state. In the tiny town of Ritzville, a store owner who hadn't read his paper couldn't figure out why the local priest was stocking up on panty hose.

He was now having the same dream over and over. Roald could see himself and Venus. They were jumping from one log to another on the river. It was not a nightmare. He felt no fear. But the number of logs seemed endless. The couple could only quit leaping from one log to the next when he woke.

When Leonty Skorohodoff's vital signs began to fail, Melissa Metcaff was caught by surprise. She had cared for him as his day nurse. He had been stable. Now she was serving as evening supervisor for the unit. Suddenly the alarm bell rang on Skorohodoff's monitor. A young physician yelled out that the Russian was going into cardiac arrest. She called the code team and then phoned Dr. Parshley. She was put on hold while his answering service tried to contact him. Listening to the *ding ding ding* of the alarm bell, she decided she could not wait. She rushed into Skorohodoff's room. The code team of medical specialists trained in emergency resuscitation arrived and started working to revive him. Metcaff finally reached Parshley, who said he would be right in.

When Parshley examined the Russian, he realized Skorohodoff was already in mass organ failure. Parshley ordered the team of physicians, nurses, and technicians to continue their attempt to save him. Then he went to the family to tell them there was little hope. At first the family members were in disbelief. Only two days before, Skorohodoff had been taken off the respirator because he had shown such great improvement. How could he be dying?

The medical team worked to keep Skorohodoff alive, but the death watch had begun. Several times the mother and wife entered his room, only to run back crying to the waiting area a few minutes later. Other relatives arrived: brothers, cousins, and children. They were allowed into the dying man's room to say goodbye. The waiting area filled with wailing and crying as the Russians tried to comfort each other. Tanya could take it no longer. She left the area.

Then word came that death was imminent. Tanya returned just in time to hear Parshley report that Skorohodoff was dead. Wails rose up from his mother as she collapsed to the floor in sobs.

Parshley was shocked when Tanya demanded that the family be allowed to take the body with them. Autopsies were not allowed by their religion. He explained that the medical examiner had ruled that in the public interest there would be autopsies of any volcano patients who died at the center. But the Russians did not understand. Security had to be called to keep them from taking the body.

Helen Scymanky had also been in the waiting room She cried and could not stop. Once the Russians left she sat in the room by herself. "It's my turn to get the news next," she thought.

Late the next morning a pickup truck backed up the driveway that led to the morgue in the hospital basement. Two men got out, slid a door out of the bed of the pickup, and walked with it to the morgue. They placed Skorohodoff's dissected body on the door and carried it back to the pickup. Then they drove off.

Peter Frenzen had not expected the phone call. He had already signed the papers to attend the Yale University graduate school of forestry. Funds were lined up for his studies and living expenses. He had informed Oregon State University of his decision to attend Yale instead of OSU.

So he was surprised to hear the voice of Jerry Franklin, an OSU professor who also worked for the Forest Service. "I've been awarded grant money to study at St. Helens when it's safe to go in there," Franklin said. "I wanted to see if you might want to reconsider and become my graduate student?"

Frenzen's head spun. But he knew immediately what his answer would be. He didn't want to blurt it out, however, and seem overanxious. He waited a whole hour before he called Franklin back with his acceptance.

Authorities had stopped looking for survivors and believed they had found most of the bodies of the dead. But stories of both survival and death on the day the volcano erupted were still coming to light. The stories of death were not easy for the reporters to write, especially when local friends and relatives of the deceased had to be called for comment. But now Stepankowsky had been assigned such a story.

Terry Crall and Karen Varner, both twenty-one and from Longview, were missing and presumed dead. Varner was the daughter of a local high school coach and teacher. She had worked for a local physician. Crall had worked for the area's largest employer, Weyerhaeuser. The couple had gone camping and fishing with other friends on the Green River. Crall and two of those friends were fishing from their campsite on the river when the mountain exploded. Crall rushed into a tent to warn the sleeping Varner.

"That was the last anyone saw of them," said Sue Ruff, one of the friends. "They were the most vivacious, outgoing, fun-loving, nature-loving people you'd ever meet," she told Stepankowsky.

The concussion of the blast, debris, and heat had swept over the camping party. Trees shielded them from the worst of the heat wave. "Trees were falling around us," Ruff said. "We were standing straight up holding each other. The sky got dark as midnight. It was incredibly hot."

Ruff and her boyfriend, Bruce Nelson, hid under a huge log for ninety minutes before emerging from the nest of trees that saved them. They found one member of the camping party severely burned and another one injured under a log. Frantically they searched through enormous piles of fallen trees for Crall and Varner but could not find them. Rescuers later found the survivors.

"I was going to Seattle," Crall's mother, Sallie Nichols, said to Stepankowsky, "and Terry told me to have a good time and I told him to have a good time. That was the last thing he told me." She began to cry. "They were out of the danger area," she said. "He was a good boy. No one should think he wasn't obeying the law." Stepankowsky confirmed that Crall's campsite had been thirteen miles out from the nearest roadblock meant to keep the public out.

Stepankowsky moved on to other stories he had been assigned that day, concerning plans to dredge volcanic debris from the rivers. The Army Corps of Engineers had upped its estimate of the amount of debris it would have to clear from the Columbia River to 22 million cubic yards. The ports of Portland and Vancouver were losing $5 million a day in revenue because ship traffic had been halted. Silt that flowed down the Toutle and Cowlitz Rivers continued to seep into the Columbia at an alarming rate. Dredges were working and more were on their way. The Corps estimated the cleanup of the Columbia would cost more than $44 million.

There was great fear that the Cowlitz would flood that winter, devastating local communities. The Corps had asked for but not yet received funds for dredging that river. Meanwhile, giant cranes had been brought into the moorage at Kelso to raise seventy-five sunken boats.

As Stepankowsky wrote, he could see that dredging and the flood threat was shaping up as a big ongoing story. He was one of those rare reporters who had both creative and mechanical-minded sides. This was a story he could get his teeth into, one that was made for the dogged nature of a reporter who did not mind poring over lines of statistics in piles of reports and who loved figuring out how things worked—things like a dredge.

Donna Parker was trying to persuade the funeral director to take the bodies of her brother and his wife out of the plastic bags. Michael Everhart talked her out of it. "Think about it, Donna. The condition that they are in. This isn't the Billy that you knew."

Everhart had tracked down the bodies the day before. Authorities had yet to officially notify family members that Billy and Jean were dead. Everhart found they were in a morgue in Portland. The bodies arrived at the Molalla mortuary just before the graveside services. A family neighbor had arranged flowers on top of the caskets. "Hold on, I don't have them in there yet," Everhart said. They took the flowers off the caskets and Everhart and an assistant put the bodies inside and closed the lids.

Afterward, standing in front of the caskets with Parker's mother, the neighbor, Loa Schultz, talked about how she had taken the flowers off for the bodies to be placed inside. Parker's mother asked her a lot of questions. Schultz told her how she could see the bodies through the obscured vinyl bags. They were like dark shadows.

Parker's mother wanted to see Billy one last time. "Think, Mom," Parker said. "Would Billy want you to see him in this condition?" The bereaved mother finally decided against viewing the body.

The funeral limousine was the only car to follow the hearse. Parker had wanted to invite at least George Gianopoulos, her brother's best friend. But relatives of her dead sister-in-law wanted no one attending outside of the immediate families.

Now the cars wound through the streets of the small town to an ancient cemetery where headstones of pioneers from a century before could be found mingled with freshly dug graves. They did not have enough able family members on hand to carry the caskets from the limousine. They had to enlist Everhart and grave diggers, along with Parker's half-brother, her son, and a son-in-law of Jean's, to take the caskets to the graves.

A minister began intoning the words of a simple graveside service. But Parker could not hear him. She just kept repeating to herself her new mantra, "Billy would understand. Billy would understand. Billy would understand. . . ."

CHAPTER 7

June 2–19

THE VASTNESS of it all, the endless gray, overwhelmed him. On Stepankowsky's plane trip during the May 18 eruption, there had been no way to gauge the destruction. There had been too much smoke and ash in the air, plus the distraction of trying to dodge other aircraft. Now as he flew in the plane above an endless plain of ash, there was an unsettling calmness and silence, while the volcano sent a few puffs of steam harmlessly into the air.

Stepankowsky was not deceived by this ash wasteland, seemingly immobile and reposed. It too represented the potential for more death and destruction. When the winter rains came, he knew the ash would have to go somewhere. That was most likely down the Toutle and Cowlitz Rivers, in waters cascading over riverbanks as the silt settled to their bottoms, cutting the rivers' capacities and bringing floods to cities in the lowlands.

As the plane flew back to Kelso, he took one more look over his shoulder at the gray blanket that covered all he could see. It was a haunting scene.

When the doctor and nurse came into his room that morning, he realized by the expression on their faces that something terrible had happened. As they told him about Skorohodoff's death, Scymanky refused to believe it at first and then couldn't understand it. Both of them had been through the same experience, but he was still alive and Skorohodoff was dead. The Russian had taught him everything he knew about working in the woods. Whenever Scymanky needed help, Skorohodoff seemed to be there for him. All the good times they had spent together around a campfire after work flashed before his mind.

Scymanky knew he was going to make it now, but the question of why his friend had died and he had not became an obsession for him. There was no answer, scientific or otherwise, that made sense to him.

Venus Dergan pulled the brush through her hair as she stood looking into the bathroom mirror. Another wad of hair came off. She was now almost bald. She rubbed the top of her head where she could feel the bumps of silica still embedded in the skin.

Already thin before the eruption, she had lost twenty-five pounds and often refused to eat because she had no appetite. Her gaunt face was scratched and torn. She worried about the scarring. Her body was wrapped with bandages, which her mother and father changed daily. Her hand was in a cast.

The Centers for Disease Control, doing a study on the effects of the ash from St. Helens, monitored her condition and called often. Then there were the phone calls from newspapers. But she no longer wanted to talk to anybody. She wasn't even sure she wanted to talk to Roald. How could he love anyone as hideous as this creature that stared at her from the mirror?

They were celebrating their homecoming after a frustrating attempt to build the radio station in Utah. Robert Rogers and his friend Gray Haertig had returned to Portland with boxes of confiscated fireworks they had found in the police storeroom in the municipal building where they were installing the radio station. Now as he watched the rockets explode in a dazzling display of shooting stars and fountains of streaming colors, Rogers experienced a twinge of introspection. He was a twenty-nine-year-old with no real job to speak of, a recent failed relationship with a woman he had left because he thought they were too "weird" together, and few ideas about his future. And here he was setting off fireworks like a kid. Where was his sense of responsibility? He smiled and bent over to light more rockets.

In the morning he got a call from a friend. His mother was looking for him. She had read an article in the Vancouver *Columbian* that mentioned his and Francisco Valenzuela's exploits on the mountain. Rogers figured Valenzuela had said something to the newspaper. But he didn't even consider calling his mother. His family didn't understand him any better than most other people did. He was estranged from them and he wanted to keep it that way.

Besides, he had better things to do. Right now he was going down to the U-Develop Darkrooms to print some of his volcano photos. At U-Develop he noticed a tiny photograph tacked to the wall next to the cash register. It showed a landscape of trees with Mount St. Helens in the background before the May 18 eruption.

"I know where that photograph was taken," he told the woman behind the cash register.

She looked at him in shock and started crying. "That's the last picture we have that Robert Landsburg took. We think he might have been in the same spot the morning the volcano erupted. But we don't know where it is." Landsburg was a close friend of hers.

"Was he kind of slender and maybe on the tall side?" Rogers asked.

"Yes."

"Did he have a green station wagon?"

"Oh, god, how did you know?"

"I saw him the morning of the eruption. I know exactly where he was. I even phoned the authorities and gave them a detailed description of where they could find him."

Rogers went home and phoned Valenzuela. Valenzuela said that in his work with the Forest Service, he had flown in a helicopter to the spot where he and Rogers had seen the green station wagon and the nearby pickup. But he saw no sign of the station wagon or the man who had been standing by it.

The pickup was still there, Valenzuela said. The three bodies that were found at the pickup had been evacuated. They were those of Day Karr, thirty-seven, and his two boys, eleven-year-old Andy and nine-year-old Mike. A photo of Andy's naked body in the bed of the pickup, taken by a *San Jose Mercury* photographer, had become an emblem of the volcano tragedy as wire services circulated it around the world.

Then Rogers phoned the emergency services agency. Once again he gave the exact location where he had seen the station wagon and the man. The next day a helicopter from the 304th Air Force Reserve Rescue Unit of Portland flew into the area with a logger who claimed that from a logging tower he had seen Landsburg taking photos two days before the eruption.

On ashy ground, down a slope at the headwaters of the South Fork of the Toutle River, they found the wreckage of Landsburg's 1969 Dodge Coronet station wagon. The vehicle was just off 4170 Road in the area where Rogers told authorities he saw it last. The site was four miles from the summit of the volcano. The car had been blasted into four large pieces and blown down the embankment. A trail of auto debris led to the spot where the pieces came to rest.

Landsburg's body was found just over a hundred yards from the road near the wreckage. There was no way to tell whether he had been inside or out of the car when the volcano's explosion ripped it apart. The searchers also found his billfold, cameras, and a camera bag that contained what looked to be a hurriedly scrawled inventory of the photos that were on his exposed film.

His body was the twenty-fourth found on the mountain. Many more people were still missing and presumed dead. The helicopter flew the body to Kelso Airport. The crew and a Cowlitz County deputy sheriff transferred the body in a green bag to a waiting hearse. "He died doing what he wanted to do—on the mountain," Landsburg's brother said. "What a tombstone."

When Donna Parker went to pick up her brother's belongings from the Cowlitz County Courthouse, she took George Gianopoulos and his son Steve with her. They were familiar with many of Billy's possessions. The coroner dropped a bag about six inches high and four inches wide on the counter in front of her. "Here's your brother's stuff." She looked inside and found his wallet, change, and a few other items.

"This is it?" Parker asked. "Isn't there something else?" George had told her that Billy packed several cameras with him. In fact George had just lent him a set of telescopic lenses. Parker also knew that the pickup and its camper were supplied with all the accouterments imaginable for comfortable living in the outdoors. Where was it all?

"Well, there are some more bags down in the basement," the coroner said. "There might be more of your brother's stuff down there. Would you like to have a look?"

"Sure."

"Okay. If you just look on top of the bags, there should be a license plate. Do you know your brother's license plate?"

"Yes."

"Well, the license plate should tell you if those are his things."

A worker led them to the dark, damp basement. The worker turned on a dim light and left them alone. They were surrounded by a multitude of huge, black plastic bags that gave off a strong, musty odor. The smell mixed with the dank aroma of wet ash. The basement was filthy with mounds of the gray grit from the mountain.

They looked at the license plates on the bags. None were Billy's. They saw several bags with no plates. They opened the bags and searched them. They found cameras but none that belonged to her brother. Were his possessions back at the truck or were they somewhere else?

As they started to leave, Parker went back to one of the bags and stared at the license plate. It was from Ohio. She couldn't remember anybody from Ohio being on the list of dead or missing people. It was a list that she had memorized.

He and the photographer had to travel over rough back roads. But he didn't care. Stepankowsky was going to interview Stan Lee, a living legend always good for a laugh and a colorful quote. Like that other great character of the mountain, the crusty old Harry Truman who had enthralled millions of TV viewers with his defiant stand against leaving his home and had died for it, you could never tell what would come out of Lee's mouth.

Lee owned the Kid Valley Store, a small general store housed in a rickety log building plastered with signs: "Hunting and Fishing Licenses, inquire here," "Say Pepsi please," "Last Stop for Groceries." But it was the sign that Lee had scrawled in big letters on a piece of cardboard that told the story: "Open Now." Lee's store was the closest business enterprise to St. Helens to have survived on the west side of the mountain.

He reopened the store in the small settlement of Kid Valley a few days after the eruption. Mudflows had swept away many of the homes. Spirit Lake Highway to the settlement was mostly buried under ten to twenty feet of mud. The eruption also wiped out the two bridges used by tourists and customers to get to his store. The only way to Kid Valley was over miles of rutted, rocky logging roads. There were few takers. His biggest day had brought in twenty dollars in the weeks since the eruption

Stepankowsky asked why he reopened the store. "Because of the fortune-teller," Lee said. More than thirty years before he had stopped to see a card-reading fortune-teller in Cornelius, Oregon. "She told me I would open a business in Washington. I laughed. Then she told me I would take out a loan. And I laughed again." She didn't tell him a volcano would blow away his business. But she did say that his future enterprise "would pay off if you hang on long enough."

"Everything she said has come true," Lee said.

Then the sixty-seven-year-old curmudgeon, who always wore military green clothing with suspenders for the pants, erupted with a string of expletives and railed against his favorite target, the government. The whole disaster was the government's fault. "Had the state built the Coal Banks Bridge twenty feet higher, it would have survived anything." And customers might still be streaming to his store.

He was going to have the last laugh, though. Lee was going to prove the fortune-teller right. He had heard the predictions that hundreds of thousands, maybe millions, of tourists would be lined up bumper to bumper on a reconstructed Spirit Lake Highway to see the volcano. Every one of them would have to pass by his store. Lee made no bones about planning to milk the tourists.

The photographer posed him in front of the store next to the "Open Now" sign. Lee stood stiff and straight with his arms crossed, the perfect defiant symbol of the businessman as survivor against even the forces of nature. The nearby restaurant advertising "Volcano Burgers" was shuttered tight. The only sound aside from the wind was the call from a rooster that led a harem of four hens past the two gas pumps in front of the store.

Everything medically possible was being done for Jose Dias. IV lines ran to his arms and medication kept his blood pressure up while a respirator breathed for him. He had been in a coma-like state for a week, but it still came as a shock to nurse Metcaff when he died. As the hours passed that day, his vital signs began to drop. Still, she couldn't bring herself to believe that he would die on her shift. She felt helpless. All that could be done was being done. At noon it was finally over.

The burn victim with a tube down his throat, the man who never recovered enough to explain his past, died a mystery. The driver's license in his wallet had given the name Raymond Casillias. In a story published a few days after the eruption, a Stockton, California, newspaper claimed that the real Raymond Casillias was in prison, serving a five-year term for armed robbery. A man fitting the description of Dias had worked for Casillias's mother for a few days cleaning her yard.

Maria said Dias was a Mexican national. Scymanky had been told by Skorohodoff that Dias was a soccer player from Honduras who had played across South America. The physician who signed the death certificate wrote in the names of both Jose Dias and Raymond Casillias.

When Maria got the news, she came into the room, clasped Dias's hand, and sobbed. After twenty minutes Metcaff had to ask Maria to leave so she could prepare the body for the autopsy. As she worked, Metcaff realized again that no matter how much dying she witnessed, she never got used to it.

The day after Dias died, Jim Scymanky began asking everyone why he was still alive while his two friends were dead. Two of the nurses, on different occasions, became upset after trying several times to answer him. They started crying and ran from the room. One of them was nurse Metcaff.

He was all over the tiny control room, wanting to know what each of the fourteen levers did on the control panel of the dredge *Oregon*, a two-story-high monstrosity with an even higher tower. Andre Stepankowsky was in his element. He was fascinated with the dredge—really a ship 265 feet long. Nothing was beyond his mechanically inquisitive mind and he peppered everybody with questions.

The lever man reveled in the reporter's attention. He moved the levers with a deft touch like a musician's to maneuver dredging equipment on the end of the huge boom that extended off the front of the ship. A thirty-inch-wide pipe was suspended from the long arm into the water, where it sucked up silt and water stirred up by an agitator. Stepankowsky listened to the slurping sound and thought he had the lead for his story: "SSSlllllurp."

The ship shuddered from the suction power provided by a 5,000-horsepower diesel engine. It sucked the debris and pumped it through pipe that ran like a snake across a series of rafts to the banks of a nearby island. The *Oregon* could suck up and deposit thirty thousand cubic yards of silt and mud a day. Stepankowsky was surprised to learn the dredge had no propulsion of its own. It maneuvered by using river currents and wire cables. The anchored cables moved the dredge in an arc, clearing an area on the Columbia River three hundred feet wide and three hundred feet long each day. The reporter wanted to know everything he could about the dredge. It would be similar to the dredges promised for the Cowlitz River to help save the region from flooding.

After the funeral Donna Parker's mother searched through Billy's jeans and found the keys to his pickup in a pocket. She and her daughter were surprised. They had assumed, along with everybody else, that Billy had been trying to escape by driving away. Photographs in the newspapers had shown his black, mummified body with one hand gripping near the top of the steering wheel. But now it looked like Billy might have realized this was the best he could do and had hoped the pickup would offer him and Jean some protection. On the other hand everything might have happened so fast that he didn't have time to get his keys out.

Later her mother received a phone call. Both government and private psychologists were recruiting subjects for studies to learn how survivors and relatives of victims were dealing with the tragedy. "How are you handling your pain?" the man on the phone wanted to know.

"I buried my mother when I was seven. I buried my Dad. I buried my brother. I buried a husband. I buried my only grandson after an auto wreck. I buried my six-year-old son after he got shot by a neighbor boy. I just now buried my other boy who got killed by the mountain. How do you think I'm handling it?"

Stepankowsky in his bullheaded manner had browbeaten the city editor to assign him what was sure to be tomorrow's lead story. The Army Corps of Engineers was going to meet with citizens to outline its dredging plans to prevent winter flooding. Stepankowsky wanted to dazzle the editors in order to snare the vital dredging story as his regular beat.

The meeting had to be moved to a gymnasium to accommodate the crowd. Nearly a thousand anxious residents packed the Castle Rock High School gym. The eruption and its mudflows had left the Cowlitz River only two feet deep, with an 85 percent loss in water-carrying capacity. If the river wasn't deepened by immediate dredging, a large area could end up under water when the rains came.

A row of officials sat facing the audience. Officials from the military side of the Corps wore neatly pressed uniforms. Other government officials and Corps civilians were impeccably dressed in suits and ties. The rather rotund mayor of Castle Rock stood up to begin the meeting. He was a baker who got up early for work and retired early. From the rumpled appearance of the mayor's clothing, Stepankowsky surmised the man had taken a nap in his clothes and didn't change them before the meeting. Then the reporter did a double take. The mayor's fly was wide open. Stepankowsky's first reaction was to laugh. Then he felt an urge to rush up to the mayor and say something or at least to gesture toward the man. But in the end he did nothing. And neither did anyone else.

Patrick Keough, a planner for the Corps, was speaking. "We'll have dredges in the Cowlitz by January 12," he said. A howl rose from the audience. Keough quickly caught his mistake. He had meant to say June 12, a week away. "I just wanted to see if you're paying attention," he joked.

The Corps had a plan, he told the audience. It involved dredging parts of the Cowlitz, building five dams to trap sediment in the Toutle River, which feeds the Cowlitz, and permanently evacuating people from parts of the area's floodplain. The cost would be $219 million. The only catch was that they didn't have the money. An aide from the office of the local congressman rose to say that the appropriation bill probably would pass the Senate. "But I don't want to give you a load of hay," he said. The bill would face stiff opposition in the House.

One man in the audience hollered out, "What are the chances of getting half of those guys up here and look? I'd pay taxes to get them up here to see this." The audience applauded.

"Will we get money as fast as the Cubans?" another man yelled out, referring to the millions of dollars the federal government had allocated to help refugees being allowed by Fidel Castro to flee Cuba. The crowd now cheered.

Stepankowsky had never seen an audience at a public hearing so worked up. Then Van Youngquist, a Cowlitz County commissioner, stood. "Ladies and gentlemen, the sheriff's office has just informed me that the volcano is acting up and some ash may be headed this way." People hurriedly began filing out.

The next morning Stepankowsky began writing. Everything came together from the lead paragraph on. He had never produced a major breaking news story so effortlessly. Then he worked on an accompanying story about a self-employed statistician in Longview who predicted that a mere inch of rain could cause the Cowlitz to flood. The man was known as a gadfly who often took on government experts and proved them wrong. The figures that Dennis Jacobsen showed to Stepankowsky challenged government calculations that normal rainfall would not cause flooding.

The reporter did not want to be accused of writing about any crackpot theory in order to get a headline. He checked Jacobsen's figures and thought the statistician was right. He called several Corps engineers and other scientists who criticized Jacobsen's calculations. Then he wrote the story. It called for the use of many statistics and arcane terminology like "run-off coefficient." Stepankowsky felt confident in interpreting these for the reader, rendering in simple language the jargon and theories of the engineers and scientists. He thought he could do it better than anyone else on the newspaper.

Stepankowsky transmitted the two stories into the city editor's computer. City editor Dan McDonough, a man much older than Stepankowsky and a consummate professional, stopped after reading the first few paragraphs of the Corps meeting story. He looked up at Stepankowsky, now standing expectantly beside the editor's desk. "Great lead," he told the reporter. "Even if it is editorializing a little bit."

"If two usually uncontrollable forces cooperate," Stepankowsky had written, "the U.S. Army Corps of Engineers has a plan to save the Cowlitz River Valley from extensive flooding this fall.

"Those forces are Mother Nature and the federal government. . .

"Mother Nature must cooperate by not sending too much rain in one dose. And the federal government, Congress in particular, must appropriate money to improve flood protection near the river."

It seemed like Scymanky was always telling them no. No, when they turned on the lights after he was asleep and said it was time to change his dressing. No, when they kept stuffing food into his mouth, saying he needed the calories. No, when they tried to draw blood from his big toe. No, when they had him in the tank and began pulling and stretching his body. It never made any difference. They went right ahead.

He screamed at them almost the whole time when they put him on the Stryker bed designed to change position, relieving the pressure on skin grafts. The bed, between two large metal hoops that surrounded it on each side, could rotate from front to back at any desired angle. The nurses placed him on the bed face up. Then they tilted the bed 180 degrees until he was face down looking at the floor.

He grew bored. He begged the nurses to turn him over. Then he screamed at them. One hour passed, then two and then three. Each time he asked a nurse to turn him over, she told him "not now." They had to follow the doctor's orders.

Finally a nurse brought him a tray of food and placed it beneath his face. His arms were stretched out from his sides. "Why don't you try eating something?" she said. He laughed. "How are you supposed to eat? You know my hands are out here."

Helen came in. A nurse told her that Jim's tray had arrived and she should try to feed him. Helen slid under the bed and raised a spoonful of food to his mouth. He took one bite. "No, this is it," he told her. "I'm not going to eat this way anymore. This is the bottom line."

Except for minor ash explosions, the mountain had been relatively quiet for several days. Seismic activity was in a flat-line pattern, leading some scientists to speculate that St. Helens' eruptive phase might be over. Everything was so quiet that the Forest Service was afraid the public might start drifting back into the restricted zone. "We don't want a replay of May 18," a spokesman said.

The USGS, too, didn't want anyone lulled into ignoring the danger. H. William Menard, director of the agency, told Congress, "There is no indication that we have seen, even in the May 18 eruption, the biggest thing that could happen." He pointed out that in the period of the previous eruptions of Mount St. Helens, in the nineteenth century, three other volcanoes in the region had been active. Mount Rainier, which shadows the Northwest's largest city, Seattle, had erupted along with Mount Hood, just sixty miles outside of the region's second-largest city, Portland. There had also been activity on Mount Baker in the northwest corner of Washington.

The mountain's continuing quiet period would have been conducive to scientists getting work done, if only the weather had cooperated. Many times when Don Swanson and his colleagues wanted to helicopter near the crater, place new tiltmeters, or establish lines for geodetic readings, the mountain was covered by clouds and steam.

Swanson was what was called a "deformation man"—an expert in how volcanoes deform their surroundings as they move toward an eruption. He did this by using tiltmeters and geodimeters. Tiltmeters are so sensitive they can measure a microscopic rise as they sit on the earth's surface. The geodimeter works with a laser beam that is bounced off a reflector. This instrument provides incredibly accurate readings by measuring a point in the wavelength on which a light wave returns as compared with a hypothetically perfect wavelength. Over the course of time, lines are reshot to measure any changes.

Using the geodimeter requires a mathematical mind that can factor out weather and other influences to establish a geodetic line. It could take some scientists hours to make the calculations. Swanson could do it in minutes. He was known to be the foremost geodimeter expert in the USGS.

While in Hawaii, in a great show of patience and fortitude, Swanson used the geodimeter to create a net of measuring lines over the entire Kilauea volcano. In follow-up measurements, other scientists marveled at how accurate the lines proved to be time after time. When he arrived in March as St. Helens started showing signs of life, he had established a similar geodetic network. The May 18 eruption blew that network away, and ever since he had been working to establish an expanded network.

However, neither Swanson nor any other scientist had figured out how to use deformation, or any other gauge, to accurately predict a volcanic eruption. Scientists had studied any number of possible indicators: temperature and gas changes in the small volcanic vents known as fumaroles, changes in the Earth's magnetic field, infrared photography, seismic activity, even the nervousness of animals. But all fell short.

The long history of volcano eruptions recorded only two instances of scientific tools and theories being useful in specific predictions. Both took place in Hawaii: on Mauna Loa in 1942 and at the Kilauea volcano in 1973. But in neither case could the predictions be formulated into a technique or theory that could repeatedly provide accurate forecasts. The fact remained that no one had yet found any reliable method to predict an eruption on an individual volcano, let alone a technique that could be used on other volcanoes.

Swanson now had more immediate concerns trying to get a geodetic measuring system at St. Helens up and running again. So far all the measuring points he had reestablished were outside the crater itself. But where Swanson liked most doing his studies was at the center of a volcano. In the case of St. Helens, this would be on the lava dome that had yet to form, where he could most accurately take the pulse of the volcano. If there was another significant eruption soon, he thought, it might promote the creation of such a dome.

That June 12 morning, a Thursday, the USGS had its hands full. The next night would have a full moon. The initial major eruption on May 18 had been predicted by a Seattle scientist, outside of the USGS, based on the fact there would be a full moon that night. A full moon, the scientist said, also played a role in the second eruption. The scientist believed the moon's gravitational pull was a major contributor.

Swanson and the other USGS geologists put little credence in the theory. They hoped there would be no eruption the next day because it would only be a coincidence that would lead the public astray. The USGS had been flooded with calls from a public concerned about the potential for an eruption brought on by Friday's full moon. It didn't help that it was Friday the thirteenth. "If it blows on Friday it will set science back ten years," USGS geologist Pete Rowley said at a press briefing Thursday morning.

However, harmonic tremors—sustained rhythmic vibrations—already had started as Rowley spoke. Two hours and forty minutes before midnight, Mount St. Helens exploded in its third major eruption since May 18. Billowing clouds of ash kept scientists away from observing the crater until Sunday, when they were awed by what they found. A dome of lava had built up from the crater floor. The newly formed dome looked like a toasted mushroom cap giving off a slight red glow. It stood fourteen stories high.

Swanson now had the lava dome he wanted. All he had to do was figure out a way to get into the crater. "We expect another explosive eruption," Rowley told the news media after the eruption. Unnecessary flights to the crater would be "extremely foolhardy," he said.

The helicopter hovered over the great gaping mouth of the volcano. "You are clear to enter," the air controller told the pilot, who happened to be Robert Rogers' brother Warren. Rogers and a geology graduate student, Chuck Meisner, were also in the tiny helicopter that had only a single bench seat. Meisner had a pass to work in the Red Zone.

"Okay, I'm going in," Warren said. What the air controller couldn't know was that the pilot wasn't talking about simply entering the Red Zone, the destination shown on his flight plan. They were heading into the crater itself. In Robert Rogers' ongoing game of infiltrating forbidden territory, he was raising the ante.

When Rogers found out about Meisner's permit, he talked the student into making the trip into the crater. Rogers persuaded his brother, a flight instructor and helicopter salesman, to fly them. But even with the Red Zone permit, it was prohibited to land in the crater or hover near its floor. That morning Warren "borrowed" a helicopter from where he worked and disconnected the meter that measured the amount of time it was flown.

Now Rogers' brother flew into the crater and circled the lava dome with the chopper. They had taken the door off the passenger's side to facilitate picture-taking. Rogers leaned out, snapping photo after photo. They decided to fly directly over the dome before landing in the crater. Wisps of smoke rose from the dome. Suddenly the helicopter started shuddering and Rogers tried to keep from falling out. The chopper began to fall, its engine sputtering. Rogers realized what was wrong. "We're right over the dome. There's probably no oxygen here," he screamed. The helicopter's high-speed, turbocharged piston engine needed air to function.

The pilot fought for control. He was preparing to crash-land on the dome when the helicopter hit a pocket of air and jolted up into the sky. They cruised away from the dome and hovered above a nearby sandy ridge. To improve chances of the helicopter landing on or near the dome, one of the passengers would have to get out. It wouldn't be Meisner, because Rogers' great inducement to get him to go on the trip along with his pass was that the student would be able to get a sample from the dome. Rogers hopped out onto the ground with three cameras dangling from his neck.

The chopper took off and circled as Meisner took photographs of Rogers below. Rogers couldn't get over how the volcano was screaming in his ears like a jet engine as steam blew from the ground. Then the helicopter descended over the dome and landed next to the steaming pile of ejected rock and hardened lava The student raced partway up the side of the dome, grabbed a rock sample, and ran back, trying to keep his balance on the hot, rough surface. He hurled himself back into the chopper, which rose quickly to the sky. Rogers, on the ridge, took photos of the operation.

The helicopter picked him up and they flew out of the crater. Rogers had little time to explore. The trip only whetted his appetite for spending more time in the crater. His brother made it plain that after this experience, he would not be coming back. Rogers knew the chance of finding another pilot was virtually nil. He would have to figure out some way to get past the roadblocks and hike in.

The rock sample from that day was believed to be one of the first secured from the dome. It was sent to a government laboratory in Washington, D.C., where it was used to help confirm the composition of the volcano. The names of Rogers and his brother and the use of the borrowed helicopter were kept out of the telling of the event.

—✳—

When Donna Parker returned to work at the Tektronix plant, she felt strangely changed. All her life she had pretty much accepted authority without challenge. Now she couldn't get out of her mind what Jimmy Carter, Dixy

Lee Ray, and others had said about the victims of the eruption—that somehow these people had knowingly put their own lives at risk. She also found herself questioning what was really important in life as she grieved for her brother.

Parker was a supervisor at the plant. A woman came to her with a complaint. It seemed the woman complained almost every day, always about the same fellow worker. Distracted, Parker took awhile to realize what the woman was saying. But when she became aware of the woman's whining voice and her complaint, Parker became enraged. The complaint seemed so inconsequential. "If she bugs you so much, why don't you just get a gun and shoot her and quit your bitchin' to me all the time," Parker said.

The woman shrank back in horror. Parker couldn't believe she had just said such a thing. Those couldn't have been her own words. Parker immediately went to her boss and told him what happened.

With his dust mask on, Peter Frenzen waded through the developing vegetation on the aged mudflow on Mount Rainier. The vegetation was covered with volcanic ash from St. Helens that blew into the air when disturbed. Frenzen had ash in his hair, on his clothes, and he could taste it. It was a constant reminder of the active volcano to the south.

The study of vegetation and soil development in the mudflow was going well. But it could not go fast enough for him. His studies at St. Helens with Jerry Franklin would not begin until September. His hope to gain more immediate access to the eruption area lay with the National Science Foundation. Then one day he was told that the addition to his original study proposal had been accepted. He would be given use of a van and enough money for a trip down to the volcano.

Authorities had the restricted zone tightly secured both in the air and on the ground. Robert Rogers thought he could get around them, but he needed some type of pass to show if he was caught inside the zone. Then he had a stroke of luck.

Rogers decided to try the USGS again to see if he could spark any interest in his sequence of eruption photographs. Rogers and his friend Gray Haertig met with a USGS scientist named Richard Waitt. The geologist was interviewing principal witnesses to the eruption to gather data for establishing the timing of events. Photographs could help, he said.

"Your impression of time between pictures can be a big help," he told Rogers. "If I get you a permit would you mind going back in to where you were to reconstruct the timing?"

Rogers tried not to look at Haertig. He was afraid if they made eye contact, one of them would burst out laughing.

"Uh, gee, I guess I can go back," Rogers said with his best poker face.

Waitt wanted Rogers to return to the ridgetop and plot out with highway cones where he had been as he took each eruption photo. Then Rogers was supposed to time everything with a stopwatch as he ran from spot to spot, taking photographs as he had done the day the mountain blew.

Rogers studied the pass that Waitt had helped him get from the Forest Service as he and Haertig walked back to the car. He had no intention of using the pass to do timing studies, at least not immediately. The pass was only good for a one-time visit into the restricted zone. To use it legitimately, he would have to have it stamped upon entrance, which would limit its use to that day, and surrender it when he left. But if he figured out a way around the roadblock at Cougar, he could think of a thousand excuses to give the patrolling authorities about what he was doing inside the Red Zone. And the pass, even unstamped, would provide him with the legitimacy to avoid arrest. It would be his safety net, he told himself.

"Take the thing. I don't care. You can have it for nothing. That thing isn't worth a dime." Jean's oldest daughter had seemed a little surprised when she asked Donna Parker what she wanted to buy from her brother's estate and Parker told her, "Billy's bike."

Parker brought the bicycle home. It was a Roadmaster that Billy had used as a child. Her brother had restored it and it was in immaculate condition. Later she took it on a back road. She rode and rode and rode. With the green countryside whizzing by, she thought back to when they were kids together. The family could only afford an old pickup when she was growing up and they would often take Sunday rides. She and Billy would ride in the back. She would sit on a blanket while her older brother stood looking above the cab. He would tell her where they were going, often with a little embellishment. "There's a limb over the road ahead and there's a cougar laying right on it."

After he got the Roadmaster he took her everywhere on it, with her riding the handlebars in front of him. Then he taught her to ride it and do tricks. She still had scars from failed attempts to emulate him. But she didn't care. He was her big brother and she loved him.

Now she rode faster. She threw both of her hands up into the air and rode with no hands just like Billy taught her to do.

CHAPTER 8
June 20–30

"IT'S TIME FOR YOU to start feeding yourself," the nurse told Scymanky as she gave him a fork and placed the tray of food before him. As the nurse left the room, he yelled after her. For a long time he stared at the fork and the food. Then he screamed for the nurse again. When he told her that he wanted her to feed him, she left the room again.

He looked at the food and grew hungry. He sat up in bed. When he picked up the fork, it fell out of his hand. His fingers were stiff. Once more he picked up the fork. He lifted food to his mouth. His arms ached. It was his longest meal in memory.

Two nurses the next day strapped him between the two large metal hoops of the Stryker bed. They rotated the bed until he was in an upright position, as if standing. With blood flowing down through his body, muscles in his legs began to ache. He realized how weak he was.

The twelve of them sat at the lunch counter at Jack's Restaurant and Sporting Goods, five miles from the little town of Cougar. The crew had been installing tiltmeters under contract with the Trojan nuclear power plant, some forty miles away. Robert Rogers was part of the crew made up mostly of Portland State University geology students. He had been recruited as a volunteer. Nobody knew the road system like he did on the southwest side of the mountain.

Jack's sat just inside the restricted zone. Cougar, a hamlet of barely a hundred permanent residents, and the surrounding area had been devastated by ashfall from the June 12 eruption and by being on the wrong side of the governor's new line for safety. Ordinarily this time of year, the town and the nearby restaurant and store would bustle with tourists, fishermen, and campers. Now there were only about twenty people left in town and no tourists. The store's owners, a husband and wife, started complaining about their plight. The woman, who stood behind the counter with her husband, pointed to Rogers. "Is it safe? Would you stay here?"

"Of course I would stay here," he said. "As a matter of fact, I'm trying to get residence in Cougar or nearby so I can beat the Red Zone."

The woman's husband, who had been battling with authorities to get the line for the restricted zone moved beyond their store, puffed out his chest. With relish he dug through a drawer and came up with a receipt book. He smiled as he filled out a rent receipt. "There. That ought to make you legal," he told Rogers. "You're now officially the renter of that old trailer we've got out back nobody lives in. Those bastards can just take that and stuff it."

Now all Rogers had to do was take the receipt to the Department of Motor Vehicles, the agency that issued Red Zone passes. The pass would get him past the checkpoints without having to give up his one-day USGS permit. With a creative cover story, he thought he could use the USGS permit to bluff his way through any confrontation.

Rogers and the rest of the crew finished eating, drove into Cougar, and got out of their vehicles. The streets were deserted. The post office was shut down and so was the liquor store and the town's only tavern. Houses sat empty, streaked with volcanic dust. Trees had broken from the weight of the ash. Wind blew up a blanket of ash from the street, enveloping Rogers and the others in a gray cloud that kept them from seeing much more than a few feet. As he leaned into the wind, he felt like a character in one of those Depression-era documentaries about the Dust Bowl. He tried to wipe the ash off his face. It felt like slivers of glass rubbing into his skin. He bit down. There was ash grit between his teeth.

They walked by the Lone Fir Motel. It was out of business. The water in its swimming pool had turned to a grayish green muck. A bulletin flapped in the wind where it was tacked up next to the locked office door. It was a yellowed USGS document titled "What to Do When a Volcano Erupts."

Stepankowsky wanted in the worst way to be assigned the beat on the dredging and flood-potential stories. He saw it as a way to continue covering Mount St. Helens for a long time. He found that he didn't have to make his case very hard with his fellow scribes. Nobody wanted it. The other reporters thought it would be boring.

The editors were another matter. They were noncommittal until they met with reporters to divide up the volcano coverage. Before now there had been little time to worry about establishing beats. Much of the reporting had been assigned as events unfolded. Now it was time to return part of the twelve-member reporting staff to normal coverage like sports, business, and everyday community events. Five reporters would be assigned to handle aspects of the volcano story.

The reporters and editors filed into the conference room. They assessed where they had been with the St. Helens story. Then they started dividing up the coverage into beats. Managing editor Bob Gaston said he thought the community was going to be struggling with the flood control issue for a long time. They agreed. Then he asked which reporter should be assigned to the flood and dredging beat. There was silence. "I'd like to have the job," Stepankowsky said.

Gaston looked down at the table for what seemed a long time. Then he looked up. "Let's go to work." As they walked out, reporter Rick Seifert, a friend of Stepankowsky, said, "Thank God you got it. I was afraid they were going to saddle me with it."

The helicopter carrying Swanson and other scientists descended into the crater, then hovered as the men jumped out. The first thing he felt was heat from the floor of ash and pumice. He had to perform a hot-foot dance as he sank into ash almost to his knees. The men pulled a fence post from the helicopter. Swanson would mount a reflector on it to bounce back laser beams for his geodimeter to measure.

They had landed on top of the rampart, a ridge running crosswise across the crater. Swanson and the men made their way along the ridge as the blades of the waiting helicopter beat the air, ready for a rapid escape. Swanson wanted to plant the post and reflector near the smaller of two domes that had developed in the crater. The larger dome, which had grown to over one hundred and forty feet high, loomed over them. It was more than twelve hundred feet across its crusty surface. In the cracks he could see the faint red glow of lava.

Steam from hundreds of small vents, or fumaroles, wafted to join the clouds that whisped above. Small avalanches of rock fell from the steep slopes of the crater, their sound a constant applause that echoed around the men. Waterfalls steamed with heat. A small boulder tumbled down, punctuating the crater slope with brown puffs of dust before splashing into the sea of gray below. Swanson picked up his pace. The heat hung heavy in the air, which stank of sulfur.

This was Swanson's first trip into the crater since May 18. At first he found himself caught between fear and awe. He wondered how he could ever run back to the helicopter through this ash if the mountain began to erupt. Then a calmness settled over him. Long ago he had told himself that to survive in this work he would have to keep his wits about him. He couldn't allow himself the luxury of fear. Working in live volcanoes he had found his senses greatly heightened. Every rock movement, every smell, every sound, every change in the wind registered with him on a visceral level that his mind intuitively processed as being a sign of danger or not. When he was in this state, he was one with the volcano.

With the help of another man he pounded the fence post into the ground with a sledgehammer while he continued to march in place in the hot ash. He attached the reflector and guyed up the post. They rushed back to the helicopter. The trip had taken twenty minutes, but it had seemed like forever.

"Venus, if you don't do these exercises and get up and move around more, your limbs are going to atrophy," the physical therapist told her in his gentle but firm voice. "The more you wait, the harder it is going to be for you to recover."

She went home and her father put her on the couch and walked out of the room. "Goddamn, everything is screwed up," she thought. "I don't have a job. I don't have any income. I'm going to lose my house. I still look like shit. And I'm going to have surgery on my hand. When is it going to end?" Those survival stories on the evening news never showed the rehabilitation of the injured. The healing could take as much effort, if not more, than the original act of survival itself, she thought.

She started to cry but caught herself. "Look, Venus," she told herself, "you've never been a lay-around person. Aren't you tired of it? Tired of calling mom or daddy every time you get thirsty or have to piss? You have to get moving here."

She stood up. Her body ached. She did a few stretching exercises. She took one step and then another and another. Then her body began to wobble. "Dad," she called out. It was going to take longer and more effort than she had expected. She was going to have to fight the battle every single day until she recovered. But now she knew she had the will.

"Gray, do you still have that motorcycle?" Robert Rogers asked his friend.

"Yeah."

"You mind if I use it?"

Haertig looked at him curiously. Haertig had taken several horrific falls from the motorcycle. "You know where it is," he said. "Take it. Get it out of here. I hate that damn thing."

Rogers pulled the Honda 90 out of the bushes by Haertig's house. It needed lower gears for the sprockets. He found Oregon Motorcycle Parts in the phone book and went to the combined garage and store. The owner was spotless in a white T-shirt and clean pants. But it was obvious he had been working on a motorcycle as he used a rag to wipe the grease from his hands. Rogers explained that he was repairing the old Honda. The owner wanted to know what of kind of use the machine would be getting so he could determine the quality and durability of the parts he would sell to Rogers.

"Well, I'm going to use it to get to a place to climb to the top of St. Helens," Rogers told the startled owner. For some reason Rogers thought he could trust this man. He explained his plan as the man became intrigued with the scheme. The owner went into the back and returned with the parts.

"How much do I owe you?" Rogers asked.

"Nothing. Just keep me informed of your progress." Rogers could see the man now felt like he was part of the conspiracy to get Rogers to the top of St. Helens. He took the parts and went to Haertig's house to repair the motorcycle. When he finished a few days later, he spray-painted it gray.

They came mainly from California, Washington, Alaska, and Oregon, but there were people from other states as well. Donna Parker was amazed to see the hundreds of people who streamed into the Western Forestry Center auditorium for the memorial service for her brother and his wife. There were the couple's fellow members from the Mazamas mountain climbing club. There were friends and coworkers from the telephone company, former military buddies of Billy's, nurses from Kaiser Hospital where Jean had worked, and neighbors and relatives.

The forestry center sat on a hill overlooking the zoo and the lawns and woods of Portland's Washington Park. The center was reflective of the mountains, forests, and the outdoors that Billy and Jean loved. Billy would not receive the funeral he wanted, but the family hoped this service might help make up for it.

Parker had helped her brother's friend George Gianopoulos select the slides he would show at the service. She helped pick out the best slides of Billy on family occasions and vacations. Memories flooded back of shared birthdays, Christmases, Thanksgivings, New Year's celebrations. She felt good about plans for the service. "Billy understands. He now knows, Donna," she told herself.

They ran out of chairs in the auditorium. People lined the walls. The service started and there were prayers, and readings, and the grasp for memories by grieving speakers. Then the lights darkened and Gianopoulos started the narrative dedicated to the life of his best friend and his friend's wife. As the projector clicked, sixteen-foot images of Billy flickered across the screen. He was laughing at a family party, he was sunbathing with Jean in Hawaii, he was clowning with George, he was standing tall against the wind on the summit of Mount St. Helens.

Parker found herself growing short of breath. It was one thing to look through tiny slides and another to see your brother's image before you larger than life. This had not been a good idea, she thought. Maybe in a year or two she might be able to take this. But not now. She shut her eyes. She found her throat and chest tightening. One of the daughters from Jean's previous marriage was in a wheelchair and she had to go to the bathroom. "Mom, I'm going to help take Debbie to the bathroom," Parker told her mother. "I'll be back."

But she knew she wasn't coming back. She stayed in the bathroom until Debbie was done. Then she found someone else to help with the girl while Parker stayed in the hallway. She waited until the lights came on, the doors opened, and the crowd came out. Then she went in and tidied up.

Swanson's trips into the crater continued to provoke argument. Some people thought that he and other scientists were risking their lives recklessly. But now Swanson and the USGS team had gotten a new boss, a man they hoped would settle the matter once and for all.

Don Peterson had been appointed scientist-in-charge of the newly created David A. Johnston Cascades Volcano Observatory. One of his first acts was to ask Swanson to become a permanent member of the observatory team whose main function was to monitor Mount St. Helens. Now, Peterson stood at Swanson's desk waiting for an answer.

Like Swanson, Peterson had received part of his training at the Hawaiian Volcano Observatory. The two men thought much alike. Peterson believed that like firefighters, volcanologists faced inherent risks in doing their jobs. He supported the work of Swanson and other scientists inside the crater.

Swanson also hoped Peterson could bring unity and direction to the team. Ever since the first days when USGS scientists gathered to study the nascent activity of St. Helens, there had been two distinct factions with differing outlooks on the study of volcanoes.

A good part of this evolved from the structure of the USGS itself. One faction came from the USGS Denver office, staffed with scientists from the agency's engineering geology branch. These were scientists who spent great amounts of time in the field, often alone, studying the geological impacts of past eruptions.

Almost all members of the faction to which Swanson belonged had worked at the observatory in Hawaii, studying *live* volcanoes. These scientists had worked more closely together and had become almost a family.

There were several ironies in their differences. The Hazards Team from the USGS was staffed with Denver scientists, who mostly studied dead or dormant volcanoes. The Monitoring Team had almost all Hawaiian Volcano graduates, who, though they worked with live volcanoes and knew their dangers intimately, were much less hesitant in entering the crater than the Denver group.

The division meant you had monitoring people gathering data without much thought to how that data would be used by the Hazards Team to evaluate the dangers the volcano posed. And you had hazards people interpreting the monitoring data without much thought of what went into the data and how much uncertainty it represented.

Swanson found the structure ridiculous. It created constant tension between the two groups that often led to office spats and inefficiency.

He tried to stay oblivious to the politics being played out around him except when it affected his work. Some scientists played the political game well and saw St. Helens as a stepping stone for their careers. Swanson was ambitious, too, but he was ambitious for his science. He now had answers to find and promises to keep. This is where he needed to focus his energies. He told Peterson he would take the job.

The nurse left Scymanky in the wheelchair in the middle of his room. In the past the nurses had always helped him get in and out. "Now it's time for you to get in and out of that chair by yourself," the nurse told him.

"Are you crazy? I'm not about to try to get out of this wheelchair without help."

"Yeah, you are," she said. "You're going to have to do it sometime." Scymanky lifted himself slightly before falling back into the chair. He convinced himself he couldn't do it and quit trying.

Day after day the nurses repeated the ritual, always ending the same way. One day he sat for a very long time in the chair, his back to the doorway. He was growing angrier by the minute. Didn't they realize he was doing the best he could, that he was not a slacker. Maybe this was as good as he would ever be, he thought.

An overwhelming feeling of helplessness fueled his anger. Leaning forward all of a sudden, he swung his body up out of the chair and was shocked to find himself standing on his own two feet. Even though his legs began to wobble, he wanted to keep standing to savor the moment. A voice behind him broke his trancelike state. "Oh, you finally decided to get up out of that thing." The nurse had been standing in the doorway behind him all along, witnessing his trial and triumph.

Hordes of townspeople waited to watch the dredge *Art Riedel* turn up Third Avenue in West Kelso to make its way to the boat launch on the Cowlitz River. Andre Stepankowsky could feel the vibrations grow beneath his feet. Streets had been closed. Traffic lights and telephone, cable TV, and electric wires had been moved. The *Art Riedel* was the first of the dredges that would be moved up the Cowlitz to remove mud and pumice from the river to reduce the flood danger.

To the accompaniment of flashing lights from escort vehicles and police cars, the *Art Riedel* turned up the street. Stepankowsky thought he might hear cheers go up from the crowd, but it was more like a collective gasp. The two-hundred-ton dredge was massive, even more so for being out of the water. It sat on the flatbed of a fifty-wheel truck. It was 86 feet long, 35 feet wide, and stood over 24 feet high on the truck.

It gave Stepankowsky a thrill to watch it. Then he looked around at the faces of the people. The dredge and its size constituted a symbol to these people who had suffered so much. It was a massive piece of machinery to provide some hope and protection against an unpredictable mountain and the forces of nature. In sharing their struggles, their fears and hopes, Stepankowsky realized he was forming a bond with these people.

In her nightmare it rained hard on Billy's truck. People were in the cab, eating popcorn and drinking. They threw trash everywhere. Their laughter echoed ever louder in Donna Parker's head. Then, in the dream, she stood looking in the distance at the truck that was parked on a huge hill of ash. It looked so very far away. She heard the voice of her brother calling her over the laughter. But she had no way to get to the truck.

Her head continued to throb with the laughter until she thought it would burst. Then she woke. It was the same terrible dream she had experienced so many times. Each time, the only difference was the weather. If it was raining when she went to bed, it rained in the nightmare. If the stars brightly lit the sky, they shone also on Billy's truck. But the trashing of the truck, the laughter, Billy's calling her, and her inability to get to the truck never changed.

Roald Reitan could get around on crutches now. Under the watchful eye of his physical therapist he was lifting weights and walking on a treadmill. As he limped along on the treadmill, he thought of Venus and the good times. Then in his mind's eye he saw himself running on the banks of a pristine river, trying to haul in a salmon with his fly rod. But soon his knee grew too painful to think of anything. He stopped. Roald wondered if he would ever be able to walk again without a limp, let alone run.

He had his best man with him and they picked up his future sister-in-law in Seattle. Stepankowsky was about to be married. As he drove the I-90 freeway heading for Spokane in Eastern Washington, the scene became bizarre. West of the Cascade Range, there had been occasional patches of uncleared land where ash still lay. But coming down off Snoqualmie Pass onto the eastern side of the range and the vast, flat farmland beyond, it looked like it had recently snowed. Mounds of grayish white ash lined the freeway. For miles the land lay in a mantle of ash. He knew the reach of the volcano had been enormous, but the reality had not sunk in until he was two hundred miles from St. Helens.

They had the ceremony in the small chapel of Fort Wright College where his fiancée, Paula LaBeck, had attended undergraduate school. She was a determined woman who was a match for Stepankowsky in many ways. He had not had much hand in the wedding plans. He had been too busy covering the volcano. When he got to the church he found he didn't know 90 percent of the people. Most were friends and family of LaBeck. From his side, his parents, two brothers, and a friend had flown across the country for the ceremony. Then there was his best man and the four staff members from the *Daily News* whom he now counted as friends.

It was a small but formal occasion with Stepankowsky in a tuxedo and LaBeck in a flowing white dress. After the vows, the couple signed the marriage certificate. When they turned around they were wearing ash masks. It was a gag they had planned, but they also wanted a reminder, putting their marriage into the context of the volcano that had come to so affect their lives. His sister-in-law snapped away with her camera.

Swanson and his crew installed eight new laser targets on the mountain. This was in addition to the other targets, tiltmeters, and seismographs already in place. Now he waited for the next blast. He knew it would again blow away much of his geodetic network, but he hoped it would leave him with valuable information about events leading up to an eruption—the all-important precursors that could be used to develop a dependable method of predicting eruptions.

Rogers lashed the motorcycle to the rear bumper of his Simca and headed toward Cougar. The deputy sheriff who manned the roadblock looked at him suspiciously. But there was nothing the officer could do. Rogers' pass was in order.

Rogers went into Jack's Sporting Goods and picked up the key to the trailer. In the trailer he plotted his trip for the next day. His photos of the May eruption were beginning to turn up in quickly published books and he was getting a little money for them. He stuffed plenty of rolls of film into the gray jacket that he hoped would blend with the ash if he came under surveillance from the air.

Before sunrise he walked the motorcycle a few miles up the main road. Then he revved it up and began climbing the paved Merrill Lake Road. He was ready to run the bike off the road if he saw a Forest Service truck. When the road became covered with the grayish black, sandy ash from the volcano, he stopped the bike. He pulled over, found the branch of a fir tree, and tied it to drag behind his rear fender to cover his tracks.

Rogers drove past Goat Mountain where he had met Francisco Valenzuela. He found a road that took him to the other side of the silt-filled Toutle River. After a while he hid the motorcycle in some bushes and walked up the South Fork of the river. He found the pickup he had seen before the volcano blew. It was here that the boy Andy Karr had died along with his brother and his father. The morning was growing clear and warmer, as it had on eruption day.

The river here was covered in a mudflow. Across the river, on the side he had been with Valenzuela the day before the blast, there was a path of devastation. Trees had been knocked down and their tips pointed toward the volcano, instead of away from it. Rogers always pumped his scientific friends for information. From them he had learned that the area may have been swept by horizontal vortexes, one from the east and another from the west. A turbulent, viscous flow of ash, traveling at times at hurricane speed, had gone down the slope below where he and Valenzuela had been and then changed direction and swept up the hill. It was easy to trace its path to where it had stopped, just before the place where he and Valenzuela had stood.

From the pickup he walked toward where he had seen the photographer Robert Landsburg and his station wagon. Rogers found a debris trail left from the destruction of the vehicle. He followed it down to the slope where the river bent. When he looked down to the river, he could see why authorities had trouble finding Landsburg's body and car. The car had been blown several hundred feet, along with the body, and had disintegrated into four large pieces and many smaller ones to rest on the bank of the river.

He stared at the wreckage for a long time. Rogers speculated that on the morning Landsburg died, the turbulent flow of ash had gone to the north out of his view and then had swung around and come over the ridge and down on top of him. Landsburg would have lived only a short while in the suffocating, intensely hot mixture of volcanic ash and rocks.

When Rogers turned to walk back, something caught his eye. It was Landsburg's tripod. He picked it up to take with him. He looked up at the volcano emitting lazy puffs of smoke. That was where he was going on his next trip, he told himself—to the top of the ridge to look once more into the heart of the volcano itself.

CHAPTER 9

July

"**DO YOU NEED** any psychological help in dealing with what has happened to you?" the psychiatrist asked Venus Dergan. She was getting ready for the surgery on her hand.

"No, I don't think so" she said.

"Well, a lot of people who have these types of experiences have traumatic aftereffects," the psychiatrist said. "Like post-traumatic stress disorder."

"I'm okay."

"Did you have any kind of special or out-of-body experiences while this was happening?"

"Like what?"

"Did you see any bright lights? Did you see your family flash in front of you or did you relive your past?"

"No. I never had any of that happen. I was worried, though, I was never going to see my family again."

"I've been told about what you and Roald went through. You know, you shouldn't be alive. How do you feel about that?"

"Listen, I don't think I'm going to have any traumatic stress over it, except for the stress I feel in getting my life back together."

A gray-haired woman stood in front of him in the supermarket line. Don Swanson guessed she was the grandmother of the young boy next to her. The boy wanted to know when the volcano would quit erupting. "Not until it erupts andesite," she said, referring to a form of dark grayish volcanic rock.

It was widely believed by the general public and many scientists that andesite would signal the end of an eruptive phase. But Swanson and other geologists had their doubts. Studies were under way to test this theory and a multitude of others, such as the possible effect of the moon on eruptions. Whatever their outcomes, Swanson thought the results would prove useful. They would help move the field of volcanology toward becoming a more definitive science than it had been throughout its history.

Three ancients made considerable contributions to volcanology before the Dark Ages sent mankind into sorcery and superstition. Seneca, who had better luck in understanding volcanoes than in tutoring Nero, was the first to describe the role of gases in an eruption. He theorized that each volcano is fed by its own reservoir of magma, or molten rock. He believed underground fires propelled gases to tighten like "springs of the blast" until they were released in an explosion

A Roman naval officer and scholar, Pliny the Elder, brought the next substantial advancement in the understanding of volcanoes. In his encyclopedia he listed every volcano known to man. There were ten of them. His fascination with volcanoes ended with his death in the eruption of Mount Vesuvius in AD 79. But his nephew Pliny the Younger was there to take over. He proceeded to describe his uncle's death and the eruption in exquisite detail that would inspire young volcanologists throughout the ages. In his written description, Pliny the Younger had it all and more: the telltale earthquakes before an eruption, the enormous vertical column, lightning zigzagging across ash clouds, sulfurous gases, the pumice responsible for the burial of buildings and ash-filled lungs, hot ash flows, and the receding of the sea with fish flopping on the shore in the moments before the tidal wave.

In medieval times, volcanology seemed to recede into a mixture of myth, religion, and superstition. This was best represented in Dante's masterpiece, *The Divine Comedy*, where a devil dwells in the bowels of the earth in a fiery hell reserved for the unrepentants' eternal damnation. The volcano was thought to connect the world of the living with the devil's subterranean domain. Lava burst forth from that hell, with its accompanying noises interpreted as the groans, shrieks, and moans of the damned.

More than four hundred years later, in the late eighteenth century, a man named Abraham Werner lent his name to a theory that all crystalline rocks were precipitated from an ancient universal ocean and that volcanoes were caused by the combustion of coal. Three of Werner's students disproved his theory. They showed that contrary to the belief that volcanoes were the result of rare phenomena occurring in recent history, volcanoes were widespread in much earlier times and in settings that could not have involved coal burning.

At the beginning of the twentieth century, instruments were developed that for the first time allowed scientists to study volcanic processes taking place beneath the earth's surface. Italian volcanologists were the first to install seismographs on the sides of a volcano in hopes of forecasting eruptions. Japanese scientists provided new understanding of the link of certain earthquakes to volcanic activity and the relation of surface deformation to the movement of magma deep beneath the surface.

Although the USGS was created in 1879, other entities did the pioneering work on volcanoes in the United States. In 1907 the Geophysical Laboratory of the Carnegie Institution of Washington was founded, soon to become the nation's leader in the study of volcanoes. Studies by the laboratory's scientists provided the foundation for the modern volcanology that Swanson and other scientists were now practicing on Mount St. Helens. But still it was a crude science.

The state of volcanology at the time of St. Helens' May 18 eruption was perhaps best summed up in the respected textbook *Volcanology*, by Howel Williams and Alexander McBirney, published only a few months before the blast.

> *The progress made since the classical scholars of Greece and Rome explained volcanoes as products of combustion and subterranean storms is more apparent than real. Geologists looking at Vesuvius today can tell far more about the volcano than could Pliny the Elder in AD 79; they can describe its composition, including the trace-element abundances and the isotopic ratios of its lavas, but they cannot say why it has this composition. They can date eruptions of ancient lavas but cannot say when or why new ones will occur. They know the temperature, viscosity, and gas content of magmas but cannot say where or how the magmas were formed or how they reached the surface, nor can they explain why one volcano has an eruptive behavior dramatically different from that of a neighbor only a few kilometers away. Geophysicists detect earthquakes and subtle differences in the earth's gravitational and magnetic field in and beneath a volcano but cannot delineate its magma chamber or even say for sure that one exists.*
>
> *The reason for our ignorance and the slow pace of volcanology is not hard to find. Few important magmatic processes can be observed directly; most can be studied only from a remote position in space or time. No less discouraging is the diversity of complex relations that must be interpreted if one is to view any individual feature in its true dimension. Indeed, it seems that the more we learn, the more difficult it becomes to integrate our knowledge into a comprehensive theory of volcanism.*

Swanson and others hoped that Mount St. Helens would advance their science. Rarely had an active volcano been so close to urban centers with their universities staffed with specialists in all areas of science. And the fact it was the first eruptive volcano on the mainland United States in more than sixty years and posed a threat to life in nearby cities spurred Congress to come up with record funding for volcanic studies.

Streams of data from Swanson and other scientists poured into USGS headquarters in Vancouver: tilt observations; geodetic monitoring of deformations; emission tests of sulfur dioxide and other gases; seismic readings; temperature measurements from cracks, steam vents, lakes, and magma pools; infrared photographs; recordings for sonic detection; maps and composition studies of avalanche debris and other eruption deposits. Enough data had been collected and analyzed to change the face of volcanology forever. Swanson and the other scientists now knew a great deal about what had happened on May 18. But predictions of future eruptions continued to be based on seismic activity that at most gave a few hours' to a day's warning and were not entirely reliable. They still had no foolproof way to predict eruptions.

"Have you seen how little they've dredged up?" the woman on the phone asked Stepankowsky. "Now how in the hell are they going to get this done before winter?" The woman wanted him to write a story about how incompetent the Army Corps of Engineers was in its dredging of the Cowlitz River.

"Ma'am, they just got started," Stepankowsky told the woman. "Don't you think we ought to give them a chance?" It was true that the dredges *Art Riedel* and *Herb Anderson* had just started work on the upper Cowlitz River while the *Missouri* and *Husky* continued work on the lower part of the river. But the woman on the phone was not the only one concerned about whether the Corps could get the job done before the winter rains. Experts both inside and outside the Corps disagreed.

The Corps originally announced that it hoped to restore flood flow capacity to the Cowlitz by October 1. Then the chief of the Corps' civil works division announced that homes along the Cowlitz and Toutle Rivers might have to be evacuated in the face of severe flooding in the fall. He said there was too little time before the beginning of the rainy season to completely clear the Cowlitz of volcanic silt.

The Corps was trying to squeeze dredging work that would normally take five to seven years into what was left of the summer. "The Corps," Stepankowsky had written, "has brushed aside environmental requirements, hired dredges without knowing exactly how much dredging will be done, and forgone detailed studies in an emergency flood control effort." The Corps official he had grown to respect and trust the most, Patrick Keough, admitted the Corps didn't have all the answers. "You have to change your game plan in the middle of the game and that will continue to happen," he told Stepankowsky. "What looks like a real good solution today looks like hell tomorrow."

Editors at the *Daily News* were discovering that Stepankowsky had that elusive quality essential to a great reporter, "a nose for news." When Stepankowsky wrote his story about the statistician who predicted the Cowlitz could flood with only an inch of rainfall, he had been criticized in some quarters for reporting questionable conclusions and sensationalizing. No other newspaper had picked up the story. But now none other than the National Weather Service believed the statistician had been correct. Bud Walsh of the National Oceanic and Atmospheric Administration said the Weather Service had calculated that a typical October rain of one to two inches in a twelve-hour period was "likely to result in appreciable flooding. . . . I would advise people in the area to listen very closely to the weather forecasts and to warnings. Warning times will be limited, and people will have to move quickly."

The changes in Scymanky were becoming noticeable to the nurses. He now joined their banter, joking with them. Gradually he was revealing more of what had happened on the volcano. Helen started to allow herself to believe her husband was going to make it. She rarely missed a day sitting by his bedside, giving him encouragement or just talking to him, getting things he needed, and even helping him to eat. One day she was told by a nurse that Jim was no longer in critical condition and that Helen could now help her husband most by not coming every day.

Then Scymanky was told he was to be moved to the hospital's rehabilitation center. "Do I have to go?" he asked. He was now walking and feeling good about himself. He had grown used to the people and routines of the burn center. The move scared him. When the nurses ribbed him about the move, he tried to smile, but it was a forced smile. "You won't get any coddling like you do here," they would say. "You're going to be so busy you won't have time to piss and moan."

Yet after the move Scymanky found he liked the rehabilitation center. Patients in the burn center always seemed to be in pain and were kept in separate rooms. It was hard to talk to them. Patients now came up to him and wanted to know about his experiences on the mountain. He was a hero.

Scymanky was put on a tough schedule. Promptly at 7:30 he was awakened by a physical therapist who stretched Scymanky's stiffened muscles in a very painful exercise. After breakfast he reported to the physical therapy room for mobility exercises that lasted most of the morning. There was more physical therapy in the afternoon, followed by occupational therapy. Only for a short period in the evening did he get a chance to relax. Then they changed his bandages before he went to bed.

One day he met another burn patient, a man who had tried to commit suicide by pouring gasoline on himself and igniting it. The man, unlike Scymanky, was badly disfigured, with a face that looked like a rubber mask that had been melted into contorted forms. They became close friends. They would joke with each other while they awkwardly tried out new exercises. When the pain of their burns sometimes intensified, they would talk each other through it. But Scymanky could never understand why anyone would intentionally burn himself. When he got to know the man better, he felt comfortable enough to kid him about it in such a way his new friend always laughed. "Man, couldn't you have thought up a better way to get rid of yourself?" Scymanky would ask.

Donna Parker sat in the office of her State Farm insurance agent. She had learned that Jean's oldest daughter had collected the insurance money on Billy's truck. "I want you to help me buy the truck back from the insurance company," she told the agent. The agent made a couple of phone calls. The second one put him in touch with the person responsible for disposing of totaled vehicles for Billy's insurance company.

"You could probably get it for two dollars," the agent told her as he covered the mouthpiece of the phone.

"I don't want to take a chance on losing the bid."

"Donna, you're the only one bidding."

"I can't bid two dollars on Billy's truck. That's too cheap. Tell them I'll bid two hundred dollars."

"What?"

"It's something I feel in my gut that I've got to do."

"Oh, Donna."

"Tell them."

Soon she owned the truck, and next she would bring it home. That's what Billy would have done. This became the single most important goal in her life. A few days later the attorney representing Jean's relatives called her. He had heard that she was going to rent a helicopter to fly the truck out of the devastated area. "If you know what's good for you, you'll leave that truck and its contents alone," he told her.

"Whatever you say," she told the attorney.

She wasn't going to tell him or anyone else that she owned the truck. Let the attorney think he had her buffaloed. From now on Parker was going to keep her mouth shut. She wasn't going to take a chance on someone thwarting her plan. She would spring the ownership papers on them after she had the truck out.

The dormant Mount Hood volcano sprang to life with a flurry of earthquakes that sent shudders through the people of nearby Portland. What concerned Swanson and other scientists was the evidence indicating that Mount St. Helens, Mount Hood, and even Mount Rainier outside of Seattle had erupted in concert in the past. Such a confluence of eruptive activity could threaten a great many lives.

Mount Hood continued to quake. USGS geologists started a volcano hazard watch on the mountain.

The picture in the newspaper shocked Scymanky. It showed Sharipoff's contorted body hanging in a hemlock tree along the North Fork of the Toutle River. The logger's body had just been discovered, fifty-two days after the eruption. The rest of the day, Scymanky was haunted by thoughts of what it must have been like for his friend. First there had to have been the pain, then the loneliness, despair, and finally death. Scymanky was the last survivor of his crew. He felt very alone.

"Roald, right before I had my hand surgery, there was this shrink who told me you and I need help, whether we know it or not," Venus told him over the phone. "He's not the only one. There's been other doctors and friends. Do you think we're going to have some repercussions over this?"

"Nope."

"So, you don't think we need to see a shrink?"

"I can get over this and move on with my life without any damn shrink."

"Yeah, I guess I can, too."

From the first day Scymanky entered the hospital, media demands to interview him were constant. Television stations sent crews to try to see him. Dozens of newspapers ranging from the *Seattle Times* and the *Oregonian* to *Long Island Newsday* and the *London Times* requested interviews. National magazines sent writers, and authors writing books wanted to see him. Each day, the hospital received more requests for interviews. But none were granted as the hospital's most famous patient struggled back to wellness.

When Gary Eisler from the hospital's public relations department approached Scymanky to recommend a news conference, Eisler discovered that the recovering man disliked publicity. Even more, Scymanky hated public speaking. Eisler told him it would be much easier to control the situation if Scymanky agreed to a news conference while he was still a patient at the hospital. Otherwise, Scymanky was going to be flooded with media calls after he was discharged and at home. Scymanky put Eisler off.

Then he started thinking more about the news conference. Scymanky had been able to let out what had happened to him on the mountain only in bits and pieces. He could feel his experience bottled up inside. Perhaps it would be better if he got it all out. The clincher, however, was when he started thinking about his friends on the work crew. Who was left to tell their story?

Scymanky, wearing an "I'm a Mount St. Helens Survivor" T-shirt given to him by another patient, was overwhelmed when he entered the room set aside for the news conference. He had thought there would be perhaps a half-dozen reporters. Instead he was facing scores of media people. Photographers rushed forward, flashes going off in his face as he sat down behind a bank of microphones. His stomach churned but it was too late to back out. Reporters peppered him with questions. Scymanky kept looking down at his hands and almost mumbled as he tried to answer.

Then a reporter asked him what had happened on the mountain. Looking up, he began to talk. He talked on and on, his narrative never missing a beat. The room grew quiet as the media people found themselves back on the mountain with him, living through the horror. There was a long silence when he finished. The reporters had been stunned by his presentation and seemed to forget they were there to ask questions.

Television stations broadcast, without interruption, Scymanky's description of that day on the mountain, though it was much longer than their usual news segments. Some newspapers printed it almost verbatim. Through the wire services, the story was told across the nation. It had been a virtuoso performance from a man whose worst subject in high school had been public speaking.

"Is anyone missing at this address?" the voice on the other end of the phone line asked Roald Reitan.

"Why are you asking?"

"This is the Cowlitz County Sheriff's office. We've found a car registered to a Roald Reitan at this address."

"That's my car. I'm Roald Reitan."

"It's mostly buried in dried mud, but the trunk is accessible. There's still valuable contents in it. When do you think you can claim your property?"

He laughed. "I'm not exactly in shape to be going anywhere, but I can send someone."

His brother and a friend went to the car. They found the trunk sprung open. Inside they recovered Venus's purse, her wallet, a cooler, the tent, cans of food and beer, a Winchester .22 rifle, and a bottle of champagne. The men dug out around the doors to the front of the car, but before they could finish, the car began sinking deeper into the mud. After they got back, Roald packed away the champagne. He would save it for when he and Venus could celebrate being together again.

Donna Parker called Portland's KATU Channel 2 television station, hoping they could help her get a helicopter to fly out her brother's truck. She thought the station would want to do it for a news story. "We've got our own truck stuck in there, we're trying to get out," the woman at the station told her and hung up.

Parker then called Columbia Helicopters, located just a mile from where she lived. She asked the receptionist how much it would cost to rent a Huey helicopter to pick up the truck. "Just a minute. We have somebody who's been working up there who can help you."

Dwight Reber was a pilot. Parker explained what she wanted and told about the county withholding permission to retrieve Billy's pickup. "Don't make any more phone calls, Donna. I will help you," Reber said. "Why don't you stop after work and we'll talk about it." She felt like she was floating on a feather the rest of the day. Somebody was going to help.

Parker bounded up the stairs to the second floor of the Columbia Helicopter building. When she met Dwight Reber, she was taken aback. He was tall, slim, dark-haired, with a full, neatly trimmed beard. He searched deeply into her eyes for a moment, as though looking into her soul, she thought. For some reason she felt the first sense of peace she had experienced since her brother's death. She knew it was silly, but she couldn't help thinking that except for the lack of long hair, the pilot reminded her of Jesus. The hair on the back of her head stood up.

Reber told her about his background as a helicopter rescue pilot in Vietnam and then launched into his plan to retrieve Billy's truck. She listened to him for a few minutes. "Now, wait a minute. How much is this going to cost me?"

"It takes three thousand dollars to just turn on the engine of a Huey."

Parker tried to muffle a gasp. She searched her memory for when she had even seen three thousand dollars in cash. Finally she said, "I guess I could borrow three thousand dollars from somebody."

The pilot looked at her and smiled. "Listen, I'm going in there soon to retrieve some film in Reid Blackburn's car for the *National Geographic*." Blackburn had been on assignment for the magazine when the mountain erupted May 18. He had suffocated in his ash-clogged automobile. "Maybe I could stop at your brother's truck and pick up some things."

"Can I go along?"

"No, you can't go."

"Why not?"

"It's a real mess out there, Donna. It's really, really bad. If I have to set down, nothing personal, but you're not going to be able to walk out of there."

Parker got the drift about her gender and she tried to ward off her feelings of offense and disappointment. "My brother's good friend can go with you," she said.

He stared at her for a few moments. "Okay. Here's what we'll do. When the weather changes and opens up again, I'll call you and you call him. Have him meet me at Pearson Air Field in Vancouver. I'll tell you when."

"What about Cowlitz County?" she asked. "They told me if they gave permission we would need to take a deputy sheriff along."

"Don't have time. Don't have room," he said. "I don't want to deal with them. I don't even need to ask their permission to fly in there. I'm already cleared. Don't ever talk to them again."

George Gianopoulos was happy to be recruited to help. Each morning Parker woke and looked out at the weather and wondered if this would be the day. Several days after her talk with Reber, she got the call just after she arrived at work at 7 a.m. "Get your friend over to Pearson Field right away," he told her. She called Gianopoulos, trying to contain her excitement. "They're ready to go, George."

For the rest of the day she kept looking through a window or going outside to scan the sky and imagining the chopper landing at Billy's truck. It began to cloud over. By quitting time at three o'clock she wondered why she hadn't received a phone call. Parker rushed to the airport. Officials there had not heard from the chopper and were getting worried. She knew she could not stand another tragedy, especially one she might have set in motion. Finally the helicopter appeared from the flannel-gray clouds and landed.

Gianopoulos told her that Blackburn's car had been filled with ash past the steering wheel. They had to dig out the ash while sifting for film and cameras. At last they found them but it was then past the time they had been allotted to work in the Red Zone. They ran to the chopper and flew to Billy's truck. It too was filled with ash.

In their rush the men flung sleeping bags, camping gear, fishing rods, and other paraphernalia out onto the ground as they dug through the ash in the canopy and then the cab.

Gianopoulos showed her what he had recovered. There were the telescopic lenses that Gianopoulos had lent to Billy and a tripod with Billy's camera on it. Parker grew excited when she heard there was still film in the camera. Gianopoulos told her he had not been able to find Billy's handgun, rifle, and the three other cameras they knew Billy had with him. They might be there, but it was going to take another trip to find them. Reber refused to take any money for the flight.

He was sure of it now. The rampart, the ridgelike feature that ran crosswise in the crater, was accelerating in lateral movement. From a reading station miles outside the crater, Swanson had been using his geodimeter to monitor a target on top of the rampart. For almost a week he had recorded the rampart's movement north at a rate of 1.2 centimeters per day. Then for the past four days he watched the ridge's movement increase until it was moving 4.2 centimeters a day. Something, which Swanson speculated was probably magma, was stressing the ground beneath the crater as it made its way to the surface. He wondered if his measurements of this movement and stress could be used as an indicator of an impending eruption.

He reported his findings to his superiors. This sparked a debate over the significance of the movement and whether to release the information to the public. Some scientists argued that Swanson's findings could indicate only the beginning of slumping or a landslide. Others pointed out that his measurements were close to what could be caused by instrument error. A decision was made to not release the information. But in the days that followed, Swanson watched the rampart increase its rate of movement. He became convinced that Mount St. Helens was on the verge of another eruption.

Scymanky was apprehensive the day he was to be discharged. He still felt periods of weakness, and the pain had not fully gone away. The doctors assured him he was ready to go home, but he was not sure he was prepared to face the world again. His room was crowded with nurses, physicians, therapists, and patients. Everybody wanted to say goodbye. At noon, Helen helped him carry his things out.

As he passed through the double doors of the rehabilitation center, Scymanky felt strange. He was happy to be going home, but he also felt alone. He thought of Skorohodoff, Dias, and Sharipoff. There was no one else left who could really understand what it was like for them the day the volcano exploded.

Scymanky had been only thirteen miles from the volcano when it erupted. Few others in the path of the eruption had been closer and survived. The mountain had taken its toll as the most lethal recorded in North American history. By the time of Scymanky's release, the bodies of twenty-eight volcano victims had been accounted for. In addition, thirty-six people had vanished and were presumed dead. While still being adjusted the official list of dead and missing thus stood at sixty-four. But it was recognized that if the volcano had erupted just a day later, on a Monday with the woods full of loggers, the official death toll would have been closer to a thousand.

"The mountain just blew up. Get back up there," the new city editor, Dave Connelly, told Stepankowsky and photographer Brian Drake. It was July 22. Mount St. Helens had erupted at 5:14 p.m. in a massive blast that sent billowing steam and ash plumes to an elevation of sixty thousand feet.

"They'll have the mountain closed off," Stepankowsky said in mild protest. He was tired, hot, and hungry. He had spent the day with Drake at a sediment dam the Corps of Engineers was building on the North Fork of the Toutle. Earthquakes shook the area most of the time he was there. Dust clung to his clothes. He could still taste the ash.

"Just go back up there, anyway," Connelly said.

The two men jumped into the company-owned, four-wheel-drive Chevy Blazer, Stepankowsky at the wheel. He decided to stop at Silver Lake where some tourist attractions had sprung up and there was a halfway decent view of St. Helens. At least here he could interview tourists and Drake could get some photographs even if it was well outside the Red Zone and more than thirty miles from the volcano.

Thick gray clouds over the mountain wiped away at the brilliant blue sky. As they drove, Stepankowsky and Drake saw the plumes of steam and ash head northeast. Debris sifted down as a hazy layer over the volcano.

At Silver Lake a few sightseers still had their cameras aimed at St. Helens and the remnants of the eruption. Others gazed at the dissipating eruption clouds and drank soft drinks. Stepankowsky and Drake walked toward the tourists.

"My wife told me it would blow for me," George Adams told Stepankowsky. Adams, a retired Los Angeles naturalist, had arrived only an hour before. "It was glorious," he said.

His wife chimed in, "If you promise to make it go again, he'll never leave."

Just when it looked like everything was settling down, a second eruption jetted in black cauliflower balls up through the hazy layer to catch the first eruption cloud now zigzagging to the east. All around him Stepankowsky heard the *click* of cameras and the *whirr* of movie cameras. The photographers jostled for position amid exclamations of awe. The volcano was putting on its best show since May 18.

A kneeling woman posed her son fifteen feet in front of her, with the erupting volcano behind him. Drake caught the scene with his camera. "Kodak, you're making a mint today," a bystander yelled.

The blast was spectacular, but weather clouds and eruption smoke obscured the top of St. Helens. Drake thought they might get a better photo closer to the mountain. Stepankowsky and Drake continued up Spirit Lake Highway and crossed over to the South Toutle Road. A few miles farther on they encountered a roadblock manned by a Cowlitz County Sheriff's deputy.

"You can't go in there," the deputy said. But Stepankowsky knew they were still far outside the Red Zone.

"Look," he said, "we're from the newspaper. We're just trying to get a picture. We're not going to try to get into the Red Zone." He knew the newspaper's Red Zone permit was not valid during actual eruptions.

The deputy hesitated and looked over both men. "All right. Go on through. If you're not out of there in fifteen minutes, I'm coming in to get you."

Ten minutes later, just as Stepankowsky and Drake had found a good location to take photographs of the fading eruption, the deputy reappeared. "You got to get out of here, now," he told them.

"But we're miles outside of the Red Zone."

"Now, I said. I want you out of here."

As they drove back past the roadblock, Stepankowsky's stubbornness kicked in. "You know we're not in the Red Zone here," he said to Drake. "I want to try to find a way to get up to where we were, so we can get a picture."

He found a logging road with the gate swung open. Stepankowsky turned up the road and smiled to himself. For several miles he drove the Blazer with clouds of ash dust trailing them. The setting sun threw hues of red through the sky and the settling veil of the eruption cloud. The moon was rising over the volcano. For a few miles more, he drove toward the mountain. Then he began to get worried. He could tell they were now in the Red Zone.

Stepankowsky stopped the Blazer and they hiked up a hill of ash. The eruption cloud had dissipated too much to be distinguishable. There were no great photographs to be had. "We got to get out of here," Stepankowsky said. He gunned the Blazer back down to the gate they had entered earlier. It was now closed, secured with a heavy padlock.

"We're going to have to find a way across the North Toutle," Stepankowsky told Drake. The only problem was, they would have to drive back toward the mountain to cross over the North Fork of the river. Stepankowsky drove toward the still smoking volcano. The sun was setting in the west behind them. They tried several roads that dead-ended into the roaring North Toutle. Then they found a temporary logging bridge. The wooden bridge creaked and groaned as Stepankowsky slowly drove across it with the swirling river below. On the other side Stepankowsky looked up at the mountain, whose dissipating eruption cloud was now over them, blocking out the last rays of daylight. They had been inside the restricted zone for hours and it might be hours more before they got out.

He found a road that appeared to be going in the right direction, away from the mountain. Time seemed to slow to a crawl even though he floored the accelerator. They were now going somewhere very fast, but where they didn't exactly know. Then he heard the *whop, whop, whop* overhead. It was a Weyerhaeuser helicopter. The helicopter swooped down, keeping pace with the Blazer, a few dozen feet in front of the windshield. The copter tilted its body from side to side and slowed further, until Stepankowsky thought for an instant the Blazer might hit it. Then it swooped almost straight up into the air. He listened to the noise of the whirling blades fade away.

The Blazer shook and rattled as he drove it at top speed, with total darkness now descending. Unseen and unheard by the two men, St. Helens was now erupting for the third time that day. Finally Stepankowsky slowed down. He couldn't believe his luck. The gate to the main road was unlocked and there were no police nearby. He sped through. The reporter breathed a sigh of relief.

Then the headlights of a car behind him reflected off his windshield mirror. They grew larger until the car was right on his bumper. He turned to see the car pulling up alongside of him. It was a Weyerhaeuser security vehicle, with the guard indicating to him to pull over. He put on the brakes and pulled the Blazer to the side of the road.

"You guys were in the Red Zone," the guard told them. "We're going to need to straighten this out with the sheriff's department."

"You've got to be kidding me," Stepankowsky said. Then he looked into the guard's stone face and realized he wasn't. "But we got into the Red Zone by accident."

"Right." The guard went back to his car to ask the dispatcher to send out a deputy.

The deputy was one Stepankowsky recognized: a cop's cop, hard-nosed with no patience for nonsense. The officer was angry. They had gotten him out of bed. "You almost got that helicopter crashed, too," the deputy yelled at them. He cited them but did not haul them to jail.

It was 1 a.m. by the time Stepankowsky dropped off Drake and the Blazer at the office and arrived home. When he got to the office the next morning he was still numb from lack of sleep. Managing editor Bob Gaston let him know he had placed the paper's Red Zone pass in jeopardy. That afternoon he opened the *Daily News* and found the story he had written about the tourists at Silver Lake on page three, along with one of Drake's photos. Above that story in the prominent top, right-hand corner, a headline said, "Reporters cited in Red Zone."

"A reporter and photographer from the *Daily News* were charged Tuesday evening with unauthorized entry of the restricted area around Mount St. Helens," the article said. "Reporter Andre Stepankowsky, 24, of 1215 20th Ave., Longview, and photographer Brian Drake, 25, of 3120 Maryland Street, Longview, were cited and released by a Cowlitz County sheriff's deputy. . . . 'I was in there because I wanted to get a story,' Stepankowsky said."

The volcano, in the series of three eruptions on July 22, had jetted plumes eleven miles into the sky. The eruptions annihilated the crater's twenty-story dome, leaving a huge glowing pit in the crater, and sent an avalanche of superheated ash, gas, and rock—a pyroclastic flow—five miles down a valley almost to Spirit Lake. Witnesses reported hot pumice rock flowing off of St. Helens "like a waterfall."

At a press briefing a USGS spokesman described Swanson's measurements of the rampart movement as a "hint" that someday the volcano's eruptions might be predictable. The spokesman, geologist Tim Hait, cautioned against excessive optimism. "Perhaps these movements tell us there was a slight expansion beneath the surface, but we haven't fit them into the overall picture," he told reporters. "There was nothing we could base a prediction on." But in reality the link between the movement of the rampart and eruptive activity was becoming indisputable.

Although there were three distinct blasts on July 22, they were counted by the USGS as only a single eruption, starting when lava and gas first broke the surface and ceasing when the molten rock stopped flowing. Swanson made preparations to land in the crater again. He had to replace the target he had planted on the rampart in June. The blasts had charred the target beyond use.

In the dark Rogers rode the motorcycle up the west approach to Mount St. Helens. He hid the cycle and began his climb to the rim of the volcano. It was six days after the mountain's one-day series of three eruptions. When he got to the top the sun was rising over the rim across from him. He looked far down at the floor of the crater. It was glowing red. "That's what I came to see," he told himself. He sat down and dug into his backpack to grab his camera and a roll of film. By now, though, the red glow from the crater below had washed out.

"I'll just stay here all day," he thought. "This is going to come back tonight, clearly. I'm not in any hurry." He sat on top of the rim. Steep slopes fell away on both sides. Suddenly from the floor of the crater an explosion produced by steaming underground water spewed a geyser into the sky. The rising sun silhouetted and backlit a large dark cloud beyond the geyser. Through the breach of the crater, he could see Spirit Lake and Mount Rainier. He took several photos as the cloud mushroomed out into a dragon's head. Then he looked back toward Longview. Two triangular shadows met across the forest like two cones balancing point to point. A plume went up and drifted toward a nearly full moon about ready to set. He clicked away.

He spent the rest of the day hiding from airplanes, hunkering down within a crack on the rim of the crater. He mostly drank water and slept. Every once in a while he awakened to the sound of the crater shedding rocks off its inner walls.

A glorious sunset made up for the long and boring day. The sky stratified into bands of orange, purple, blue, and black hanging just above the horizon, with a ball of reddish yellow sinking into them. He shot almost a whole roll of film. As it grew dark he lay on his belly, peering through the cockeyed binoculars he had traded some of his photos for. He became fascinated with watching a huge crack that lay across the crater floor over a mile beneath him. The crack grew redder as he watched and darkness descended. He couldn't think of enough words to describe the streak of brilliant red that now zigzagged across the darkness below.

It was spectacular. It was phenomenal. It was astounding. He grabbed his camera and found he couldn't stop shooting. The full moon rose over the crater. Rogers shot roll after roll. At almost midnight, he quit. His next trip, he told himself, would be to the floor of the crater.

He walked down the backside of the mountain. Suddenly out of the ash, in the volcanic wasteland hundreds of feet above timberline where it was thought no prey could exist, an owl took to the sky right in front of Rogers. In the moonlight he watched the owl's effortless glide across the mountainside. He wasn't sure, but he thought he saw a rabbit scampering across the ash in front of the owl.

Swanson had reestablished most of his geodetic monitoring stations and targets into a network that surrounded the mountain. Initial measurements had been taken from stations he named Butte Camp, South Fork Toutle, Clearcut, Muddy, and Road 100. Because of poor weather and other factors, he had not been able to establish the most important measuring post, a northwest position looking into the gaping breach of the crater to observe developing domes and ramparts.

At the end of July he got his chance. The helicopter set him and assistants down five miles from the crater on a prominence not far from where Harry Truman and his lodge were buried under three hundred feet of mud, ash, and debris. Scientists had dubbed it Harrys Ridge in honor of the fallen folk hero. He set up the geodimeter on a portable tripod and focused it. First the crater came into view, then his target reflector sitting on top of the post on the rampart. This was going to be an excellent station, providing a clear sight line into the crater.

Then the wind began blowing. He had to brace his body and hold down the geodimeter. This instrument was even more sensitive and finicky than earlier models. Minor temperature and pressure differences between the reading station and the reflector could throw off the calculations. If the wind was this bad now as it almost knocked him over, he wondered how bad it would be in the winter when fierce storms hit.

The hazard watch at Mount Hood outside of Portland had been called off. The earthquakes had stopped and the mountain returned to its slumber.

CHAPTER 10

August

ON THE MORNING of August 6, scientists monitoring seismic activity predicted an eruption. Logging crews were evacuated a little after noon. On the following day the mountain erupted at 4:23 p.m. and again at 10:23 p.m., with flows of rocks, debris, and dust settling halfway to Spirit Lake. The eruptions were much smaller than any previous ones.

To celebrate the successful prediction and unwind from weeks of grueling work, the scientists staged a pig roast. Steaming vents at the base of the mountain provided a natural oven. The venture was given the code name "FPP temperature experiment" after a professor from Portland State University who was nicknamed Front Page Palmer for the way he grabbed headlines associating himself with the volcano. A crew cooked the pig at 575 degrees Fahrenheit at a shallow vent precariously close to the ramp at the mouth of the volcano, then radioed back to headquarters: "Vancouver, this is Five Six Yankee. We've completed the FPP temperature experiment and we're bringing the bacon home."

The scientists partied at a Vancouver park late into the night. It was the first break they had taken since May 18 to relax as a group. But Don Swanson was not among them. He was busy at work.

The world was now getting a more precise scientific account of what happened that horrific day in May. The theory of plate tectonics gave Swanson and other volcanologists a structure they could use to generally account for what led up to the May 18 eruption.

Scientists believe the earth is broken into approximately a dozen fragments known as plates. The plates, up to about sixty miles thick, float on a roiling molten core. When the plates bump against each other or pull apart, they cause earthquakes along faults. But when the edge of a plate slides beneath a neighboring plate, another ominous threat results.

The Juan de Fuca plate off the coast of Washington state has long been expanding eastward and diving under the much larger North American plate, in a process known as subduction. The process produces pressure so intense that the crustal edge of the Juan de Fuca plate melts into magma at a depth between sixty and one hundred miles. Some of this magma snakes its way through fractures in the plate, searching for a weakness in the earth's crust. The magma has found these weaknesses repeatedly over time, bursting through the surface in eruptions that formed and re-formed Mount Rainier and other Cascade Range volcanoes including Mount St. Helens.

In 1980 the magma exploited one of these weaknesses beneath St. Helens. It pushed high into the cone, where it was blocked. The edge of the magma crusted. The mountain above and the hardened crust pushed down on the magma with a force of 31,000 pounds per square inch. But the magma below was still buoyant and produced a more powerful counter-thrust.

This thrust pushed out the northwest side of the mountain in a bulge called a cryptodome. Enormous pressure built up. On May 18, as it had many times before in the history of the Cascade Range, the magma erupted in a huge explosion. But there was a difference. This time the explosion was released by an earthquake, a belief held by Swanson since eruption day and later confirmed by other scientists.

The 5.1 magnitude earthquake was centered a mile below St. Helens. The mountain already was weakened by the intrusion of the pressurized gases and magma that had pushed the bulge out more than four hundred feet. When the earthquake hit, the whole north flank of the mountain started rippling like a wave. Within seconds the side of St. Helens collapsed into the greatest avalanche in recorded history, traveling at 150 miles per hour.

Pent-up pressure with the force of twenty-four million tons of TNT sent a hurricane filled with rock and ash, along with glacial ice, jetting laterally to the north at the speed of sound in a thick, black plume. A vertical column of pulverized rock and ash shot to the sky, rising at a speed of over sixty miles per hour.

Flows of gray fumes boiled over the mountain. A shock wave led the lateral explosion, toppling trees like toothpicks. Rock churned in the air to become pulverized as the blast hurricane ripped up entire forests.

By the time this fury reached David Johnston at five miles out, it was traveling at 650 miles per hour. Purple-black clouds filled with lightning reached down to cover valleys and rip up the earth to bedrock. The avalanche now behind it, the hurricane grew to become a wall more than a mile high with entire forests in its grip.

When the avalanche hit Spirit Lake, an 850-foot-high wave rose into the sky, leaped out to grab trees, and then sloshed back into place taking a whole shredded forest with it to rest more than two hundred feet above its former level. The avalanche continued its journey, moving not only down into valleys, but up over hills, seeming to defy gravity.

At fifteen miles out from the summit, the blast hurricane suddenly stopped, reversed course, and rushed back toward the mountain at about fifty miles per hour, leaving forest fires in its wake.

All of this took place in less than ten minutes. Within five minutes more, the vertical column had risen to eighty thousand feet, where it flattened out into a mushroom top more than thirty-five miles wide. The cloud drifted toward the northeast where it would deposit an estimated 2.4 billion cubic feet of ash and material as it circled the globe. Within twenty minutes of the eruption, the sky rained hail that had been taken up from the mountain's glaciers, pellets of ash as big around as a quarter, and even larger mud balls.

By now more than three-quarters of the mountain's glaciers and snow had started melting from the intense heat. Over the hours, billions of gallons of water gathered force and rushed down slopes to combine with earth, ash, and debris into mudflows traveling more than eighty miles per hour. The mudflows, with temperatures of nearly a hundred degrees, picked up houses, vehicles, trees, and boulders, and rolled over hills and up the sides of valleys to a height of 360 feet.

The region and beyond was now feeling the first effects of the eruption. It caused barometric pressures to differ over a wide area. Curtains were suddenly sucked out of windows as far away as Las Vegas, Nevada. Doors slammed or popped open in homes in many locations.

Continuous tremors rocked the area a hundred miles around. Avalanches rumbled down the slopes of several North Cascades mountains, narrowly missing climbers or campers. To people in the Puget Sound area around Seattle, it sounded like someone was taking a wrecking ball to their homes.

Reports told of domestic and wild animals from Canada to Northern California becoming quiet right before the eruption and then acting disoriented. Antelope and deer stampeded east of Redding, California, some even running into vehicles as they tried to escape "the spook" from the north. Songbirds that had grown quiet before the eruption now engaged in erratic behavior, flying into themselves, buildings, and landscape. Carp by the dozens threw themselves onto the shore at the Pot Holes Reservoir in Central Washington, well over a hundred miles from St. Helens. When the volcano's sound waves hit eastern Washington, even the insects became abruptly still.

While survivors in the immediate vicinity heard the avalanche, none reported hearing a sound from the initial blast, even though people seven hundred miles away in Canada heard it. Scientists discovered there was a twenty-mile zone of silence where the eruption cloud was moving so fast and was so dense that it actually captured sound.

For nine hours the volcano continued to erupt its vertical plume with a force equivalent to the detonation of a Hiroshima-size atomic bomb, or twenty thousand tons of TNT, every second.

By the end of the day, when the last helicopter flew away, the whole area was left with the cemetery sound of silence and death except for the thousands of acres of forest fires that crackled in the wind and lit the sky. St. Helens had left a swath of destruction measuring twenty-three miles from east to west and out to the north for eighteen miles. In the inner blast zone extending out five miles to where David Johnston had been at Coldwater II, scientists believed that most living things had been vaporized.

The blowdown zone extended beyond the inner blast zone to a distance of seventeen miles from the crater. A total of 4.7 billion board feet of timber were blown down on 86,600 acres of land. Some of the blown-down trees fell in perfect swirls following the path of the blast cloud. Almost all the trees were stripped of branches and bark. Millions of stumps rose to the sky in fingered splays of wood, all that was left where the hurricane cloud snapped the trees off. Debris from the fifteen-mile-long avalanche extended into this area. Next came the scorch zone where trees still stood, though dead, with branches shriveled and curled.

Ash blanketed everything in the swath of destruction. More than 11 million fish were killed in the lakes, streams, and rivers of the area. More than 200 black bears, 300 bobcats, 1,400 coyotes, 1,500 elk, 5,000 black-tailed deer, 11,000 hares, and 27,000 grouse lay dead or dying on the smoldering ground. Scores of people had been injured and at least fifty-seven killed or presumed dead.

The fight over proposed dikes to be built by the Army Corps of Engineers now pitted neighbor against neighbor. The clock was running out for Lexington and other communities along the Cowlitz River as twenty-six riverbank landowners debated whether to donate or sell land for the dikes and easements. Unless work started immediately, the dikes would not be completed before winter rains set in. Building only some of the dikes would be no more effective than building none. All the riverbank landowners would have to join in an agreement.

"I'm the first to help the community," Ron Brown, a volunteer Lexington firefighter, told Stepankowsky. "But I don't want to be the guy to lose thirty thousand dollars just because I happen to have land there."

That was the amount Brown estimated he would lose if he donated his land for a dike easement. Seven of his ten acres would be needed for the project. He was asking the Corps to fill the area behind the dikes with dredging spoils—soil and mud sucked from the river's bottom. Properly deposited, it could help restore land mass to his property. Without the spoils, the land would be useless to him, he said. "I'd be paying taxes on a dead horse. I'm waiting for a commitment from the Corps."

Stepankowsky found Brown's neighbor willing to donate land for the cause. The neighbor could understand landowners wanting to be paid for the loss of buildings, but his land would be destroyed anyway without the dikes, in the event of flooding. "To charge for an easement would be like a drowning man asking for a hundred dollars to be thrown a life preserver," Vern Adams said.

"I can get you real money for these," the man told him as he looked through Rogers' photo file. The man was a photo agent from Earth Images in Bainbridge Island, Washington.

Rogers wasn't much of a businessman. He had priced his early volcano photographs at six hundred dollars a copy on the advice of a professor who told him what he had was valuable. He had received one inquiry from a geology teacher whose interest quickly cooled when he found out Rogers' price. Some of his photographs were beginning to appear in books on the volcano, but the pay had been minimal.

He had higher hopes for his latest series of photographs, but he didn't know how to market them. Then he bumped into the agent at a conference dealing with St. Helens. The man signed Rogers on the spot.

About a week later the agent called him. "Good news, Robert, I've sold one of your photographs."

"Oh really? Who to?"

"*Life.*"

Rogers smiled. "You're putting me on, aren't you?"

"No, Robert, I've actually sold a photograph of yours to *Life* magazine."

"Jesus! Which one?"

"The nighttime shot showing the lava glow in the crater's cracks. They want to use it for a special edition they're putting out on the volcano."

"How much are they going to pay?"

"Over a couple grand."

Jim Scymanky had lost strength since leaving the hospital. Maybe it was because he couldn't sleep. He was no longer on morphine and found his body bathed in a sweat. He wondered if he had become addicted and was suffering from withdrawal.

He thought about the two youngest of his four boys. They were staying with relatives so Helen would have enough time and energy to help him. He missed his boys and he missed the way it had been before the eruption.

When the two boys finally came home, they seemed a little hesitant at first, as though they were afraid they would hurt their father if they touched him. Scymanky grabbed them and held them close.

What impressed Peter Frenzen first was the amount of security they found getting into the restricted zone. There were the barricades manned by state police and the sheriff's office. Then as the eight students in the University of Washington van made their way over temporary bridges and up what was left of the Spirit Lake Highway, they were stopped several more times not only by law enforcement officials, but also by Weyerhaeuser security people and forest rangers. They would be stopped and interrogated, even after they showed their pass, and warned about the dangers.

Frenzen and the others were excited and impatient. They hoped to not only see fresh mudflows, but also an up-close view of the mountain itself. The restricted area was like a giant construction project. Dump trucks lined up to receive earth, rocks, asphalt, and other debris from the jaws of huge steam shovels trying to clear the highway. Logging trucks brought down scorched and stripped trees from the salvage operation farther up the road. Other trucks hauled building material needed for sediment dams being built high up on the Toutle River.

The day was hot and the students often found themselves in swirls of dust kicked up by the work. They discovered an approach to the Toutle River and the mudflows they sought to study. As they hiked down to the river they noticed marks on the trees where high water had come through and left a rind of mudflow material. From the marks they could tell that initially the flow here had been more liquid and had come down in waves reaching over twenty feet, a greater height than the denser water and debris that followed. The river was chocolate brown and thick with sediment. The mudflows looked like long, flat islands of fresh brown cement.

Frenzen and the others waded to one of the mudflows. They searched for signs of returning plant life. In his studies on the mudflow on Mount Rainier, he developed a theory that standing dead trees play a major role in plant recovery. The branches, bark, and needles drop off the dead trees, providing a nice early mulch in which a seed can sprout. Now, there were trees standing everywhere that had been strangled by the mudflow and they were dying and shedding dead material. He searched beneath several of them. Then he found it. A seedling a few inches tall was growing in decomposing material beneath a dead fir tree.

The van made its way several more miles up toward the mountain. Once again they hiked to a mudflow at the river. On first glance only a few dead trees rose from the flow. But then Frenzen spotted green tree seedlings sprouting from the gray ground near the banks of the river. Life was coming back and taking root in the devastated land.

Beyond Camp Baker they found another mudflow. They were aghast at the sight. A helicopter was spreading a slurry of garden bark, water, seed, and fertilizer from a bucket onto the mudflow. For the group of fledgling forest scientists, this was almost sacrilegious. Nature should be allowed to do the job in her own way.

They still wanted to see the volcano. They finally spotted it above a ridge. About a quarter of the mountain stuck up beyond the ridge like a little gray flattop. The volcano wasn't even puffing out steam. Frenzen remembered how powerful St. Helens had looked before the eruption. They had a hard time believing the puny mountain they now saw had caused all this destruction.

Frenzen and the others didn't have much time to dwell on their disappointment. Their two-way radio crackled alive. "This is the Department of Natural Resources. A wall of water at least five feet high is now traveling down the Toutle. Floods are expected. You are ordered to evacuate. Repeat, evacuate the area. Do you copy?"

The students were incredulous. It was a hot day. There were no clouds in the sky. The mountain was peaceful. And they had just been down on the river's banks and had seen no signs of rising water. A flood warning station on top of a ridge had reported the water's onslaught. "Get out," the radio crackled. "Evacuate."

The van sped up. Frenzen and the others were not particularly panicked. They had spotted high ground they could escape to. But the van buzzed with the excited talk of the students. As they traveled down the highway, they kept in contact with the DNR. They thought they were traveling ahead of the water, but there was no way to tell. Frenzen wondered if perhaps one of the debris dams created by the May 18 eruption had been breached, releasing the waters behind it. If so, they could indeed be in trouble.

As they saw the tiny town of Toutle on the other side of the river, they breathed a sigh of relief. They felt safe. They were stopped by a sheriff's deputy who was standing by his car in front of the temporary bridge over the river. "The bridge is closed. It's not safe. There's a flood coming. You'll have to find another way out."

Frenzen looked down at the river. The silt-filled water had risen dramatically. It swirled around the wooden pilings. Just then a logger drove up in his pickup. The deputy explained the situation.

"I can take you out," the logger told the students. "Follow me closely."

As they made their way around hairpin turns, up steep ridges, and over bumpy roads, Frenzen tried to fathom what had just happened. He had suddenly developed a very healthy respect for how fast things could change around Mount St. Helens.

"We do a show on people caught in life-threatening disasters," the man said. He told Roald Reitan he was with a weekly television program hosted by movie stars Charles Bronson and Jill Ireland, who were married to each other. "We're sending a crew up there to film the aftermath of the eruption. We've heard about your ordeal and we want to know if we can take you back to where it happened and do an on-camera interview?"

"Sure. The only problem is that place is now buried under thirty feet of mud."

"Maybe some place close by, then."

"I'm feeling well enough to do something like that now," Roald said. But Venus was not, he told the caller.

"Well, maybe we can just get your interview, then. We've got some timelines to meet."

Roald was more curious than anything else when he went back to where he and Venus were swept away in the mudflow. He had wondered what he would feel coming back here. He was surprised at how numb he was to it all. He gave the film crew a vivid description of his and Venus's life-and-death struggle. But it felt like he was talking about somebody else caught in an event that happened long ago.

He hiked downstream and found the twisted remains of his car. A piece of heavy equipment used in cleaning up flood debris had crushed the car in its powerful jaws and thrown it aside above the riverbank.

Three days after the interview he got a call from an excited producer. "We've just seen your tape. We didn't realize you went through all that. Would you like to come down to Hollywood at our expense and be interviewed by Charles and Jill?"

"Well, there's someone I have to talk to first." He called Venus. He was not going to Hollywood without her.

"You don't mean it, Roald, you're kidding me. Us on TV with Charles Bronson and Jill Ireland? You know how I love her. You're kidding me aren't you?"

"They want us to come pretty soon. Are you going to be well enough?"

"You bet."

Whenever he could spare the time, Swanson returned to where he believed David Johnston had perished. The site was on a ridge that overlooked the gray and barren valley below and the mountain beyond. The mountain appeared close and overpowering in the cool clear air, though it was five miles away. Lazy plumes of steam rose from the crater to dissipate high in the sky. It had rained and the air was musty and tinged with the smell of sulfur. Swanson hoped the rain had worn away at the ash, perhaps revealing something that belonged to Johnston, some reminder of his dead friend.

Swanson had always been able to separate his emotions from the work at hand. But now there were quiet times when he reflected on his own life, his role as a scientist, and Johnston's death. Now when he worked and faced critical decisions, he found himself asking, "What would Dave want? What would Dave do?"

That summer Johnston's spirit motivated many of the scientists. Some, like Johnston's understudy Harry Glicken, let everyone know they were dedicating their work on St. Helens to him. Others were not so demonstrative, but Johnston in death had given meaning and drive to their mission to unravel the secrets of St. Helens in the hope of preventing future loss of life. Swanson was not the only one who sneaked away after work to look for Johnston's effects.

Swanson was poking in the ash with a stick when he spotted part of a strap. He reached down and pulled. It was a backpack, with a name tag. It was Johnston's. He began poking around again. Nearby he found a piece of cloth that looked like part of a sleeve. He began digging with both hands. It was Johnston's parka. He decided he would reveal the discovery only to a select few people. Scavengers were raiding sites, even stealing entire engine blocks out of victims' vehicles and selling them to souvenir hunters. Swanson and the other scientists thought they knew the culprits—some of the same helicopter pilots who were contracted to fly USGS personnel around the volcano.

The scavengers and souvenir hunters would like nothing better than to get hold of an item of Johnston's. The dead scientist had become a folk hero of almost mythic proportions. Hollywood was going to make a movie about him and the old innkeeper Harry Truman, who had refused to leave St. Helens before the eruption. Swanson went on digging. But he could not find the body.

Chapter 11

September

"DONNA, IT'S GEORGE. Why don't you, your son, and your mother come over to our place for dinner? I've got Billy's spotting scope, his tackle box, and some other things from the truck that I've cleaned up." He paused. "And by the way, I've got a surprise for you and your mother."

After dinner George Gianopoulos brought out her brother's belongings that he had retrieved from the truck. She ran her fingers over the cold steel of the scope. Then she opened the banged-up tackle box. Inside were the flies her brother had so meticulously tied, miniature pieces of art.

Parker looked up to see Gianopoulos setting up a screen. The film in the camera retrieved from the truck had proven worthless. It had been cooked. Maybe George wanted to show some slides he had taken on the trip. "This is the surprise," he said as the first slide flashed onto the screen.

She gasped. It was Billy. He was sitting on a stump. Hundreds of wildflowers surrounded him. Behind him, looking majestic and whole, stood the mountain that would soon kill him.

Gianopoulos had found a yellow film canister in the ash by the truck, he told his stunned audience. The exposed film inside contained Billy's last photographs. Close-ups of wildflowers flashed onto the screen. The mountain was alone in several shots. Then there were the photos of Billy and Jean, arms around each other, smiling with St. Helens as backdrop, which Billy had taken by using a delay timer. The same stump showed in several photos. Then a jarring view of the stump appeared. It took Parker a second to realize what was wrong. It had been taken at the same angle and distance as the other photos, but the stump was blackened. Behind it a sea of gray extended to the mountain now missing its whole top. Unknowingly at the time, Gianopoulos had taken a photo from the same spot.

She talked with Gianopoulos about the canister he found and the mystery it suggested. Billy usually shot lots of photos and kept the canisters with exposed film in the glove compartment. Was it possible there had been other canisters in the glove compartment and that someone had taken them, accidentally dropping that one canister into the ash? And what about the three cameras that were missing? Billy was an avid photographer who would take some photographs with one camera fitted with a certain lens or certain film, switch to another while leaving the roll in the first camera, and so on. It was not unusual for him to have all four cameras loaded with film. But no more film or cameras had shown up when Parker retrieved her brother's things from the courthouse or when Gianopoulos searched the truck.

They speculated on who could have Billy's cameras and film. The only ones let in immediately after the blast were law enforcement officials, geologists, and helicopter pilots from the armed services. Supposedly everything they found was handed over to the county. But why would anyone keep the film and cameras when they would have to know how much they would mean to a grieving family?

A few days later the Vancouver *Columbian* used a full page to print Billy's last photographs. Many people considered them a miracle. The canister had lain in the ash for over two months. It had survived not only the raw weather, but several apparent searches and the eruptions of the volcano.

Rogers left Jack's Sporting Goods and walked up to the sheriff's deputy who was standing by the nearby roadblock. "Keeping pretty busy?" Rogers asked.

"Yeah, between the mountain and tourists, I keep pretty busy."

"Is it still pretty dangerous up on the mountain?"

"Yeah. We saw some idiot up there the other day. We've seen him before in his blue sleeping bag." Rogers could feel the hair on the back of his neck stand up. He knew that Forest Service rangers, state police, and sheriff's deputies were alerted to his escapades. He had made the volcano and its crater his personal playground. While he usually tried to climb at night, he had grown more brazen and sometimes sneaked into the crater during the day. His friend Valenzuela told him the authorities had spotted him several times and now knew him as "the man in gray."

Rogers looked down for a moment and then looked up. "I wonder what makes people take chances like that?" he asked the deputy.

"Because he's managed to get away with it, so far. We haven't even figured out which side he's coming up on." Rogers relaxed.

"I hope you get him," Rogers said.

"We dispatched a helicopter for him the other day, but he was gone by the time it got there."

"I'll bet you'll nail him."

"Yeah, well, we have a heli-attack crew ready for him this time."

Rogers looked up at the mountain. Heli-attack crews were made up of elite smoke jumpers from the Forest Service. They were tougher than he would ever be, he thought. They were going to have to take him, though. He wouldn't hurt anybody, but he didn't plan to cooperate, either.

The eighty-seven-year-old man didn't sound too concerned about the flood threat on the Cowlitz as Stepankowsky interviewed him. He had seen it all before. Roy Mohr had lived on this piece of riverfront property since 1907. He built this very house with his own hands forty-two years earlier. It replaced his first home, which was destroyed in a flood, and he took care to build it several feet higher than the first one. "That's when I made my flood preparations," he told Stepankowsky.

The old man recalled what his buddies had done when the river flooded in 1933. "They just drank moonshine while their chickens flooded out," he said with a laugh.

Many other low-landers were not as nonchalant. Much of the pastoral farmland south of Castle Rock was still pocked gray with mud and debris from flooding during the May 18 eruption. In some places it was even worse. Stepankowsky met an older couple at a place where volcanic silt still covered all but 20 acres on their 110-acre cattle ranch. The ranch had been the couple's dream when they came out of retirement to buy it. They now lived in a trailer parked in the driveway of their flood-devastated home. "I have too much money invested. I have to keep going," Dutch-born Sam Hornstra told the reporter.

The Hornstras faced an uncertain future. Sam Hornstra didn't know if grass and corn would grow well in the silt. The twenty acres that had escaped the flood were not enough to support his herd, which consumed half a ton of grain a month. Already he had spent over fifteen thousand dollars on feed he hadn't anticipated on needing. Even if the government came through with a loan, he didn't know how he would pay it back. "I've had lots of ups and downs but this is like standing in the corner not knowing which way to go," he said. "I think I'll make it back . . . somehow . . . maybe."

Stepankowsky found much the same situation up and down the river—devastated families putting up brave fronts to face even more possible devastation. The Army Corps of Engineers still wanted to build dikes to help protect the six-hundred-family community of Lexington. But of the twenty-six landowners involved, only four were not holding out for sizable monetary settlements for easements on part of their land. "I can see that the dikes have to be built," one landowner told Stepankowsky. "But I can't see building them at my expense."

The next day, Stepankowsky's story about the plight of the Hornstras and other river families made the front page. His stories were regularly appearing there now. He had learned how to make dry statistics come alive while relating them to his readers' daily lives. He knew how to give a human face to the story.

Her friends were coming for a visit. It was part of the old gang that she and Roald used to spend so much time with. Venus Dergan primped before the bathroom mirror. She marveled at the power her body had to heal. Most of her wounds had vanished and left little scarring. The miracle of her recovery felt almost spiritual.

At first the four girls wanted to make sure she was okay. They wanted to know if she needed anything. But then the teasing started. "Did you have to find a volcano to get a little attention, Venus?" "You could have found a better way to try killing yourself." "You always just have to do something to make life exciting." She didn't mind. Since the day of the eruption she had been touched by the concern and support of her friends. These were people she wanted to keep in her life forever.

"How is Roald?" Venus asked.

"He misses you," one of them told her. "He seems a little lost."

Peter Frenzen hadn't expected the phone call. "We have an extra seat on a chopper to do a flyover of St. Helens. Would you like to come?" Jerry Franklin asked him.

"Sure."

"How soon can you get here?"

Frenzen didn't have a car. "I'll catch a bus right away."

Franklin told him he would be needed immediately to help with a project funded by the National Science Foundation and the Forest Service. Franklin was coordinating a multidisciplinary team of leading scientists to study how the volcano had affected plant and animal life and the land itself. Basically the scientists would take the pulse of the environment during an intense two-week effort to provide baseline data for future studies. Franklin called it just that: the Pulse Study.

Frenzen filled his backpack and a duffel bag. As he boarded the bus to Vancouver, he found it hard to contain his excitement. One day he had been a lowly recent graduate and the next he was going to help unlock the secrets of Mount St. Helens.

From the bus station he drove with Franklin to Pearson Field in Vancouver to board a waiting helicopter. It was the first time Frenzen had seen the volcano from the air. He was amazed at how monochromatic everything looked—a uniform gray and an occasional brown spot. He had trouble figuring out the scale of things. Fallen logs looked no bigger than twigs. As he viewed the steaming volcano, he felt the awe he had missed in his glimpse of the mountain during his earlier drive into the area. To see the volcano and its hundreds of square miles of destruction in one glance humbled him. It all looked so dead and barren.

For days now, a web of cracks radiating out on the crater floor like broken spokes had caught his eye. They weren't unusual. They had been noticed in volcanoes before. The cracks ranged in size from hairline fractures to fissures up to ten feet wide containing incandescent rock. Swanson stood staring down into one of the cracks. "What the hell," he thought. He had been measuring everything else to do with volcanic deformation. He might as well measure the crack.

Some force of inquiry and simple curiosity led him to pound a short length of metal reinforcing bar into the ground on one side of the crack. He pounded another piece of rebar into the ground on the other side. Then he took out a store-bought steel tape measure, measured the distance between the two bars, and recorded the result in a notebook.

Some of his colleagues might laugh at his crude methods if they could see him now, but he didn't care. He was the one who pushed the USGS into buying the most sophisticated and costly measuring device they now used, an advanced model geodimeter. But he didn't believe in using high-tech tools just to be using them. He had found that often he could get the job done more quickly and at lower cost with simple tools. Besides, his preferred tools rarely gave false readings and never broke down. Over and over he had learned in taking volcano measurements to "do what you can as fast you can" or take a chance on losing vital data.

Still, truth be told, he took a lot of satisfaction in displaying his Yankee ingenuity and seeing the shocked surprise in colleagues' and reporters' faces when he told them about the outlandishly simple tools he sometimes used to solve complex scientific problems. In Hawaii he had pioneered use of fifty-cent highway reflectors when there was not enough money to pay for the reflecting prisms he needed for his laser-beam measurements. The prisms cost several hundred dollars each.

In the days before the May 18 eruption, he and a colleague wanted to find out if the area around Mount St. Helens was bulging and tipping like the mountain's peak. They nailed wooden yardsticks to stumps and docks on the shores of Spirit Lake. They figured the lake would move up or down with the earth's movement. Swanson called it the "world's greatest carpenter's level." He got measurements accurate to one part in a million. The yardsticks were handed out as free promotions from a lumberyard.

Now he enlisted the help of another member of his team to measure more of the cracks in the crater floor. They moved from crack to crack, one on each side, pounding in rebar and then measuring widths with the tape measure Swanson had kept with him for years. He didn't want to spend too much time on what could turn out to be an exercise in futility. Swanson decided they would just follow the progress of four cracks.

Rogers was hit hard when *Life* notified his agent that the magazine had changed its volcano layout and would not be using his dramatic photo of the red lava vein at night. But he had recovered by the time his kill fee arrived in the mail—a check for sixteen hundred dollars. This was more money than he ever had at one time in his life. And he had a plan for the prized photo.

He showed it to a group of graduate students and professors at Portland State University. A young professor asked, "Can I get a copy?"

"Yeah, I sell copies of that," he replied. "But what I would like to do is give you a slide show."

"What's involved in that?"

"Have you got twenty friends with jobs?"

"What?"

"Have you got twenty friends with jobs?"

"Yeah, I think so."

"Well, invite them over to your place and I'll give them a slide show."

He placed the dramatic photo at the end of the slide show. When it appeared on the screen, gasps rose from his audience. One person asked, "Can I get a copy of that?"

"Sure," he said as he turned on the lights. He grabbed a stack of photographs. "The 8-by-10s are ten bucks. The 11-by-14s go for twenty. And if you really want something to put on the wall to get your friends' attention, there's the 16-by-20s I can let you have for forty bucks."

He sold out of the 16-by-20s. The photographs had cost him $1.30 apiece to make.

Roald drove to Venus's rented house in Tacoma an hour early for her homecoming. He sat in his car in front of the house and thought about himself and Venus. What had happened to them wasn't going to change their love. They would pick up where they left off. But as he told himself that, he wasn't exactly sure he believed it.

Venus and her mother were on their way to the house. All the way up the freeway, her mother kept saying, "Venus, I don't know about this. Are you sure you want to do this?" But with each mile the young woman's anticipation only grew. She was going back to her own house. Her life was going to start again.

When they pulled into the driveway, Roald jumped out of his car and began hobbling toward them. Venus still had her arm in a cast and Roald noticed how much thinner she was. He took her in his arms and they both tried to hold back their tears.

Venus felt relieved to be home, but strange too. After all these months it was eerie to find everything exactly as she left it. It was like she had never left, like all of this had been a bad dream.

Swanson had almost no free time. His wife, Barbara, had stayed in Cupertino, California, completing graduate work. His only social contacts were with other scientists, and there was no escaping the shop talk. Still, he kept quiet about his work on monitoring the cracks on the crater floor. They were gradually widening at an accelerated pace, but he didn't know yet what that meant.

They drove to the house where Roald lived with his parents. Venus wasn't exactly sure what was going to happen. Her parents and Roald's parents had decided they wanted to get together in a joint meeting.

They all sat in the living room. Venus stared at Roald, whose leg was bothering him. He hadn't gotten up when she and her parents entered the room. The silence was full and heavy with expectation, as though in a search for words none were ripe and expressive enough in light of all that had happened.

"Damn kids," her father finally said as he looked up from inspecting his hands. "Just when you think you got them out on their own, they're back again raising hell. I just don't have that much more hair to lose."

With his words it was like a dam had broken as both sets of parents shared experiences in coping with what happened. Venus discovered just how much the parents had been through. Her mother and father had never expressed all their fears and thoughts while she was recovering. She now realized they had hidden their emotions well. The discussion was therapeutic as the parents exchanged stories.

It seemed the whole world had been after their children. Both families were inundated with calls from reporters, geologists, doctors, and an assortment of other people, some of whom had no legitimate reason for their inquiries. They shared ideas about how the two families and their children should respond in the future. Both sets of parents related problems in getting the state of Washington to help with medical bills.

Venus smiled at Roald. Except for an occasional comment that they managed to slip into the conversation, it was like the two of them weren't even there. The parents seemed to be making decisions as if their offspring were young children again.

The parents wanted a say in their children's plans to appear on the TV show hosted by Charles Bronson and Jill Ireland. Venus and Roald were excited about the prospect of a Hollywood trip. It seemed like it took forever for their parents to examine all the potential pitfalls and ask questions about the logistics as they reassured themselves their children could handle it. Finally both sets of parents said yes.

Late that night, when she was alone, Helen Scymanky sat down on the couch and bawled. She was exhausted. Her mornings began with helping Jim into the shower. Afterward she dabbed medicine onto his wounds with Q-tips. Then white bandages had to go over the wounds, followed by Ace bandages wrapped around his arms. She helped him put on the Jobst stockings that covered his body from the waist to the ankles. These tight, long, elastic socks were designed to prevent clotting and promote circulation. Then she helped him with stretching exercises for an hour. Later in the day she drove him on the half-hour trip to the town of Mount Angel where he received physical therapy for an hour and a half, five days a week.

All this was in addition to keeping house and caring for her children. In the evening there was another shower and more bandages to be changed. And she worried about whether she was doing it all right and what would happen if Jim got an infection and why a wound on his head did not seem to be healing as fast as the other wounds.

This was his first helicopter trip to the volcano for the Pulse Study. They landed on the south flank of St. Helens at daybreak. Peter Frenzen was with a University of Washington botany professor interested in plant succession in the meadows and on the mudflows at higher elevations. Frenzen began moving down the steep slope, pounding rebar at specified intervals into the ground and then stretching tape around them to establish plot boundaries. The professor came behind him, recording plant species and their relative abundance.

As he worked, Frenzen kept looking back at the lip of the volcano that loomed over his shoulder. He remembered the flood from his earlier trip. It had taught him just how fast things could change around the volcano. Frenzen thought he heard something. He jumped up and spun around. The volcano was puffing out thick streams of steam. "What's that?" he asked as he braced his body. The botanist said it was probably harmless.

By the end of the day Frenzen found himself more exhausted than fearful. They returned to the former Job Corps training camp that served as headquarters for the small army of scientists and graduate students working on the Pulse Study. The camp by the Cispus River stood amid a towering forest of firs, untouched by the violence that struck the blast zone thirty-five miles away. After he showered and ate, he found his energy returning. This was the time of day he cherished most, when scientists assembled in the library to compare notes.

That night he listened in fascination to a team of scientists from Utah. It was believed that plants and animals in the immediate vicinity of the May 18 eruption had been fatally scorched by waves of superheated air or blown to their deaths by winds with a velocity many times that of a hurricane. But the Utah team made what was being called a miraculous discovery. Above Spirit Lake, on a ridge facing the volcano, they found pocket gophers. The gophers were known for burrowing mazes of deep tunnels and living off the roots of plants. They were also known to hide caches of roots for long-term survival. The scientists speculated that the gophers had hidden deep in their tunnels during the eruption and were now living off either their caches or the roots that still existed from plants sheared off during the blast.

The implications of the discovery for the possibilities of rebirth on the mountain were not lost on Frenzen and the other listening scientists. Mounds of earth built by the gophers had been discovered. The mounds should be made up of a mixture of ash and the organically rich undersoil the gophers brought to the surface. And these mounds should be filled with spores of fungi that would facilitate the absorption of nutrients by plants whose seeds hopefully would blow onto the fertilized soil and take root. The plants would then attract insects that in turn would attract birds that would bring preying animals. And the cycle of life would have begun again.

Venus still couldn't believe it. It was like one day they were nobodies and the next they were flying to Hollywood to do a national television show with movie stars. At the airport in Los Angeles they waited for the limousine to pick them up. Instead a young man appeared in a car. "Are you Roald and Venus?" he asked. He was the producer's son. They all hopped into his 1960s' Chevy, a car in immaculate shape. The air in the car was thick with lingering marijuana smoke. The man's face was flushed. Well, after all, this is Hollywood, Roald thought.

The producer's son took them to a luxury hotel where they had been assigned separate rooms. "Hey, we're going to save you some money," Roald said to him. "We only need one room." Up in their suite Roald parted the drapes. "Come here, Venus. You're not going to believe this." Through the window he could see the name "Hollywood" spelled out in huge letters on a nearby hillside.

Across the street they found Grauman's Chinese Theatre. They held hands as they looked at the actors' names on the walk of stars in front of the theater—Cary Grant, Gregory Peck, Rita Hayworth, Marilyn Monroe, John Wayne, Bette Davis, and others. They viewed the names in the golden stars on the sidewalk, checked out the signatures, and laughed as they measured up their own hands and feet with the imprints the actors had left in concrete. Mount St. Helens was the farthest thing from their minds.

The next morning they got up early for the taping. At the studio they met other guests. There was the announcer whose radio and newsreel broadcast of the exploding German airship *Hindenburg* at Lakehurst, New Jersey, in 1937 had become world famous. And there were survivors of other assorted calamities: a couple who had lived through a gigantic explosion in a Texas port, and a man who had survived a flash flood when a dam broke in Idaho.

Finally it was their turn to be prepped for the interview. They were taken to one of the multiple stage sets designed so that several guests could be ready for their interviews at once and Charles Bronson and Jill Ireland could hop from interview to interview. Stylists checked the guests' hair, the makeup artist padded a little rouge on their faces, and technicians hooked tiny microphones on them. As they watched other people roaming the stage area, Roald heard Venus gasp. "There he is, Roald, in real life," she said, pointing to a distinguished-looking man walking across the huge room. "It's Luke."

"Who in the hell is he?"

"You know. Luke from *General Hospital*." During her long recovery the soap opera had been the highlight of her day. "Look, there he goes. Can you believe it?"

A technician came up to tell them that Ireland and Bronson would be with them shortly. They looked at each other. "Are we ready for this, Roald?"

"We'll be fine."

She couldn't keep her eyes off Ireland as she approached with Bronson. The actress was everything Venus thought a movie star should be, graceful and glamorous. Roald found that Bronson was short but carried himself like a much larger man with his broad shoulders and rugged good looks. The bright lights came on. The cameras moved in and the director called out, "Action."

As they answered questions about their ordeal, their jitters faded away. They seemed to work well together, one of them chiming in with information at the right time just when the other hesitated, never interrupting one another.

After the filming they had the rest of the afternoon to do whatever they wanted. They took their two-hundred-dollar appearance checks and went straight to a bank to cash them. Then they headed for shopping on Rodeo Drive.

That night they shared dinner with the announcer from the *Hindenburg* disaster and his wife. The announcer was full of stories about major events from the past. Roald found it hard to believe. Here he was talking to a news celebrity who had seen with his own eyes many of the events Roald had only read about in history books.

That evening in their room, Venus and Roald stood looking out the window at the colored bulbs of racing neon on nearby theater marquees, the lights that sparkled like fuzzy white diamonds from the hillside homes of movie stars, and the lit-up "Hollywood" sign with its huge white letters outlined in a halo of haze in the smog-filled night.

Tomorrow they would go home.

As he lay there, he heard the *crack* that usually preceded earthquakes. Then Rogers' body shook with the ground. He woke up. But he thought his body continued to shake. It took a moment for him to realize that he had not been sleeping in the crater, but in his own bed. He had now spent so much time on the mountain with its continual earthquakes that he found himself often waking up this way.

They were working in a forest northeast of the mountain that had received considerable ashfall. Frenzen and other graduate students tagged trees in various plots and recorded which ones were alive and which were dead. At lunchtime, as he sat on a downed log eating a sandwich, he had the feeling he was being watched. He looked up. Through the trees he could see a pair of eyes. Then he could make out the outline of what looked to be a coyote. The animal gradually moved closer. The coyote moved out of the tree line, stood staring for a while, and then sat down thirty feet from Frenzen.

The animal was emaciated. He was tempted to feed it, but then he had the thought that he might be inadvertently, in some small way, influencing another scientist's study. Feeding a single animal might not be a big thing, but the idea went against the grain of everything he and the other scientists had been fighting for—the chance to study this unique environment with as little influence from the hand of man as possible. But it didn't help when the coyote the very next day cautiously emerged from the forest to sit down and stare at Frenzen again as he ate lunch.

A few days later he and other researchers entered a blowdown area. As if they were on an Easter egg hunt, they searched for signs of plant life as they scurried over and between fallen logs. Frenzen jumped off one of the logs, his feet sinking into the ash as he turned to look under the trunk. "I found one," he yelled out. There, sheltered by the log that must have fallen before the eruption, a tiny fern sprouted from the ash.

When work was done for the day, he jumped on top of a log and began running its length. In his spiked logging boots, he leaped to the next log without slowing down, and then to another and another. As he gained speed, it was like he was a schoolboy again. His fellow team members watched as the silhouette of his body grew smaller against the late afternoon sky.

Stepankowsky was listening to the emergency radio scanner when he heard about the bodies. Weyerhaeuser workers had found the remains of a man and woman near Elk Rock. These were the first volcano victims' bodies found in over two months. No photographer was available, so he found another reporter, Jay McIntosh, and had him bring a camera.

They raced up Spirit Lake Highway, across a road over a sediment retention dam the Corps was building, and onto another road that ran behind the impoundment. This led to a series of logging roads, some cleared and some not. The way to the road marker mentioned in the radio dispatch had not been cleared. They jumped out of the car and climbed a hill, their feet sinking into the ash. Now they were high above the Toutle River Valley. Down below was a tiny quarry where they spotted the remnants of a camp. At the camp they found a tattered pillow, half buried in a layer of damp ash a few feet from a tarnished coffee pot. The site also revealed a dirty barbecue grill, a gas stove, a crushed water jug, and a melted sleeping bag.

About a hundred yards down the hill they could see a smashed pickup truck. They split up, trying to find the bodies. Stepankowsky climbed over a fallen log in a nest of downed trees near the road that wound down the hill. He stopped in his tracks, startled. There lying in the ash were human remains. The body was propped up slightly by a clump of gnarled roots. Its arms were flung open as though in a welcoming embrace, but the legs were bent in a contorted position. Feeding birds had left only a few tiny pieces of skin to dangle from the skull, its jaw hung open. He could see the bones of the skeleton clearly beneath what little skin was left on the torso. Strips of tattered clothing blew in the wind. He could tell from the rib cage that this had been a man.

This was the first dead volcano victim he had seen. He had never been around a decaying corpse. This body had been exposed to the elements for more than four and a half months. He found himself filled with an odd mixture of attraction and revulsion. This was different than taking down reports over the phone and writing up the latest statistics on the number of people killed. This had been a human being who had once lived, laughed, and loved as he did and simple statistics could not begin to tell this person's story. He looked up. Ash clung to everything, including the corpse, like a vast mantle of death. There was no sound except for the wind. It hit him that this was a graveyard, sacred in its own right. He felt like a trespasser.

In the distance he could hear helicopters approaching. The sound of his own voice almost startled him. "Jay, I found one," he screamed out.

About fifty feet away they found the twisted, decaying corpse of the woman. The only way they could judge it was a woman was by her slender feet. The battered 1979 Chevy Luv pickup was farther down the hill. Its plastic parts had been melted.

Stepankowsky tried to figure out what had happened. By the position of the truck and its distance from the campsite, he thought the pair had been riding down the road in the pickup. If they had been at the campsite, the truck would have been blown in the opposite direction by the blast. They probably jumped out of the truck when they saw the ash cloud billowing toward them. The man got a little farther than the woman. Both,

Stepankowsky thought, died from ash inhalation, like most victims of the volcano.

The sound of helicopters grew louder. Stepankowsky waved them in. The two helicopters from Unit 304 of the Air Force Reserve in Portland carried a six-member recovery team. They found a wallet in the pants pocket of the male victim. Harold Kirkpatrick, according to his Oregon driver's license, was thirty-three years old.

"They must have had a good view of the mountain," someone said as they all turned to look at the volcano. Ten miles away St. Helens steamed peacefully. McIntosh took a photograph of the recovery workers dragging the victims' remains in black plastic bags to one of the helicopters.

Back at the office Stepankowsky found the name of Joyce Kirkpatrick on the list of missing people. He started making phone calls. Harold Kirkpatrick was a laid-off welder at a shipyard, his aunt Clara Kirkpatrick told the reporter. Joyce and Harold were cousins by marriage and both were thirty-three. He had never been married. She had three children by a marriage that had been failing.

"Joyce was an adventurer," Clara Kirkpatrick said. "She'd like to go into the mountains to be alone. She liked to go fishing and hunting. She was pretty. Had short brown hair."

Chapter 12
October

DONNA PARKER wasn't daunted when the forty-page special tabloid section on the volcano dropped out of her Monday *Oregonian*. It was called "A Terrible Beauty." Parker waded into the section with a voracious appetite. Buried within its pages was investigative reporting that, if it had been relayed in a front-page series in a more traditional story format, might have won a Pulitzer Prize for the *Oregonian*.

A team of reporters had worked for months on the section. They conducted hundreds of interviews and pored over thousands of documents including U.S. Forest Service and state Department of Emergency Services logs, private notes and diaries, meeting transcripts, and memos. The reporters had listened to recordings of rescue missions and reviewed hundreds of photographs taken by survivors and others.

Parker's pulse quickened as she discovered revelation after revelation.

- Billy was not killed in a restricted zone as she and her family were led to believe. Only three victims met their deaths in the Red Zone, and these "had permission to be where they were," Parker read.

- Despite what state officials led the public to believe, the restricted zone was not a twenty-mile ring around the volcano with variances of up to five miles either way at certain points. The *Oregonian* reported that its boundary ebbed and flowed, with national forest land on the off-limits side and with lands owned by the state and Weyerhaeuser on the other side of the line, which was kept open. On the northwest side of the mountain, where Weyerhaeuser had its holdings, there was in essence no zone at all. The property line for the logging giant butted up against the mountain. The line for the supposedly outer Blue Zone lay directly on top of the Red Zone line on the northwest side of the mountain. Billy and almost all the victims had been killed on Weyerhaeuser land in the area outside the restricted zone.

- Officials of both government and industry ignored clear warnings of danger as they set up the restricted zones. "The Red Zone grew out of a series of meetings the Forest Service held," the newspaper said. "Among those present were representatives of the Forest Service, Geological Survey, sheriff's offices, Emergency Services and the governor's office. So were people representing private interests, such as the Weyerhaeuser Co., International Paper Co. and Pacific Power and Light."

In formulating the restricted zones that were recommended to the governor and then established by executive order, this group chose to ignore the warnings of the U.S. Geological Survey's top hazards scientist, a geologist who knew the eruptive history of Mount St. Helens well. "Dwight 'Rocky' Crandell, of the Geological Survey, warned government and industry officials that mudflows and pyroclastic materials could spread as far as the confluence of the Cowlitz and Columbia Rivers, near Longview 40 miles southwest of the mountain," the *Oregonian* reported.

"Crandell would recall that his meetings with Forest Service officials and others dealt not only with volcano hazards, but also with timber harvest considerations, political subdivision boundaries, 'and of course, people who wanted the zone to extend as far out as possible, as long as it didn't include their land.'"

Crandell told the newspaper he had even drawn hazard maps and provided them to officials. "We also showed that in the past, ash deposits had been three feet thick up to 20 miles north of the mountain. . . . They chose to ignore that. I'm not sure why."

• Several members of the group at the Forest Service meetings remembered how Weyerhaeuser, International Paper, and Washington state agencies were all pressuring to have no restricted zones or only extremely limited ones. "Cowlitz County Deputy Ben Bena recalls that Weyerhaeuser officials and representatives of the Washington Department of Fish and Wildlife argued at some meetings against a restrictive closure zone, saying it would affect not only timber harvest but also fishing and hunting access to the country west of the mountain," the *Oregonian* reported.

"Leonard Bacon, a public relations man for Pacific Power & Light, remembers that Weyerhaeuser and International wanted to make certain the zone infringed on neither their property rights nor their ability to keep logging," the newspaper said.

Representatives of the state Department of Natural Resources were blunt in arguing for continued logging with no restrictive zone. "Timber sales had meant $175 million to public education in Washington the year before," the *Oregonian* pointed out.

"'We were anxious to continue to harvest timber,' State Land Commissioner Bert Cole . . . would say weeks later. 'We were concerned about the safety of lives, too. But we had timber sales going on in there that had to be taken care of.'"

"Cole said neither his agency nor private timber harvesters like the Weyerhaeuser Co. believed the volcanic hazard analyses in April. They wanted to keep working to bring out timber."

- All records of the meetings the Forest Service hosted disappeared after the May 18 eruption. The *Oregonian* said it was told by Ed Osmond, Forest Service disaster coordinator and leader of the meetings, "that only notes had been taken at the March 26 meeting, and that no detailed record had been made of any other. What records he did have were misplaced, Osmond said.

"In fact Maggi Courville, a Forest Service stenographer, made a verbatim transcript of the March 26 session from tape recordings. Minutes were taken at a half dozen others by Penny Hiatt, secretary to Gifford Pinchot National Forest Supervisor Robert Tokarcyzk. Ms. Hiatt and a Forest Service file clerk said all records of volcano-related meetings before May 20 were gone from the central filing system."

- Loggers made numerous protests to the state about Weyerhaeuser making them work in dangerous areas close to the volcano. But the men's fears were ignored by representatives of the state. Callers to the Washington Department of Industry and Development were referred to Les Ludwig, the department's senior safety inspector in Longview, home to Weyerhaeuser's southwest regional headquarters. The concerned loggers asked Ludwig "to prevent the company from sending crews into areas behind the roadblocks where the public was being advised not to go for safety reasons."

In explaining his lack of action, the newspaper quoted Ludwig as saying, "It wasn't my jurisdiction. You don't go in and tell a company like Weyerhaeuser not to go in and work. . . . I thought that most of the complaints were from people who wanted a way to get out of work. Frankly, I thought they were out to get unemployment [compensation]. Now I wouldn't want this to get out in the press, but that's what I thought."

The barricades the loggers cited as designating danger zones in which they were forced to work were moved "under pressure" after their complaints, from twenty-six miles from the mountain to less than twelve miles. Captain Richard Bullock, commander of the State Patrol's southwest Washington district, said he first established a barricade on the northwest side of the mountain twenty-six miles from the summit at the intersection of Washington Roads 504 and 505. The newspaper said he called this "'the most logical choice' because it would give tourists access to several scenic viewpoints along the highway while denying easy entry to logging roads" that could lead dangerously close to the mountain.

Fifteen days later Bullock's decision was countermanded and the barricades were brought in closer to the volcano not once but twice, as "pressures built up to move the barricades closer again." The State Patrol could no longer control people who could now even unknowingly enter logging roads leading to the volcano. For this reason, and out of concern for the safety of its own officers, the order was given to discontinue staffing the barricades continuously, checking them only at dawn and dusk.

- Coordination of rescue efforts was badly bungled. This situation had been reported before, but the *Oregonian* went into much more detail. "No one thought to plan for air search and rescue operations before the day the mountain blew apart," the newspaper said. "No one thought to take charge after it happened. No one thought to establish a central command center from which search and rescue teams could be dispatched. No one thought to assign a single radio frequency so helicopter crews from different military units could communicate with one another. No one thought to make certain that aviation fuel supplies would last more than a few hours."

There was a great deal of infighting as to who was in charge. Sheriff's departments, the State Police, Forest Service, Army National Guard, Air Force Reserve, and Washington Department of Emergency Services all vied for power or sought to protect the right to control their own personnel without regard to the overall effort.

Relatives of the missing who could pinpoint the last whereabouts of their loved ones found their information disregarded or lost before it got to rescue pilots. In frustration some mounted their own searches. The newspaper took pains to point out the courageous efforts of pilots and crews who continually risked their own lives in rescuing victims, often ignoring their own commanders.

Only two days after the eruption, officials decided there were probably no more surviving victims to be found and efforts slacked off. The *Oregonian* said that coroner reports indicated at least two men had been alive at the end of the first day of the eruption, and perhaps longer. One of them was Jim Scymanky's friend and boss of his tree-thinning crew, Evlanty Sharipoff.

- The Washington Department of Emergency Services exhibited callous indifference to the victims' families. The *Oregonian* said the department "refused at first to fully identify the dead and missing, releasing only their names and omitting their ages and hometowns. It was reported at the time that this policy was an effort to spare relatives the added grief of news media inquiries. But it went beyond that.

"Emergency Services shift supervisor Bud Lien said during the first week after May 18 that his agency also was refusing to verify identifications to private citizens who had friends with the same names as those missing or killed."

Department representatives, the newspaper said, "freely acknowledged that dozens of people in the Pacific Northwest alone had names identical to or similar to those of the missing and dead. Hometowns without specific addresses eventually were disclosed, but ages were not. William Parker of Portland was one of those killed by the volcano. Officials refused to give Parker's street address even though there were 12 William Parkers listed in the Portland telephone book.

"Federal Emergency Management Agency representatives who worked with Emergency Services were dismayed at the state agency's attitude and privately expressed concern about the unnecessary grief it would cause friends and families of living people who had similar names."

Donna Parker's anger grew as she dug information out of the section. Billy was named twice in the report, first in a part that dealt with his death and then in connection with the bungling of officials in identifying victims. When she read her brother's name, all the grief and anxiety swept over her again as she remembered her frantic efforts to find out whether he had been a victim of the volcano.

He pulled the "Terrible Beauty" section from the *Oregonian* and flung it aside to throw out as trash. Stepankowsky thought it was just another of those routine special sections that inundated the public after the eruption and that subscribers kept as souvenirs if they kept them at all. The publications helped a publisher's bottom line but Stepankowsky found they contained little in the way of news.

The *Oregonian* report presented fifty-six thousand words in column after column of gray type, occasionally broken up by photographs. There were no headlines summarizing the information. Instead the tabloid was broken into sections with plain-Jane titles like "The Tree"; "The Land"; "The Ring"; "The Warning." The executives of the *Oregonian*, part of the Newhouse media empire, planned to turn this section into a book. It read like a book. The writing was beautiful, but facts and significant revelations got lost.

For most readers it was a publication you would save to digest over several long winter nights but then end up throwing out unread. It had been issued not on Sunday, the highest circulation day of the week when readers have more time, but on a Monday, one of the lowest readership days. It didn't help that news about the presidential election to take place in a few days dominated the newspaper and overshadowed any impact the *Oregonian* hoped to make with the special section.

In the days ahead, no other newspaper picked up on the important stories the *Oregonian* told in its special section. There were no calls for investigations, no calls for accountability about what happened, no calls for resignations. The *Oregonian*, long the venerated and dominant newspaper in Oregon, had a relatively small circulation in Washington, concentrated in the very southern part of the state. Its clout in the rest of Washington, including the halls of the capitol building in Olympia, was virtually nil.

Something in the *Oregonian* special section puzzled Parker. She reread one part. "Many sightseers who heeded the Spirit Lake Highway closure either parked their cars along the pavement or pulled a short distance off the asphalt onto the 3500 Road. It widened there into a large gravel-covered area which state troopers manning the highway barricade had come to call the 'gravel turnaround.'

"Souvenir hawkers and food vendors had joined the crowd that gathered in the turnaround that Saturday. Smells of food and campfire smoke filled the air. Radios blared music and people mingled in small groups. There were hitchhikers of several nationalities, photographers, painters, amateur geologists and vacationers of wide description."

She flipped a few more pages. There was a photograph by a Weyerhaeuser photographer named Jessie Wilt taken late on that day before the eruption, showing recreational vehicles and clusters of cars filling the turnaround. This was the area that she later learned was known as International Camp because a lot of foreigners had been camping overnight there.

The newspaper mentioned a couple in a car racing through the International Camp as they tried to escape the eruption cloud and seeing at least four other vehicles. In its epilogue, The *Oregonian* report almost casually stated that the people parked at the International Camp probably had "vanished" in the eruption. The statement surprised Parker. She knew the authorities had reported no one dead in the area, and only two presumed missing.

Not all readers of the special *Oregonian* section had a full appreciation of the role Weyerhaeuser played in the state of Washington, but such an awareness was essential to understanding the report. The company and the state government were powerful entities whose fortunes were often intertwined. From her research Parker knew well the history of Weyerhaeuser.

The company first came to the state in 1900 when founder Frederick Weyerhaeuser purchased nine hundred thousand acres of prime timberland at six dollars an acre from railroad baron James J. Hill. Two years later the company shifted its emphasis from Puget Sound to southwest Washington. In 1927, one year after Longview incorporated, the company built its first lumber mill in that city, soon followed by a shingle mill and paper mills.

The company grew until it became the largest timber company in the world, with nearly six million acres of highly productive land. And none of the company's other eleven regions could compare with the business its southwest Washington region generated. If that regional district, with its five thousand employees, was separated from the company, by itself it would still have rated well within the Fortune 500, *Fortune* magazine's list of top U.S. corporations. The district and its production was the dynamo that drove the company through good times and bad.

With the growth of its economic base, Weyerhaeuser's political power grew apace. Unlike many other large corporations, Weyerhaeuser actually encouraged its employees to enter politics, including top executives who served in the state legislature. Members of the Weyerhaeuser family also were elected to high office in the state. The support of the company helped put several governors into office over the years.

When Peter Frenzen returned with his team of scientists late in the afternoon, the camp at Cispus was abuzz. The Centers for Disease Control had flown in a special team to draw blood samples from aquatic scientists. A number of them had turned up with mysterious flu-like symptoms. Researchers had drawn water samples from lakes next to the volcano. They discovered the water contained extremely high concentrations of Legionella bacteria, the cause of Legionnaires' disease that was known to lead to serious sickness or death.

On Wednesday October 8 at 4:30 in the afternoon, Charlotte King called the *Daily News* office from her home in Oregon's capital city, Salem. King had successfully predicted two past volcanic eruptions. She said dizziness and pressure behind her eyes led her to believe there was going to be an earthquake in southwest Washington in the next twenty-four hours.

At 9:11 a.m. the next day, after a weeks-long slumber, St. Helens came alive with a moderate earthquake. Within ten minutes a plume of steam and old ash shot sixteen thousand feet into the sky.

Later that morning King called the newspaper office again. She said her head was about to burst. She had been to the hospital but doctors had not been able to find anything wrong. She said she believed something big was going to happen but she didn't know exactly when.

The gritty taste in his mouth seemed all too familiar to Peter Frenzen as he bicycled across campus to the research laboratory at Oregon State University. Then it hit him. This was ash from Mount St. Helens. When he reached the lab he confirmed that the mountain, more than a hundred miles to the north, had erupted again.

It was the morning of Friday, October 17. The eruption had occurred the night before at 9:58 p.m. spewing ash forty thousand feet into the air. Ash swept across southwest Washington, where twenty-mile-per-hour speed limits were imposed in the cities because of poor visibility from falling ash. An air quality alert was called in Portland.

Not long after Frenzen got to the lab Friday morning, the mountain erupted for a second time, at 9:28 a.m. At that moment, Robert Rogers was sitting in the Portland office of a geological consultant for Portland General Electric. Rogers was impressed as the consultant, looking out his office window, used power poles as a gauge to determine the height of the plume. Rogers listened as the geologist called out the increasing altitudes of the eruption cloud just ahead of the radio reports of a pilot with an altimeter flying over the volcano. Ash rose to forty-seven thousand feet, this time accompanied by superheated flows of pumice and gas that shot out two miles north from the volcano.

At 9:12 that night the mountain unleashed a massive third blast. The volcano shot steam and ash nearly five miles into the sky. The eruption triggered another flow of superheated gas and debris that "just seemed to fall off the north slope and travel two miles," according to a forester who witnessed it. Cities as far as two hundred miles away in Oregon, Washington, and Idaho reported ash dustings, which residents said looked like "fine, white snow." The three eruptions were all part of one event, the longest eruptive period since May 18. The USGS officially categorized it as a single eruption.

A Forest Service pilot flying over the crater after the Friday night eruption reported seeing through the clouds that the volcano's twenty-story dome had been blown away.

In previous eruptions there had been a deep quake, about six miles beneath the surface. But by the day after the Friday night eruption, no similar quake had materialized. Scientists were finding that St. Helens could not necessarily be scripted. "We'll watch and see how this one winds down," Don Peterson, Swanson's USGS boss, told a *Daily News* reporter. "It's giving us ambiguous signals."

Scientists hoped to be able to use seismic activity and harmonic tremors in predicting St. Helens' eruptions. In some ways they had been successful. Seismic activity usually consisted of rock inside the mountain cracking and breaking as pressure built up from magma, or molten rock. For the week before the eruption on Thursday night, the mountain had jiggled. By as late as Thursday afternoon, however, the scientists had not detected the rate of buildup they had seen before previous eruptions. Then, all of a sudden at about 7 p.m., a larger quake measuring 3.0 on the Richter scale was detected directly below the volcano. Seismic events began building in intensity. An eruption alert was called and the eruption followed. Seismic activity had helped predict the eruption, but the pattern was dramatically different than in previous eruptions.

Almost immediately after the Thursday night eruption, seismic activity fell into a lull that held until about 9 a.m. Friday when scientists picked up signals so slight they were discounted. But suddenly at 9:28 the mountain erupted. Prediction through the use of seismic activity was proving to still be a crude tool.

Harmonic tremors, the sustained rhythmic vibrations used to determine that magma is on the move, had not worked much better in predicting the eruptions. The Friday night eruption had been heralded by sixteen minutes of continuous harmonic tremors. The morning eruption had been preceded by only five seconds of discernible tremors.

The final, deep earthquake signaling the end of the eruptive phase never came as it had in past eruptions, further confounding scientists. Don Swanson could hardly wait to get into the crater again, but poor weather and lack of some type of confirmation that the volcano had ended this eruptive cycle stymied him. The cracks on the crater had increased their rate of widening as the volcano neared this latest series of eruptions. He was not yet sure what that meant. It could just be an anomaly. As the volcano had shown, signals or clues were not always consistent. The current or popular theory held by scientists today might very well be blown away with the next eruption.

He was out partying with friends when it happened. Roald Reitan made a sudden move and his knee gave out. The next morning he was in excruciating pain. "My god, what have I done?" he asked himself. He limped to his parents' room and woke his father, who called the doctor.

After looking at the knee, the doctor recommended immediate surgery. At the hospital they found that Roald's original injuries from fleeing the eruption flood had been much worse than anyone thought. Some ligaments had been completely torn and others were on the brink of separating. Through his exercise therapy, Roald had developed muscles that compensated for the torn and damaged ligaments. But it was only a matter of time before he put enough pressure on them and they gave out.

After the surgery in which his kneecap was removed and ligaments reattached, his physician told Roald what was in store for him. He would spend the next ten days in the hospital. The straight-legged cast he now wore down his leg and around his foot would stay on for eight weeks. Six months of physical therapy would follow. Perhaps after a couple of years he might be able to run again, but there was no guarantee.

Andre Stepankowsky and Rick Seifert had been a given a whole week to investigate the extent of the flood danger facing the region. The two reporters were going to try to separate fact from fiction for their readers.

Stepankowsky started by plowing through government reports. He loved to find the story hidden within the dry statistics. When he picked up the Forest Service report on the Mount St. Helens watershed, he didn't expect to find much. But as he ran a finger down a column, a figure jumped out at him. He found other figures and began doing the math.

The Forest Service projected that 113 million cubic yards of debris would erode that winter off the eighteen-mile mudflow lodged in the North Fork of the Toutle River. And that was only part of what was expected to be eroded. He read it again. Then he straightened up in his chair. The Army Corps of Engineers had dredged what was considered to be a monumental amount of sediment from the area's rivers—but that had only been 24 million cubic yards. And the two silt-trapping dams the Corps had built on the Toutle River would only hold about 6 million cubic yards of sediment combined.

Stepankowsky cross-checked the 113-million figure with other figures in the report to make sure what he had read was not a typographical error. It was not.

He grabbed the phone to call reporter Seifert, who was in Tacoma. He reached him just as Seifert was coming into the building that housed the USGS water resources office. Stepankowsky told Seifert the startling information he had just found. He asked him to find out from Phil Carpenter, associate district chief of the water resources division, what the USGS knew. While Seifert was at work in Tacoma, Stepankowsky called Forest Service sources who confirmed his find. When Seifert returned to the newsroom and Stepankowsky saw the grin on his face, it was clear they were onto something.

USGS researchers, with a study area much broader than the mudflow the Forest Service had studied, concluded that massive erosion would occur if winter rains followed their normal patterns. Between 400 and 500 million cubic yards of eroded debris would reach the Cowlitz Valley from the entire Mount St. Helens watershed. This was over ten times the amount that had clogged rivers and brought flooding after the May 18 eruption. It was over twenty-seven times the amount of debris predicted by the Army Corps of Engineers in what was called a "quick and dirty" study months earlier. The region could expect to be inundated, the USGS concluded.

The USGS scientists, startled by what they had found, had worked their numbers over and over. Even the most conservative figures foretold serious flooding with normal rainfall. The scientists had gained an early clue to what the figures might indicate when a debris-dammed pond near Elk Creek had unexpectedly broken loose earlier in the summer, producing the flood that Peter Frenzen and the University of Washington graduate students had fled. The four- to five-foot wall of water that surged down the river that day had consisted of 50 per cent sediment, the scientists discovered. This was an astronomical percentage of sediment. One per cent was considered normal.

In his calls to Forest Service personnel, Stepankowsky found the service used different methodology than the USGS had used. Nevertheless the total cubic yards of erosion projected by the Forest Service was in the same order of magnitude as that of the USGS. John Pruitt, leader of the task force that produced the Forest Service study, said, "We're not trying to scare anyone. But it is possible we will get tremendous volumes of sediment."

Now the reporters knew they had a very big story. Two separate government agencies had conducted studies independent of each other, using different methodologies, and had arrived at the same result: with only normal rainfall this winter there could be massive flooding.

That night they tracked down Patrick Keough, chief of the Corps' restoration project. "That's mind-boggling," he said of the projections.

"That's catastrophic. If anything catastrophic happens, I've always said all bets are off." He said if the figures proved accurate, residents of the area might have to evacuate for months.

"Studies predict massive erosion, winter floods," the headline proclaimed in large type across the *Daily News* front page the next day. A photo showed giant dredges working to clear the Cowlitz River above Castle Rock. "Scientists predict massive erosion will nullify dredgers' efforts," the caption said. Related stories by Stepankowsky, Seifert, and other reporters filled the front page and subsequent pages.

Stepankowsky and Seifert had scored a tremendous news exclusive. The story went national. The state's two powerful U.S. senators, Warren Magnuson and Henry "Scoop" Jackson, hastily called for a meeting where the Corps of Engineers, USGS, and Forest Service could explain to them and to the public what was going on. The importance of the meeting was not lost on the agencies involved. Magnuson was chairman of the Senate Appropriations Committee. He held the key to these agencies' funding and very existence.

When the hearing was held a few days later, more than five hundred people crammed the conference room at Kelso's Thunderbird Motor Inn, with many more overflowing into the hallways. Magnuson and Jackson sat with other Congressional, state, and local politicians at a long table lined with microphones. Magnuson read a prepared statement. Jackson told the crowd this was a "fact-finding session."

It soon became apparent to Stepankowsky and other reporters that this was to be a well-choreographed dog and pony show where the disturbing facts were to become comfortably fuzzed. Colonel Terry Connell of the Corps of Engineers said the Corps expected to be able to "prevent serious flooding." However, he acknowledged that if storms were grouped so that dredging areas and dams were inundated with silt, the Corps would not be able to keep up. A spectator, Donna Malone, marched out of the room at this point, saying for all to hear, "I've had it. All they've done is talk double-talk."

Phil Carpenter, the USGS official who had provided the key erosion study results to Seifert, sat quietly in a corner taking notes. His boss, Charles Collier, squirmed in trying to give the politicians the answers they wanted, which would placate the public while not totally abandoning his agency's study. The adjective "serious" was conspicuously dropped in describing predicted flooding. "Under normal conditions we can expect some flooding. The degree of it is the question," he said.

Now the senators laid into Collier and the USGS. Magnuson wanted to know if the agency had contacted other federal officials before talking to the press. The senator knew the answer before he had asked the question.

Collier hesitated. "No," he finally said. His agency called the Corps of Engineers only after the information had been given to a reporter for the *Daily News*. The senators chastised USGS officials for releasing their information to the press before informing other agencies. The "federal family" should speak with a single voice, Jackson admonished.

The politicians could not attack the Forest Service. Its report had been made available to other agencies, though few people realized its significance or had taken the time to scour its figures as Stepankowsky had.

During public comment a woman stood, flanked by her two young children. Her name was Connie Long, she said, and she was a local resident. She was glad the information on flood potential had been released directly to area residents by the USGS through the *Daily News* report without being "sterilized by committee." The crowd applauded.

In ending the meeting, Magnuson read a prepared statement. That very day he had persuaded President Jimmy Carter to sign an executive order creating an interagency task force to coordinate Mount St. Helens activity. The meaning was clear to journalists in the room. There would be no more speaking out of turn to the press by the USGS or any other federal agency.

Then the senator astonished almost everyone in the room. He declared that all the agencies' varying conclusions were "compatible" and that, although the public should be "prepared," the Corps seemed to have things in hand. He was going to help ensure the Corps' success by securing another $80 million for dredging that the agency had requested.

Stepankowsky had now seen some of the nation's most powerful politicians at work. He was not impressed with them nor with the bureaucrats they had arm-twisted to skew what the reporter considered to be the facts. In a story filled with documenting details, but labeled "analysis" by the editors, Seifert and Stepankowsky wrote: "The bureaucrats forged agreement by omitting and changing important numbers, by making broad generalizations, and by qualifying their predictions with a flurry of ifs and buts."

The snow would soon be here. If he did not get his plots planted before the first snowfall, Peter Frenzen's study for his master's degree would be in trouble. His study was on how conifers reestablish themselves on freshly fallen volcanic ash, or tephra. The idea was to replicate nature's seeding as much as possible. Typically, seeds scatter through wind and animals in the fall, before the first blanket of snow, and then germinate in the spring.

Frenzen and a group of forestry students now scattered seed in a plot on an ash-filled clear-cut about twelve miles northeast of the volcano. The site was chosen because it was flat, meaning there would not be temperature extremes, was easily accessible, and had experienced a heavy ashfall typical of other areas around St. Helens.

Part of the study would look at which species of conifers grow best in the ash. They planted seeds from Douglas fir, Pacific silver fir, western white pine, lodgepole pine, and western hemlock. Frenzen had already planted three other nearby sites that were chosen because of the varying depth of ashfall in each location, from two to seven inches. He wanted to find out what effect the depth would have on germination.

When the planting was finished, Frenzen surveyed the work. In his mind he saw the seedlings sprouting up, hundreds to a plot. He would have his work cut out for him.

This would be her last chance until the rainy season was over. Donna Parker had persuaded Dwight Reber to fly her in to see her brother's truck. As they flew into the blast zone, she craned her neck to get a good look from the back of the helicopter where she rode with George Gianopoulos's son, Steve. George sat in the front by Reber. The miles of barrenness awed Parker. How could anyone have survived this? She thought of all the dead. This was sacred ground. Then George Gianopoulos pointed to the ground. There was Billy's truck, covered in ash.

Reber battled an updraft to land the helicopter. They got out and walked toward the truck. She had seen photographs, but somehow they had never quite seemed real. The actual battered truck, its paint chipping in the ash-filled wind, caught her by surprise.

Then she saw the sixty-pound rock that had burst through the back of the camper to kill her brother. On his first trip to the site, Gianopoulos had found the rock and placed it on a stump in front of the truck. For an instant she tried to imagine what it must have been like for her brother. Then she banned the thought.

Reber and Gianopoulos had left the site in a shambles in their rush to look for cameras and guns during their earlier visit. Sleeping bags and camping gear littered the ground. The place was a mess, she thought. She started picking it up.

"What are you doing, Donna?" George asked.

"I'm cleaning this place up."

"It doesn't matter now," George said.

"Billy wouldn't have wanted it this way," she said as she bent over to pick up a plate.

"Donna, we've only got so much time."

Parker went on for a few minutes more. She opened a cooler sitting in the ash. An egg carton inside contained a dozen eggs that Parker knew Billy had gotten fresh from their mother. She opened the carton. The eggs had been cooked solid by the blast.

She looked in the back of the camper. Some of Billy's things were there, mixed in the ash. She hesitated for a moment, then looked inside the cab. The upholstery was damaged and the plastic around the mirror had melted. The glove box was open but empty. Parker walked around the vehicle. It was buried in ash up to the bottom of its frame. She looked out over the desert of gray. She couldn't see any roads. They were all covered in ash, too. It would not be easy to get the truck out.

CHAPTER 13
November and December

HIS PHOTOGRAPHS of the volcano were selling well. Now Robert Rogers had enough money to make a trip he had thought about for a long time. Three years before, he had joined an organization called the San Francisco Suicide Club. Their activities often involved illegal trespass, daring, and danger—right up Rogers' alley. He knew he had to meet these people. Rogers especially wanted to meet their leader with the name that fascinated him, John Law.

At first he wasn't sure what they actually did. Their newsletter was a little vague about scheduled events, like "bridge tour," "sewer walk," "Tux and Gown Dinner at an exclusive club." Then he learned that the Tux and Gown Dinner was at the abandoned penthouse restaurant of a shuttered skyscraper. The bridge tour was an unsanctioned climb to the top of one of the city's illustrious bridges. And the sewer walk was a test to see who could navigate the maze of tunnels running through the bowels of San Francisco.

Rogers drove his Simca to the city by the bay. That night he showed up unannounced on the upper floor of the old building where the San Francisco Suicide Club met. He met John Law and found that was the club leader's real name. Then he got an okay to set up his projector and show volcano slides. The twenty or so men and women had never seen anything like this. They grew excited and started asking him what he had gone through to get the pictures. When he was done, the members crowded around him, some inquiring, "When are you going to move to San Francisco?"

After most of the crowd left, Law said, "Let's go climb something, like the Golden Gate." Two other club members joined them. "Remember, cops are part of the playing field," Law told him on their way to the bridge. "There's only been a couple of times we've been caught and we talked our way out of those. Just remember, you never fight cops. It's 'No, sir. Yes, sir. Right away, sir. I was just leaving, sir.'"

The deep croak of foghorns blared through the damp air as they climbed the bridge. The fog hung heavy around him. Rogers was in the middle of the men as they crawled along pipelines and over giant rivets made slick by the fog. His footing slipped once, but he recovered quickly. It was a matter of pride to keep up with Law, who led the group.

They found an access door in one of the cross braces beneath the roadway on the Marin tower. It was about half the size of a standard door. The heavy door creaked as they opened it with a key a locksmith had made on an earlier club trip. The bridge towers were made up of three-and-a-half-foot cells. Ladders ran up the cells. All but one ladder terminated before reaching the top. It took the club nearly two years to find that one ladder.

Rogers was amazed at how fast they made it to the top. The fog had started to clear. Cars whizzed across the wet roadway far below. The lights in the city and on other bridges twinkled as clumps of fog moved past them. He set up his camera for a delayed-time self-portrait that allowed for several minutes of exposure. He turned on a flashlight he had brought and began drawing what would appear as streaks of light on the film with the nightscape of San Francisco behind him. Rogers liked these people from the Suicide Club. For one of the few times in his life, he felt he belonged.

At his desk at USGS headquarters, Don Swanson pulled out the notebooks showing his measurements of the cracks on the crater floor prior to the October eruption. The cracks had started to appear four weeks before the eruption. Now they had appeared again and he had resumed measuring them. He stared at the measurements from both periods for a long time. How could he transform this data into something meaningful, something reliable that would help him predict a volcanic eruption—and with enough time for warning to save lives and property?

This had been part of the problem before the May 18 eruption. Some scientists and others promoted crackpot theories that only contributed to confusion and added to the air of levity surrounding the volcano before it exploded. St. Helens had been one big joke before the eruption. Songs, "I'm a survivor" T-shirts, plane rides, volcano burgers, souvenir trinkets, and a constant stream of volcano jokes had all contributed to the carnival atmosphere. But it wasn't only the crackpots, thrill seekers, comedians, and misguided scientists and journalists that contributed to the low level of respect for volcano predictions. It was Swanson himself, and other scientists like him. They had a poor record of success.

Swanson and his fellow scientists had gotten a firsthand taste five years before of the consequences of poor predicting. Mount Baker, 180 miles north of St. Helens and part of the same Cascade Mountains chain, jetted a black plume into the air on March 10, 1975. New steam vents opened, indicating a fresh heat source. The glacier deformed. For the next three weeks, Swanson and the other scientists saw heat from the mountain multiply tenfold. Hydrogen sulfide, which they thought indicated the presence of magma, rose to high levels.

Geologists predicted that even a small eruption would result in muddy waters cascading below, breaching a dam, and burying the region in floods. Upon the advice of the USGS, officials ordered that reservoirs be drawn down, recreation areas closed, and access to the mountain restricted. Then everybody waited, and waited, and waited. Nothing happened. Finally, over a year later, the USGS issued a public finding that there was "now no clear evidence of a forthcoming eruption."

The consequence of the faulty predicting was immense. Millions of dollars had been lost in the tourist trade, power sales, logging, and other business operations, sending the area's economy into a tailspin. Criticism of the USGS from citizens, politicians, and other scientists lasted for years. Some even wanted to disband the agency. The experience left Swanson thinking it was almost as harmful to issue a poor prediction as no prediction at all when the threat proved real. But it was hard to blame the scientists. Techniques for making accurate predictions simply did not yet exist.

Swanson grabbed some plotting paper. On one piece he plotted the acceleration and distance that cracks had widened in the crater before the October eruption. On another piece he mapped the same information for cracks since then. He could just barely see what he thought was a trend developing on the second graph that might prove comparable to the first plotting.

Now that Donna Parker had seen the truck and talked to the coroner, she thought she had a pretty good idea of how her brother had died. The front of the truck had been parked facing away from the volcano. She thought that in the minutes before the eruption, Billy was taking photographs of the mountain with his camera on a tripod. Jean was still asleep or resting in her nylon pajamas on the mattress inside the canopy on the back of the truck. Then Billy saw the side of the mountain collapse and the first part of the black cloud emerge. He was eight miles from the volcano, so he would have had a few minutes.

He got Jean out of the back of the truck and threw his camera and tripod onto the mattress. Then he and Jean ran to get into the cab. Parker figured from the keys her mother had discovered in Billy's pants pocket that he wasn't trying to start the truck. He probably knew that getting protection in the cab from the searing black cloud blasting toward them was his best option. There wouldn't be time for much else.

Billy sat there looking crosswise out the back window, watching the cloud grow larger. Then the sixty-pound rock crashed through a side window of the canopy. The rock came through the back window of the cab and into the side of her brother's head, killing him instantly. At about the same time, the black cloud, heated to 1,200 degrees Fahrenheit, overcame them. It shriveled and hardened their internal organs and carbonized their bodies like black mummies, caught in the middle of their last acts. Billy died with his left hand on the steering wheel, looking over his shoulder at the mountain. Jean was found with her hands in her lap as she looked at Billy. Ash clogged her airway, but she had not struggled long before death.

That night Parker's old dream of people trashing her brother's truck returned, only now it seemed more real and vivid.

The first heavy storm of the season hit the area. It lasted for two days. Rains resulted in heavy erosion along the Toutle River, ripping out banks and threatening the few homes left standing along the river. Three million cubic yards of silt washed into the Cowlitz River from the Toutle.

The Cowlitz experienced little flooding. The Portland forecast center of the National Weather Service said the river had responded as expected during the rain. The Weather Service in conjunction with the USGS had installed thirty rainfall and river gauges on the Cowlitz and Toutle Rivers. Stepankowsky tracked down how well they were working. He found only thirteen of them operational.

It was very cold when they flew to Harrys Ridge to set up the geodimeter and take readings. Swanson braced his body against a wind that bit into his face. Just getting to the point where he could do his work and take measurements now required hours of drudgery. He and his assistants had to continually scrape thick coats of rime ice off the new steel monitoring and reflector towers. The towers were eight-hundred-pound behemoths cemented into place. At ten feet tall they rose above any anticipated snowpack. Although they were designed to withstand 150-mile-per-hour winds, one of them in a recent eruption had been sheared off and hurled a thousand feet down the mountain. Many attempts to take measurements had to be aborted when the winds grew so fierce no one could stand on the tower platforms or the clouds were so thick that targets could no longer be seen.

Today he could see the targets, though dark, ominous clouds hung overhead. Swanson twisted the dial on the geodimeter and shouted out his readings as an assistant wrote them down. He found there had been a slight expansion of the volcano's cone, just as he had seen before the October eruption. Swanson then turned the geodimeter to the rampart. Just as before both the July and October eruptions, the rampart had moved many inches northward.

He checked his notes for the measurements he had taken on the cracks since the October eruption. There seemed to be a continuing correlation between movement of the rampart and widening of the cracks. He compared his plot graph with the previous eruption he had plotted. There was the same rate of ever-increasing widening, at first doubling and then quadrupling and so on within a specified time, multiplying by factors like a dividing amoeba. Both the flat surface cracks and the cracks from thrust faults that he was now monitoring appeared to be increasing in acceleration at the same rate.

Swanson now had a theory about what was going on. Both the widening of the cracks and the movement of the rampart were related to changing magma pressure in a narrow conduit beneath the crater. The pressure changes were so small and shallow that they only affected the crater itself, as shown in the cracks and rampart movement. Much larger and deeper pressure changes were also occurring, and these could account for the expansions and contractions of the entire volcanic cone. The two pressure sources at times appeared to work independently with the systematic movement of the rampart, indicating the shallower one was the most consistent.

Swanson thought that perhaps the two pressure sources were actually part of one reservoir system. The shallower source reflected changes, especially increases, of volatile pressure in the upper part of the magma. The accelerated rate of shallow-level indicators, such as the movement of the rampart before an eruption, could be reflecting a rapid increase in volatile pressure resulting from magma changing, convective overturn, or some other process going on in the reservoir.

Swanson also had an allied theory to help account for the northward movement of the rampart. He and other scientists were coming to believe that the volcano was sitting on top of a northwest-trending fault. It was more important than ever that he be able to measure every change in the cracks, the rampart, and the mountain overall. It was becoming apparent to him that an eruption was imminent.

Snow began to fall. With winter approaching, Swanson knew the snow would begin building up on the crater floor and dome. He wondered how he was going to monitor the cracks. The USGS water resources division had issued a report forecasting "extreme danger" from snow and ash avalanches on Mount St. Helens for the coming winter. The report said it would be "particularly dangerous for scientists to work on or near the volcano."

The fog shrouded the houses as he walked the streets of the subdivisions near his home in Woodburn, Oregon. Jim Scymanky could barely make out the outlines of the houses, and then the fog grew so thick he couldn't see them at all. It didn't matter. They all had begun to look the same to him. Scymanky now walked in the morning, in the afternoon, and at night. For hours at a time. He found he could nod off for a few minutes of sleep after he had exhausted himself. He was sleeping a few hours per day. Through the walking, he found his body growing stronger. But as he walked with the fog enveloping him, he found he could not shake the feeling that still haunted him, of being alone.

With winter near, there was now another flooding threat: rapid snowmelt from an eruption. The USGS released a report that said any pyroclastic flows of rock, ash, and gas would instantly melt the snowpack, releasing a surge of water and causing mudflows. "As we move into the winter the risks of mudflows and floods increases substantially," Chris Newhall, the new USGS coordinator for Mount St. Helens volcanic hazards, wrote in the report. At a news conference, Newhall warned that more eruptions and pyroclastic flows were almost inevitable.

A small pyroclastic flow that measured just 2.5 square miles under only a foot of snow could release 1,600 acre feet of water, according to Newhall. For comparison the Elk Creek dam break that caused so much destruction along the Toutle River and sent Peter Frenzen and his friends scurrying to get out of the area had contained only 236 acre feet.

The geologists at the news briefing acknowledged they were no closer to predicting eruptions than they were almost three months earlier when their last report was released. "It is generally impossible to predict eruptions more than several hours in advance and some eruptions can occur without any recognized signals," Newhall said in his report.

Weak harmonic tremors were now rumbling beneath St. Helens. The tremors, indicating the movement of molten rock, or gas changes, had begun just prior to midnight the night before. Volcano experts remained noncommittal about the odds of an imminent eruption.

The man was elfish, with a mop of black hair and a face that hinted at Asian ancestry. He looked to be middle-aged, was quite animated, and seemed sympathetic as he listened to Donna Parker's mother tell about what happened to Billy. Tears even glazed his eyes at one point, Parker noticed. They were meeting at the Portland home of a friend of Parker's. The man said he was writing a book about the eruption, the victims, and the relatives who had survived. Otto Sieber was his name, he said.

Snow filled the crater. Swanson and his assistant had to be careful. Hot gases from the widening cracks had melted snow above and produced snow caves. A misstep onto the roof of a snow cave that caused it to give way could mean instant death with a fall into the crack and then into the incandescent rock and gases below. Swanson lowered himself cautiously into an opening he had dug to gain access to a snow cave so he could reach one of the cracks. He began crawling along a ledge on one side of the five-foot-wide crack while his assistant crawled on the other side.

The ledges were slippery. Swanson got ahead of his assistant, who couldn't take his eyes off the rock glowing below him. The ground shook from one of the dozens of tiny quakes that often hit the area in a day. The assistant looked at Swanson, who kept crawling, seemingly undisturbed. He started sweating in the muggy, sulfuric air.

The man knew he would not do this for just anybody. But Swanson had impressed him with his dedication. The assistant knew of Swanson's theory and how important these measurements were. He knew that Swanson would do the work by himself if need be.

The men came to the pieces of rebar on either side of the crack. The crack had widened by three inches since Swanson had measured it last. And the movement was accelerating, according to his calculations. If his theory was right, it would be only a matter of days before the next eruption.

The man pointed to a dragline dredging the Cowlitz River near Castle Rock. "I would have to say," he told Stepankowsky, "that more silt passes by than he takes out every time he puts the bucket into the water." The man was Clare Cranston, part of a ten-member USGS team assigned to measure the river at eighty-four locations. Cranston's twenty-four-foot aluminum jet boat, the *Queen Mary*, idled in the water. An unrelenting, cold, and biting wind buffeted the two men.

Although 35 million cubic yards of volcanic mud had already been removed from the river, Cranston was skeptical about the dredging process. He was a veteran of five Cowlitz River surveys since May 18. "I don't think they'll find much difference between this survey and the last one," Cranston said, with a Midwestern accent muted by the cigarette that dangled from his lips.

Cranston nodded toward the screen for the boat's sonar river sounder. The picture was fuzzy as sonar deflected off silt suspended in the river. "You see what he's done with that bucket," Cranston said, referring back to the dredger. "Nothing."

The Army Corps of Engineers had announced with much aplomb that it had reached its goal to dredge a channel on the river fifteen feet deep and three hundred feet wide to carry fifty thousand cubic feet of water a second—71 percent of the Cowlitz River's preeruption capacity. Stepankowsky had hitched a ride with the USGS measuring team in order to test the Corps' claims.

Cranston revved up the boat's engine. The *Queen Mary* bounced with the river's swells on its way to the next measuring spot. The boat slowed and Cranston put it into idle as it bobbed in the water. Another worker with a long measuring pole balanced himself like a tightrope walker as he stepped along the boat's edge to the bow. The man pushed the pole into the water. Here the river was only seven feet deep. A nearby reading showed the channel was fourteen feet deep where a dragline had just finished working, but was much shallower in the center and toward the east bank.

Further readings confirmed Stepankowsky's hunch: the Corps' claims had misportrayed the actual situation on the Cowlitz. Silt was pouring back into the river channel, undoing their work. As the boat sped back down the Cowlitz toward Kelso, huge chunks of cliffs along the riverbank crashed into the water almost continuously.

On December 16 the volcano shot a plume of steam and ash two thousand feet above the rim. Steam explosions fractured the lava dome formed by the October eruption. The dome now stood sixteen stories high and was almost as wide as a football field. A little over a week later there was another eruption but nobody knew it for sure at the time. There had been seismic activity. However, there had been no explosion of gas and ash as in previous eruptions.

When Swanson and his team were able to enter the crater, a couple of days later, they found St. Helens in the midst of erupting magma, building a second dome on the side of the old one. The new dome was about 40 feet high and 250 feet wide and growing. This was the first time the volcano had erupted without an explosive blast. Swanson theorized that there had not been enough buildup of gas this time to cause such an explosion. The new magma was being squeezed like toothpaste to the surface rather than blasted. It was more than the new dome that caught the scientists' attention, however. Blocks of thick, hot rock rose up from the crater floor in spectacular pillars towering as much as a hundred feet high.

Within three days the new dome had grown to the same height as the old one. The eruption cycle had not been damaging enough to melt sufficient snow on the sides of St. Helens to cause flooding. It also helped that the snowpack to this point had been significantly less than in past years.

As Swanson pored over his data on the cracks before the new eruption, he grew excited. There was a correlation between an acceleration of crack widening and the new eruption, as there had been before the October eruption. Swanson decided that if new cracks started accelerating in the same pattern as they did before the last two eruption events, he would test his theory by making a prediction before the next eruption. The question was, would he feel secure enough to tell his superiors about the prediction and how he had arrived at it?

On Christmas day heavy rains hit the state of Washington. The downpour continued without letup the following day. That night Stepankowsky watched the television news from his in-laws' home in Spokane, where he and his wife had gone to celebrate the holidays. The rains were coupled with springlike temperatures. The storm was driving hundreds of people from their homes, washing away houses, and closing roads throughout the western part of the state.

The National Weather Service said major flooding was occurring or expected to occur along nine rivers, including the Cowlitz. Only one thing could save much of the western part of the state from catastrophic flooding. That was if the rains slackened after midnight. Early the next morning the rains did stop. The governor declared Cowlitz County, among others, a disaster area, eligible to receive emergency funds to repair over $4 million in county flood damage. But Stepankowsky and officials knew the flooding was relatively minor compared with what could potentially happen. A dam miles upstream had been put into emergency use to virtually cut off the flow of the Cowlitz River. But the limited capacity of the reservoir behind Mossyrock Dam had been taxed almost to its maximum.

Photo gallery

Top: The infamous bulge on the north slope of Mount St. Helens prior to the May, 1980 eruption. *R. Tilling, U.S. Geological Survey.* **Middle**: A scientist watches the mountain a few days before the eruption. *U.S. Geological Survey.* **Bottom**: Taken at Coldwater II observation station the evening before the eruption, this was the last photo taken of David Johnston. *Harry Glicken / USGS.*

Top:
Robert Rogers poses in front of the eruption cloud.
Francisco Valenzuela.

Bottom:
This eruption cloud bore down on Rogers and Francisco Valenzuela as they fled the mountain.
Robert Rogers.

Top:
An aerial view of the massive eruption.
U.S. Geological Survey.

Middle:
A mudslide dislodged this bridge, carrying it a quarter-mile downriver on the north fork of the Toutle River.
R.L. Schuster / USGS.

Bottom:
A helicopter crew inspects a car belonging to photographer Reid Blackburn, who died about six miles from the mountain.
U.S. Geological Survey.

Top:
A mudflow carrying thousands of logs tears loose the Coal Bank Bridge on the Toutle River. This and other dramatic photos helped win a Pulitzer Prize for the *Longview Daily News*. *Roger Werth / Longview Daily News.*

Middle:
A Weyerhaeuser Company bus was damaged by mudflow. The eruption occurred on a Sunday, a day off for loggers who worked in the area. *U.S. Geological Survey.*

Bottom:
Devastation at a logging camp on the south fork of the Toutle River. *U.S. Geological Survey.*

Top left: Snowplows removed ash from Interstate 90 in eastern Washington. *Washington State Department of Transportation.* **Top right**: An eastern Washington homeowner sweeps ash from his roof in the days after the eruption. *U.S. Geological Survey.* **Bottom**: Zero visibility due to billowing ash caused chain collisions, such as this one on Interstate 5 near Castle Rock, Washington. *Barry Wong / Seattle Times.*

Top:
Emergency workers crafted elaborate, homemade air-filtration systems to keep the ash from disabling their vehicles.
U.S. Geological Survey.

Middle:
Don Swanson worked on the volcano in 1983 as part of a scientific effort to improve the reliability of eruption forecasts.
Don Swanson.

Bottom:
After the eruption, the summit was approximately 1,300 feet lower.
U.S. Geological Survey.

Top: Swanson holds up a piece of pumice to demonstrate its weight. *U.S. Geological Survey.* **Middle**: Mud and debris from avalances left hummocks and other geological formations hundreds of feet high. See Swanson circled at lower right. *U.S. Geological Survey.* **Bottom**: Steam seeps from radial cracks extending from the crater dome in November, 1980. *U.S. Geological Survey.*

Top: Don Swanson aims a laser beam at a reflected target within the crater, seeking clues from ground movement to an impending eruption. *Cascades Volcano Observation / USGS*. **Bottom**: Swanson, left, and an assistant measure a radial crack in the crater a year after the eruption. *U.S. Geological Survey*.

Top:
The same mudflow that swept Roald Reitan and Venus Dergan from their campsite pushed this boulder downriver. *U.S. Geological Survey.*

Middle:
Helen and Jim Scymanky place crosses honoring members of his tree-thinning crew who died. *Global Net Productions.*

Bottom:
Andre and Paula Stepankowsky pose in ash masks on their wedding day a month after the eruption. *Andre Stepankowsky.*

Top:
Roald Reitan and Venus Dergan were carefree young lovers when this photo was taken shortly before the 1980 eruption. *Venus Dergan.*

Middle:
Reitan offers encouragement to Dergan in a hospital after their improbable escape from a fast-moving mudflow. *Tom Ryll / Vancouver Columbian.*

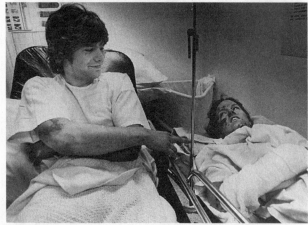

Bottom:
Friends and relatives excavate a portion of Reitan's Oldsmobile from the mudflow debris, recovering Dergan's purse. *Roald Reitan.*

Top:
Billy and Jean Parker died in this truck. The rock that flew through the back window, killing Billy, is seen near the front tire.
Donna Parker.

Bottom:
Friends try to push Billy Parker's truck out of several feet of ash to save it from souvenir hunters. *Donna Parker.*

Top: A visitor prays before crosses placed by Donna Parker in memory of Ed and Eleanor Murphy, whose bodies never were found. *Donna Parker*. **Bottom**: In 1994, Donna Parker posed at the Coldwater Visitors Center with Ray Murphy, brother of Ed Murphy. *Helen Scymanky*.

Top: A year after the eruption, Robert Rogers stands at the lip of the crater during one of his many secret treks to the volcano. *Robert Rogers.* **Bottom**: In 1997, Venus Dergan poses beside her photo on the Survivor's Wall at the Johnston Ridge Observatory. *Venus Dergan.*

Top left: Peter Frenzen, a new forestry graduate, poses in front of the steaming volcano. *U.S. Forest Service.* **Top right**: Frenzen surveys widespread devastation caused by the eruption. *Jerry Franklin / USGS.* **Bottom**: A grim view of once-pristine Hanaford Lake, north of the mountain. *U.S. Geological Survey.*

Top: Frenzen oversees a replanting site, part of a study of nature's recovery. *U.S. Forest Service.* **Bottom**: Ash and silt left in riverbeds contributed to this February, 1986 flood near Longview. *Roger Werth / Longview Daily News.*

Top: The Corps of Engineers dredged mud and silt along the north fork of the Toutle River. *U.S. Geological Survey.* **Bottom**: A deformed Mount St. Helens is reflected in Spirit Lake, two years after the eruption. *U.S. Geological Survey.*

Top:
A dramatic vista of Mount St. Helens is seen from the Johnston Ridge Observatory.
J. Quiring / USFS.

Middle:
An avalanche breaks loose from the lava dome within the crater in May, 1986.
Lyn Topinka / USGS.

Bottom:
Nature recovers, even inside the crater of America's most dangerous volcano.
Peter Frenzen / USFS.

PART 2
1981

Aftershock

What the hell is there to be worried about?

—Stan Lee

CHAPTER 14
January and February 1981

THE HEAVY WIND ballooned his parka as Don Swanson stood near the top of the thirty-story dome. He and his team had permission to make the USGS's first trip to the top of the dome, the very center of the volcano with its throat to the magma chamber below. The wind did not sweep away the ever-present smell of sulfur. Down below, steam vented from cracks on the crater floor in wispy blossoms, merged, and rose in a cloud that blotted out the sun just peeking over the south rim.

The heat of the volcano along with recent rains and warm temperatures had melted most of the snow in the crater. Boulders, flocked a transparent yellow from sulfur, peppered the crater floor. He turned to look around him. Streaks of black veined the gray crater walls, evidence of the splitting and cracking caused by magma on the move in ancient eruptions. Through the gaping breach of the north side he could see the highest volcano of the Cascade range, Mount Rainier, rising majestically with coned crown into the clear, blue sky as St. Helens once had done.

It was time to get to work. A web of cracks spoked out in all directions from the dome's center, cracks just like the ones he had been monitoring on the crater floor. The team drove pieces of rebar into the dome's surface to use as reference points for measuring the width of the cracks. When they finished the measurements, they pounded posts into the dome and mounted reflector targets for geodetic surveying.

A helicopter hovered just above the dome, ready for a quick exit. If there was going to be a surprise eruption, Swanson believed they would have at least a few minutes of seismic activity to warn them. He tried listening above the *whack* of the helicopter blades for the telltale cracking sound of an earthquake. He looked over at the smaller dome next to the one he was on. A piece of the other dome had broken off into a small avalanche that settled near the base. A furious cloud of steam, tinged a yellow-brown, surged up from where the piece had been dislodged. Swanson was not worried. The crater walls and the domes often shed rocks.

Jim Scymanky leaned back in the chair as he talked to the psychologist he was now seeing weekly. The man had suggested walks and reading. Recently he had given Scymanky soothing tapes of ocean music.

"How are they working?" the psychologist asked.

"They're helping quite a bit."

"How many hours sleep are you up to now."

"Oh, about three, off and on."

"It's going to be a very gradual recovery for you."

"I know. I've been at it for almost nine months."

Robert Rogers found a copy of the January *National Geographic* with its cover photo of eruption clouds from the May 18 blast. As he thumbed through page after page of writing and photographs, he stopped at an article titled "Robert Landsburg's brave final shots." On a two page-spread the magazine displayed three smaller photographs and one larger one, all of which had been on film retrieved from the photographer's camera. Rogers studied the photos. They had been taken at the same area where he last saw Landsburg and his car. Rogers knew the location well.

"As the all-engulfing cloud of ash," Rogers read, "climbed the sky toward him, four miles from the summit, he desperately cranked frames across his lens . . . then rewound the film into its cassette inside the camera, wrenched his camera from its tripod, and stowed it in a pack. . . .

"Seventeen days later his body was found in the ash, together with the film that he bought with his life. It contained not only telling images of the killing edge of the blast but also the scratches, bubbles, warpings, and light leaks caused by heat and ash, the very thumbprint of the holocaust."

Rogers looked at the photographs again and shook his head. The *National Geographic* had it wrong. The position of the trees was not right for a blast coming straight at Landsburg. He studied the sequence of the eruption Landsburg caught on film. It showed the plume of the initial lateral blast expanding out to the northwest and not directed at the photographer. The last shot, which was represented as the blast overcoming him, was just another photograph of that expansion, now filling up almost the full frame. Landsburg had several more minutes to live before the ash hurricane, released with the lateral blast, changed direction and overcame him. During that time he not only stuffed his camera into his knapsack, but also made notes with short details about the shots he had taken.

Because of his technical background, Rogers recognized that what the *National Geographic* was calling light leaks was actually vertical streaking caused by electrostatic discharges inside the camera body, resulting from the lightning strikes he had observed while on the ridge above the photographer that day.

The Corps of Engineers took credit at a news conference for saving the Longview-Kelso area from catastrophic flooding while minimizing damage from the Christmas rains. Andre Stepankowsky thought they were right. Without the dredging by the Corps and the boosting of emergency storage capacity behind Mossyrock Dam, the results could have been tragic.

But Stepankowsky had noticed there had not yet been any sustained series of storms. "We haven't seen much real winter weather," said Cowlitz County Deputy Sheriff Ben Bena, coordinator of emergency services. Bena went on to say what the Corps had told him but had not bothered to tell the public. At the very most the Corps could only hold back water at Mossyrock Dam for five days. Bena credited the dam for saving the area thus far, more than the dredging. The Cowlitz had been able to handle the Christmas runoff from the raging Toutle River because the Corps had virtually shut off the Cowlitz way up the river at the dam. "The crucial factor is how long can you hold back water? We have been fortunate in that our weather systems have been spaced out with breaks in between," Bena said. "The river can only handle so much."

Stepankowsky was getting mixed signals from the National Weather Service, the Corps of Engineers, and the USGS on the critical issue of flood danger. The Weather Service in Seattle announced that the peak flood season had passed, because the six largest floods in the area's history had been in November and December. Phil Carpenter of the USGS water resources division, in reviewing the same historical record, said it showed "You can still get a big rainfall event and still have problems." Then the Corps announced that because of a lack of new funds, operations were being cut back, but the announcement said the most threatened portions of the Cowlitz River had already been adequately dredged. Stepankowsky continued to wonder what role politics was playing in flood prediction. For the sake of the area's people, he hoped the mild weather and the spaced-out storm pattern persisted.

Robert Rogers was not the only one to take the *National Geographic* to task. Donna Parker turned to a double-page spread and there it was, her brother's battered truck in the sea of ash with the helicopter getting ready to land for a futile rescue. She couldn't escape the image. In one form or another it had appeared in countless newspapers, magazines, books, newscasts, and even documentary films. The image reinforced her determination to get the truck out. She feared that with all the publicity, others now recognized its significance—perhaps even imagined it as a money-making tourist attraction. She had to get to it before it fell into such hands.

What really bothered her, though, was what she found when she turned to the map in the magazine and read its caption: "Fearing the worst, state officials barred entry to a 20-mile 'red zone,' its heart mapped at right, except for critical personnel such as geologists, who sought to sniff, feel, and listen to the mountain through sensitive devices. Without restrictions, thousands of sightseers would have died."

The map and caption astonished her. She had seen similar misinformation in other publications since the *Oregonian* special section had revealed how there had been no real Red Zone on the northwest side—the side where all the volcano victims including her brother had been killed—and that there had been a zone much smaller than twenty miles on the other sides. Seeing this material in such a revered publication as the *National Geographic* only deepened her distress. The "big lie" was growing bigger, Parker thought, feeding on itself. It would take something much more than a newspaper special section to bring out the truth.

Stepankowsky had long been interested in how the land around St. Helens would be used in the future. All the players were there: local, state and national governments, timber interests, scientists, area residents, and environmentalists. Stepankowsky had grown up with a deep appreciation of the environment as his parents hauled the family all over the country to camp in national parks, teaching their children about the bounty of nature. So when his friend and fellow reporter Rick Seifert, who usually covered the beat, was busy with another story, Stepankowsky was only too happy to step in.

Environmentalists had proposed creation of a 216,000-acre national monument at Mount St. Helens. But a committee made up of logging companies and the state Department of Natural Resources, which had a vested interest in timber revenues, instead wanted a "study area" of only 50,000 acres where salvaging and tree harvesting would be excluded. Much of this proposed area had already been salvaged and harvested or else destroyed by the May 18 eruption.

The area forest manager for Burlington Northern told Stepankowsky that prohibiting salvage or timber harvesting within the larger area would mean a loss of two billion board feet of timber over an eighty-year forest cycle. A Weyerhaeuser spokesman warned of the loss of timber, jobs, and government tax income.

On the other side a representative of the Sierra Club said only about 4 percent of the 500,000 acres of commercial timberland in Cowlitz County would be within the national monument. "The potential for tourism would certainly offset any significant changes in the timber economy," he said.

Stepankowsky wrote his monument story and then rushed to another assignment. Much of the region had been saved from the Christmas flooding, but several families on Tower Road along the Toutle River had not been so lucky. He found bridges and large sections of the road missing. Where the pavement suddenly dropped off, sometimes in sixty-foot sheer walls down to the river, makeshift, rocky detour roads had been opened.

On the approach to the ravaged houses, someone had tacked flyers on tree after tree declaring, "Jesus died for your sins," as though homeowners had somehow brought on their own woes. A green span of bridge stuck up from the mud five hundred yards downstream from where it had once been connected to the road. Some toilets, a few with raised seats, sat in the mud. Parts of buried roofs attested to houses swept away.

This was the second major catastrophe for residents in the past year. The area had first been hard-hit by floods after the May 18 eruption. Half of the people had moved away. The ones who remained had to contend with looters and nosy tourists. Signs that said "Private property, violators will be shot" decorated the front lawns of the neighborhood. One hand-scrawled sign proclaimed "Looter Shooters Local No. 3644." Another sign read, "All sightseers and reporters keep out. Looters will be shot."

Stepankowsky found people who were gnawed by uncertainty. "I'm not going to rebuild my house until I see what happens to that river," one man said. His house had been swept away by mudflows. He now lived at the back of his property in a trailer. Other families had been forced to evacuate their homes, only to move back in and then have to leave once again.

Terrified and nervous neighbors had helped Jim and Jerri Hastings move their furniture out on the night of December 26 as the Toutle River cut away the couple's front lawn. They had gone to bed earlier that night, thinking they were safe. "We had tried to sleep," Jerri said. "But the bank just kept caving in. It made a thundering noise and shook the whole house."

Stepankowsky heard the incessant roar of gasoline-driven electricity generators drive out even the sound of the nearby river. Many residents had paid over a thousand dollars each for generators that didn't produce enough power to heat their homes and cost them hundreds of dollars in fuel expense in just a few weeks. Some people were going broke trying to feed their families and keep the generators going at the same time.

Families had been consumed with the daily challenge of living without electric company power, telephone, and other services. "You spend 90 per cent of your time just doing domestic things like heating water for the dishes," Joe Alongi told the reporter. It was the second time around for most of them. After the May 18 eruption the area lost power for more than six weeks.

"I'm frozen stiff and I'm congested," Joe's wife, Hazel Alongi, told Stepankowsky as she walked across her living room, which had been emptied of pictures, china, and almost all other personal items. Then the older woman surprised him.

"Do you drink?" she asked. Stepankowsky admitted to liking a beer now and then.

"Do you know how many millions of brain cells you're killing with each drink, young man?"

"I think we're going to have an eruption within the next few days," Swanson told his boss, Don Peterson, and fellow USGS scientists at an evening meeting in late January.

"Oh yeah. Why is that?" Peterson asked.

"The cracks are widening again."

"What?"

"The cracks. They're widening at an accelerated pace, like they did before."

Swanson explained his theory and what had happened prior to the past two eruptions. Peterson and other scientists looked over the data. They thought Swanson was on to something. Swanson and Peterson thought it would be best to not go public yet. There were only what scientists would consider to be two incomplete past examples to go on—convincing to Swanson, but not enough yet for the greater scientific community.

Peterson believed enough in Swanson's work, however, that he decided they ought to tell the Forest Service there might be an eruption. But before issuing any public alert, they would rely on precursors that had been useful in the past, like seismic activity.

If Swanson's theory on the cracks was found to be correct, he thought he could provide warnings up to weeks ahead of time. But one wrong prediction and this promising tool would be relegated to the second tier of precursors—interesting and useful in combination with others, but without the reliability scientists wanted and the public demanded.

Don Swanson and thirty-five other scientists who had known David Johnston signed the letter. It protested the way Johnston was depicted in a soon-to be-released Hollywood move titled *St. Helens*. Johnston's parents were also protesting.

The scientists and parents charged that the film grossly distorted the late geologist's character. "Dave's life was too meritorious to require fictional embellishments," the scientists wrote to the film's producer. The movie depicted Johnston as a reckless and careless scientist when just the opposite was true. "Dave was a superbly conscientious and creative scientist," the scientists said. The movie had invented a love interest with a waitress. "We protest any portrayal of romantic involvements by Dave while working on the volcano."

A movie based on the real life and exploits of Johnston would be a hit without stretching the truth, Swanson thought. After all, his friend did have a certain flair.

On February 5 the USGS predicted that a nonviolent eruption would occur within hours. Stepankowsky wrote:

> *Mount St. Helens spouted steam a mile above the crater this morning and scientists predicted the volcano would erupt non-violently by 4 p.m. today.*
>
> *Scientists from the U.S. Geological Survey and the University of Washington Geophysics lab issued their eruption alert at 4 a.m. after six to eight shallow earthquakes an hour rumbled through the mountain after midnight.*
>
> *Earthquake activity leveled off to about two quakes an hour later this morning, and scientists read the seismic lull as calm before the storm. . . .*
>
> *Scientists expected today's eruption would follow a dome-building pattern of the non-explosive eruptions in December and January when the dome grew the size of the Kingdome in Seattle.*
>
> *A large, spectacularly explosive eruption that would shatter the dome is not expected but still possible.*

Some residents near the mountain kept their sense of humor when informed the mountain was about to erupt. Jean Ragsdale, at Jack's Sporting

Goods near Cougar, told a reporter that every time she made lasagna for customers, the mountain erupted. She had just started making lasagna again. "My lasagna is fantastic, but maybe I'll stop making it."

Stan Lee at the Kid Valley Store said he had heard it all before. "What the hell is there to be worried about?" he said. "Some of those guys [scientists] have to keep yakking to keep their jobs."

Swanson was right. St. Helens was about to experience a nonexplosive, dome-building eruption. The last significant seismic activity of any strength was recorded at 5 p.m. that day and Swanson tried to get into the crater in a helicopter. He and his team were kept away by the massive steam pumping from the volcano's vents. Sometime early the next morning, super-hot molten rock started oozing into the crater. At dawn a spotter plane saw a glow toward the southwest side of the dome. The new dome from the December eruption was now taller than the old dome. A steam plume filled with sulfur but not ash rocketed three thousand feet above the crater lip.

That morning Swanson and other scientists landed in the crater. The new dome had grown considerably overnight. When Swanson returned to the David A. Johnston Cascades Volcano Observatory, his team and other scientists were euphoric. Swanson had predicted the eruption days before there had been any seismic precursors and had called it to within hours of the event. Stepankowsky, along with most of the rest of the world, knew nothing of Swanson's accomplishment.

On the Monday after the eruption, Longview honored the last of its pioneering fathers, John McClelland, founder of the *Daily News*. McClelland had died the previous Thursday at age ninety-five. Flags across the city flew at half mast. The newspaper went to press early so that its employees could attend the memorial service. Hundreds of people crammed into the city's largest church, Longview Community Church, which the newspaper publisher had helped build, with hundreds more spilling out in front.

Stepankowsky had known McClelland as the elderly gentleman, always impeccably dressed in suit and tie, who almost until the day he died entered the newsroom and buttonholed a reporter or editor with a news tip or cajoled them about a lapse in coverage. They all affectionately called him Mr. Mac. Stepankowsky also knew him as a newspaper legend, but he didn't have a full sense of his accomplishments until he read the obituaries and heard the eulogies.

Back in 1923, when R. A. Long began developing Longview as his company town and capital of his logging empire, he decided he wanted a daily newspaper to serve its residents. McClelland heard about the planned city and its opportunities while working as a newspaperman in Arkansas. When he arrived in Longview he found an area filled with muddy fields, sloughs, and tar-paper shacks and watched laborers at work on the city's first buildings. He persuaded Long to let him publish the planned newspaper that was to supplant a fledgling weekly publication. The *Longview News* was born in a shack, on an ancient flatbed press. Within months the enterprising McClelland made the newspaper financially independent and he owned most of it, with Long holding a minority interest. Starting as a twice-weekly newspaper, he soon published it as a daily.

He and his newspaper became a booster for the community. McClelland was a founder of the YMCA, Community Church, library, Chamber of Commerce, Lower Columbia College, Longview Elks, Masonic Lodge, Longview Rotary Club, and many more. He accomplished the almost impossible feat of being involved in these activities while keeping himself and his newspaper fiercely independent. He was a newspaperman's newspaperman who earned his employees' respect. When executives from Long's company, Long-Bell Lumber, sent employees in with stories and even demanded front-page placement, they were sent scurrying back to their bosses with McClelland's message: "You tell him he doesn't run the Longview *Daily News*."

In later years he liked to say he was the only man alive who had seen the construction of every single building in Longview. One of those buildings was an arts center, for which he was the major financial backer. A grateful city insisted on naming the center after him, an honor he did not seek. When, in his last days, he was given his own obituary for his approval, he glanced over it and flung it back at the editor, ordering, "Cut it. It's too long."

By the time he died, his city had grown to over thirty thousand residents. His newspaper, with a circulation of twenty-seven thousand and now headed by his grandson Ted Natt, was the leading publication of the region with a reach and influence that went far beyond its circulation area. It was known as one of the best small-city newspapers in America.

As Stepankowsky listened in the pew, he was proud to be part of it. He could think of nothing he would rather be than a newspaperman.

He was in search of something in his life, but he wasn't quite sure what. Robert Rogers decided to attend a spirituality workshop in the mountains above Detroit, Oregon, at Breitenbush Hot Springs, a magnet for people seeking guidance from New Age gurus and educators. That night Rogers walked through the snow and found a hot tub enclosed by a building that acted as a sauna. No one else was there. He took off his clothes, lowered himself into the hot tub, and started singing. When he was young he had wanted to become a singer, but he was afraid to tell his mother for fear she would make fun of him. But now as he soaked, he found his voice opening up in the steamy air. He practiced singing the scales. Then he sang different songs, some from operas he loved. Then he made up his own songs. For one of the rare times in his life he felt the weight of the pain he always seemed to carry with him begin to lift.

The next day he was in a group that was about to join hands for a meditation exercise. He positioned himself next to a shapely masseuse, closed his eyes, and stuck out his hand. With the strong, firm grip that now held his hand, he felt warm and safe. Suddenly he heard a loud primal scream bouncing off the walls. Then he realized it was his own voice. As he screamed, he could feel the press of bodies around him. He opened his eyes. People were hugging and pushing together. The instructor looked at him with approval. By opening himself to them and releasing his feelings, he had made the session a success. Then he looked next to him. The masseuse was not there. But who was holding his hand? He looked down. There was a young girl who looked to be in her early teens, if that.

Rogers masked his disappointment as he struck up a conversation with the girl. She was smart and intuitive. He enjoyed talking to her and sensed they had a bond. They were both outsiders. The girl's classmates had nicknamed her Yoda, after the *Star Wars* character, because she was so different from them. She used the nickname like a badge declaring her uniqueness.

For the rest of the retreat he spent a great deal of time with her as they developed a close friendship. He could talk to her and for once he had someone who seemed to understand him, a soulmate. They were oblivious to the stares they were drawing. But when she took him home to meet her parents, her mother and father couldn't understand why a thirty-year-old man had struck up a relationship with their fourteen-year-old daughter. They couldn't see that he and the girl were just buddies, and they forbade her to see him again.

Roald Reitan heard the words he had longed for during the past six months. "Your leg is now rehabilitated. You no longer need therapy," the doctor told him. He was warned to only gradually increase his activities. He still hobbled, but sometime in the future he would again be able to run. Roald's mind flooded with all the things he wanted to do, like hunting, fishing, hiking, and camping. He hoped Venus would go with him.

Scymanky found a lawyer who agreed to file a $1 million claim against Cowlitz County. The claim stated that the logger had suffered third-degree burns over 80 percent of his body. It accused the county of being negligent in allowing him to get near the mountain without warning.

The *Daily News* contacted Ben Bena, the county's emergency services coordinator. "He's barking up the wrong tree here," Bena said. "We had no responsibility for that."

On the night of February 13, a Friday, an earthquake measuring 5.5 on the Richter scale rattled the Longview-Kelso area and was felt as far south as Northern California and as far north as British Columbia. The quake was centered at Elk Lake, twelve miles northwest of Mount St. Helens. USGS scientists blamed the quake on the volcano. They theorized that the May 18 eruption changed the stresses on faults in the area.

There were no injuries and minimal damage. However, if the quake had been slightly larger or centered only a few miles closer to Longview, the result could have been much more grave. As it was, this was another blow to the psyche of people who had now discovered one more threat to their lives and property.

Rains that weekend brought more flooding. High winds knocked out power to thousands of people. The Cowlitz River quickly rose five feet. The Toutle River caused more damage. "The Toutle looks like the Colorado River. There are some rollers fifteen to twenty feet high," Jeff Hastings told Stepankowsky. The young man lived in the devastated neighborhood on Tower Road. He was the son of Jim and Jerri Hastings, who had been interviewed then by Stepankowsky. The Toutle River washed out another three-hundred-yard section of the road. Bridges were knocked out. A natural gas pipeline supply serving the region was threatened. Approaches to Interstate 5 were eroded by the Toutle at the bridge at Castle Rock.

Residents were vocal in their criticism of the Corps of Engineers. Jerri Hastings said, "I wish the Corps of Engineers could go in and reroute that river. But they said, 'We're out of money.'" As Ted Conrad watched the last of his twenty acres of timberland being swept away, he blamed the erosion on the way the Corps had channeled the river. When Stepankowsky checked the amount of rainfall, he was amazed to find that only 1.1 inches of rain that weekend had caused almost all the destruction.

For months she had been calling Cowlitz County authorities to seek permission to drive into the Red Zone to her brother's truck. It got so they wouldn't even talk to her. Then Parker heard you could get a pass from the Washington State Department of Motor Vehicles. When she called them she was referred to the state's Department of Emergency Services.

The man she finally reached listened patiently. She could sense his interest and empathy. Finally he said, "I think we can issue you a pass."

Chapter 15
March and April

THEY STOOD there, he tall and thin, dressed in a plaid shirt like he had worn the day of the eruption, sporting a mustache that was probably grown to make him look older but only emphasized his youth, and she clutching his arm, her eyes sparkling, smiling as she looked up at him while he answered the interviewer's questions. Roald Reitan and Venus Dergan were to be two of the stars to appear in a documentary film.

The first anniversary of the May 18 eruption was approaching and they were getting a rush of media requests. As they had done prior to the Charles Bronson TV show, the two decided beforehand who was going to say what. But Venus felt comfortable letting Roald take the lead, interjecting when his flow of words slackened or she was asked a direct question. She was coming to accept the eruption as an event she was fated to relive periodically, perhaps for the rest of her life.

It was late at night and the newsroom was quiet. Stepankowsky was alone, very tired, and his writing was not going well. He typed his lead over and over but nothing seemed right. The reporter had not slept for three days. He was working on a series of articles investigating the tens of millions of federal dollars spent on dredging and flood control since the May 18 eruption. Were taxpayers getting their money's worth?

He and his wife, Paula, now also a reporter for the *Daily News*, were scheduled to catch a plane to Chicago in a few hours for a needed vacation. In his sleepless marathon he had managed to finish all the stories in the series except this last one. The series would run while he was gone.

He had interviewed dozens of people, pored over figures in more than sixty contracts, and spent hours doing his own calculations. He found that the Corps of Engineers had initially done a cost analysis that boiled down to two options: providing flood control through dredging and dams for the Toutle and Cowlitz Rivers, or relocating the residents of risk areas. Other federal officials had weighed in and at one point had seriously considered whether Longview, Kelso, and Castle Rock should "continue to exist" given the flood danger over the next several years.

The Corps chose the flood control option because it had no choice in dredging the Columbia River, the region's major commercial waterway. It had to be done, and dredging the Cowlitz and Toutle Rivers could help reduce the amount of dredging needed in the Columbia. Taking that into account, the Corps thought it would be cheaper ultimately to provide flood control than pay many millions of dollars to buy out thousands of home and land owners.

However, by deciding to dredge the Cowlitz, the Corps had locked itself into a long-term project. The Corps would have to dredge that river until the flow of sediment from the mud-engorged Toutle River ended. That was expected to take years. The cost of the dredging was projected to be $700 million, an estimate made before the Corps realized it was firmly in the sights of the cutback-minded Reagan administration.

The two sediment-retaining dams the Corps had built on the Toutle River were not worth anywhere near their cost, outside experts told Stepankowsky. They had been built at a cost of $18 million, but the Corps had grossly underestimated the amount of silt the Toutle would pour into them. Several engineers and hydrologists claimed the dams had been useless since December when they filled with silt much faster than the Corps had anticipated.

The only way to keep the "useless" dams functional was to mount a huge dredging program for the dams alone. Within just a few months, it was projected the dams would have cost over $63 million, including their dredging. To keep them in existence was going to cost $277 million over the next six years. Meanwhile, one of the dams had burst during the heavy Christmas rains. Repairs would cost over $2 million.

As Stepankowsky sat at his desk that night, the hours ticked away but still the words would not come. He was burned out. The managing editor, Gaston, had told the reporter in the past that maybe he was doing "too thorough a job." But still Stepankowsky kept putting in the long hours. It wasn't so much that he had to prove himself. Rather, it was that this ongoing story had given structure and meaning to his work. This was why he had wanted to become a journalist in the first place, to make a difference.

The city editor arrived early that morning and still Stepankowsky had not written a single word that satisfied him. When Gaston arrived he found the reporter in a daze, frozen with fatigue. Gaston looked at Stepankowsky's notes, asked some questions, and coached his exhausted reporter on writing the story.

When Stepankowsky finally finished, he and Paula jumped into their car and rushed to the airport. He found he couldn't sleep on the plane, no matter how exhausted he was. He was still hyped up. In Chicago they were going to visit Paula's sister, a college student. She picked them up in her Volkswagen. He leaned over in the backseat and fell asleep. When they got to her college dorm, the women helped him up to the room. The dorm unit had a central living and kitchen area with bedroom bays surrounding it. They put him into a bed and soon he was snoring loudly. They tried waking him several times, but he would not stir. His snoring was keeping everybody else awake.

When he woke late the next morning, he felt more refreshed than he had in months. Then he started thinking about the mountain.

They were lost. Donna Parker and three of her brother's friends were riding in an ancient Bronco, trying to find a way to Billy's truck. There were few landmarks to go by, just mounds and plateaus of ash glimpsed through the fog. They carried axes and power saws and hoped to get the truck ready to be moved.

Through the fog they heard a distinctive grinding of gears and then the roar of a diesel truck coming at them. They met the logging truck on a curve. It was hauling a huge old-growth log salvaged from the blowdown area, tree limbs still hanging from the top. Parker sucked in her breath. The limbs whacked the Bronco's windshield.

They spotted Caterpillar tractor tracks in the ash, rising to the top of a hill. They decided the tracks might offer a shortcut to where the Caterpillar could have cleared the road above—a place where they might get a view of the way to the truck. But the ash was deep on the road on top of the hill and down into the canyon below. It drifted like snow in the wind. The Bronco crept along, following the tracks. George Gianopoulos couldn't stand the slow pace. "You guys stick together," he said to Parker and to his son, Steve, and to Roger Harris, who was driving the vehicle. "I'm going to see where these tracks lead." He set off on foot.

They waited a long time. Eventually they decided to follow George's footsteps through the ash. They put on packs filled with food, extra clothing, and emergency gear and set off. They came to a long series of gullies. The two men could just walk across them, but Parker was so short, she had to walk down into them. Each time Harris helped to pull her up.

As she walked, she picked up small rocks whose uniqueness in form and color caught her attention. She was a rock hound. "Oh, my god, look at this one," she exclaimed to the men who paid her little attention. Soon her pockets were bulging with rocks. She went down into a gully. Instead of helping her out, Harris just looked at her. "Donna, you're going to have to get rid of those rocks or you're not going to be pulled out of them holes anymore." She emptied her pockets.

Their route through the fog took them into a box canyon. They came upon several large pieces of logging equipment lying mangled in the ash, looking like robots that had just fought a losing battle. A log loader lay on its side, its crane arm outstretched in a final grasp.

Leaving the box canyon, they found a semblance of a road. "Hey, there's something buried up there," Steve said, pointing up a hill. They found a truck transfer case with a gearshift protruding from it and some tools. Over the bank they saw a wheel sticking out of the ash. A closer look revealed a pickup truck buried upside down.

Then it started to rain hard. The ash turned to black mush, sucking at their boots with each step on the return hike. George was waiting back at the Bronco. He was mad. "Where did you go? You were supposed to stay put."

The rain poured down. As they drove back on a logging road, the sky began to clear. A small man in hard hat and work clothes suddenly appeared in front of them with a raised hand outstretched, signaling them to stop. "What are you all doing up there?" the Weyerhaeuser employee asked.

"Well, I lost my brother up there," Parker told him. "The green pickup with the white canopy. We were going to get the truck ready to bring out, but we couldn't find it."

"Oh, is that your brother?" the man said. "I lost my son at Fawn Lake."

"John Killian!" she blurted out. Parker had memorized the names and last known locations of the dead and missing.

The man stared at her for a moment.

"It's your son and his wife," she said. "John and Christy Killian."

The man introduced himself as Ralph Killian. "I'll show you how to find your brother's truck," he said. "I know where it's at."

Steve told him about their discovery of the buried pickup and described the tools they had found. Killian grew excited. "That's Tom Gadwa's stuff. He was with another guy. We haven't been able to find anything of them. Where did you find the pickup and tools?" Steve described where they were.

They followed Killian, who drove his small Weyerhaeuser pickup like a race car. They drove to a spot high above Fawn Lake, where Killian's son and daughter-in-law had died. Parker could see three aligned peaks in the distance. "You'll find the truck on the third peak over," Killian said, pointing toward the horizon.

They stood and talked for a while. Killian told them the authorities wouldn't give him a pass to look for the body of his son. He could only search when his work crew was in the area and he could sneak away.

"I have someone who will help you," Parker said. "I haven't even met him face to face, but I can tell he's a very nice man." She gave Killian the name and number of the man from Emergency Services who had issued her pass.

They left Killian on top of the road, staring down into the vastness below where someplace his son's body lay. A feeling of shame overcame Parker. Here she was trying to retrieve a wrecked truck, and Killian couldn't even get a pass to find his son's body.

It was getting late and they drove home. She called the man at Emergency Services the next day to ask for an extension on her pass. She learned that Killian had already contacted him.

"Donna, you got to quit giving out my name," he said.

"But Dave, he was up there for his son. And I was just getting a truck out."

He told her he had approved Killian's pass. And he okayed her request for an extension.

At the roadblock, Rogers pulled out his pass for the sheriff's deputy. It had been months since he had come to the Cougar area. He breathed a sigh of relief when the officer waved him through. The pass was still good.

Rogers drove on in the Simca with his motorcycle tied to the rear bumper. On the far side of Cougar he stopped at the last house on the old road to Merrill Lake—a ramshackle wooden cottage with a huge woodpile behind it. The house was owned by a man named Gearhard Pierson.

The men had met at Jack's Sporting Goods. Rogers learned that Pierson's father, a minister in Seattle, knew Rogers' close friend Gray Haertig. Pierson was basically a hippie woodcutter. When he told Rogers that he owned the last house on the road leading to the volcano, the possibilities leaped out at the adventurer. Jack's was frequented by law enforcement officers. When Rogers stayed at the trailer behind Jack's, he felt like he had to keep looking over his shoulder all the time. Pierson's cottage would be a much better base of operations. Pierson, who had his own rebellious streak, was happy to help.

Now Rogers drove the Simca behind the woodpile to hide it from the road. He walked the motorcycle up the Old Merrill Lake Road far enough that he could start it without anyone hearing the noise. He wouldn't be going to the volcano today. This would be a test run to check out routes and try to spot any new surveillance posts that might have been set up by the authorities. With spring the weather would improve, along with his chances to get into the crater. Soon it would be time to revisit his own personal playground.

It was a clear day, two days after their attempt to reach Billy's truck. Parker and the three men bounced along in the Bronco, following logging roads toward the peak where the truck lay. They encountered a Weyerhaeuser worker clearing the road of ash with a Caterpillar tractor. Parker told him what they were up to.

"Well," the man said, "if you find any abandoned vehicles, check for cameras and film. If you find any, you bring them to me. And if I'm not here, take them to the Weyerhaeuser office in Longview. Every time we come across an abandoned vehicle, that's what we've been doing. That's the rules."

"Sure, okay. That will be fine," Roger Harris said. He was driving the Bronco.

Parker was suspicious. "None of this makes any sense," she said as they drove off. She knew they were on Weyerhaeuser land, but what was a private company doing collecting cameras and exposed film that could be used to help identify victims? Shouldn't that be going to the sheriff's office?

Finally they found the road leading to the truck. But Parker couldn't believe it. She had been given wrong information. The road was still closed a few miles before the truck. She tried not to be too disappointed. There was no way they could get the pickup out today. They decided to hike in, anyway.

As George Gianopoulos led the way, Parker realized she could feel the wind but couldn't hear it. There wasn't anything for it to blow against—no trees, no bushes, nothing. A gray, muddy desert surrounded them for miles. The three men moved quickly. All were fit outdoorsmen. Parker stopped several times to take photographs, a good excuse so she could catch her breath. She was too proud to let the men know she couldn't keep up.

From nowhere a fog rolled in. Soon she could no longer see the men in front of her. They began climbing over and under piles of shredded trees. The ash grew thicker. They sank with each step into the gray blanket that rose above their ankles and onto their calves. A couple of the hikers asked about turning back. "We are close, I know it," George said. "We're not going to turn around now."

Suddenly the outline of the pickup appeared, like a ghost ship rising from the thick, gray sea of mist. They stood in front of the truck, in silence, as if it was a shrine. The men waited for her to make a move or to tell them what to do. Finally somebody said, "Well, I brought some coffee."

As they drank their coffee, they checked inside the cab. They also opened the hood and peered at the engine. Parker started looking for her brother's cameras and other effects. The search only confirmed what she had thought during her first visit. A great many things of Billy's and Jean's were missing. Parker walked over to the rock that had killed her brother, which still lay on a stump near the truck. She lightly rubbed the rock, as though she didn't quite know how to respond, whether to be repulsed or attracted.

Parker began cleaning up the site while the men inspected the truck to figure how to haul it out. Parker decided that if it became too difficult to get the truck out, she would have the men push it over the nearby ridge and down into the lake below—anything to keep it out of the hands of sightseers and souvenir hunters.

Parker spotted a lone coyote's tracks leading up to her brother's cooler. She opened the cooler and found empty eggshells in the carton where she had found the blast-cooked eggs during her October visit. She started laughing. Billy would have gotten a chuckle out of knowing his eggs had fed a hungry coyote.

Steve Gianopoulos bent over the cooler and pulled out a bottle of Yukon Jack whiskey liqueur. "That's good stuff," he said. He held the bottle up. "I'll tell you what, Donna, we're going to get this truck out of here, I'm going to fix it up, and we're going to drive it back here and spend the night, toasting Billy and Jean." But even as she agreed, she knew she wouldn't do it. She was not going to spend a night next to this monster of a mountain that had killed her brother and caused so much grief.

"It's time to go, Donna," George said.

"Wait. I have one more thing to do."

She wrote out a note and took it along with some dried flowers she had brought and tied them with a ribbon to the steering wheel. "Please, do not disturb," the note said. "We are coming back."

Don Swanson stared at the cracks he had just measured on the crater floor. He had noticed that the cracks appeared two to four weeks before an eruption. He wondered how many times scientists in the past had looked at the same sort of cracks on other volcanoes but had failed to realize their significance. He wasn't even sure why he had started measuring them. Swanson decided it must be serendipity at work again—"the faculty of finding valuable or agreeable things not sought for," as he read long ago in *Webster's*. From the discovery of penicillin, electric current, and photography to vulcanized rubber, x-rays, and radioactivity, he knew the role serendipity had played in scientific inquiry.

Swanson wrote down the latest measurements. He checked his figures. The cracks on the crater floor and on the dome were widening at an accelerated pace, as they had before previous eruptions. The rampart also was moving at an accelerated pace. He brought this latest information to his boss, Don Peterson.

On Monday, March 30, Peterson issued a public eruption alert. The alert said an eruption could occur sometime in the next two weeks. It was the first time the USGS had made a long-term prediction of an eruption. And it was the first alert that was based almost entirely on Swanson's discoveries. If Swanson was wrong, it could seriously undermine the public's confidence in eruption warnings.

This was to be the biggest presentation he had ever given. Rogers was excited. A librarian from the University of Miami had seen Rogers' slide show about the volcano during a Christmas vacation in Oregon. When he returned to Miami, he persuaded the geology department to offer the adventurer a thousand-dollar honorarium to speak at a presentation open to the student body.

Rogers flew in early to see his sister, who lived in Miami. The geology professor responsible for sponsoring his visit invited him to lunch, along with some of the professor's students. The students, all women, peppered him with questions. Rogers answered the questions, but he couldn't bring himself to chat with any of the students. He was extremely shy around women he didn't know.

As he dressed for his presentation that night, he turned on the TV. A national anchorman was talking excitedly and Rogers perked up his ears. President Reagan had been shot and wounded. When Rogers arrived to give his slide show, he found only three people there. Then he got word the university had canceled all activities for the evening.

The following day the librarian asked Rogers to visit his daughter's third-grade class. After he showed his slides he was amazed at the questions from the thirty students. He had given more than fifty presentations about St. Helens, and these were the most thoughtful and penetrating questions he had received. They wanted to know about everything and weren't afraid to ask. Simplistic answers wouldn't suffice.

Rogers felt it must be because of their time in life. The system had not yet destroyed their natural curiosity and spontaneity. Rogers loved being with the children, felt comfortable with them, and he seemed to hold an attraction for them. Like a Peter Pan who remains forever young and adventurous in Never-Never Land, Rogers hoped he would never grow up.

It would be his first inspection since fall of the experimental plots he had planted with fir, pine, and hemlock seeds. As Peter Frenzen drove up the freeway he could see in his mind's eye all the seedlings sprouting through the ash.

Reports were filtering back to him of scientists astounded at the recovery of the mountain, which the spring season was now revealing to them after the mild winter. Nature was proving more resilient than anyone could have predicted. Scientists had identified more than fifty types of surviving plants. They included hardy pioneer plants—fireweed, Canada thistle, bracken fern, salal, and pearly everlasting—that presaged the rebirth of a forest. The plants were perennials with live root systems. The ash that covered the area was devoid of the nitrogen essential to plant life. But scientists believed the plants had sunk their roots deep into the ground before the May 18 eruption and had been able to send their stems up through the crust of ash.

With the presence of plants, the chain of life could begin again. Insects and animals, some survivors and others emigrants, now appeared. Honeybees, butterflies, and ladybugs swarmed over foliage. Over two dozen species of birds and animals had been spotted. Porcupines, raccoons, mice, gophers, hares, beaver, mountain goats, elk, deer, and cougars had been seen. But over half the species known to exist in the area before the eruption were gone.

The recovery was spotty. Scientists were amazed to find fish in lakes as close as six miles to the volcano while discovering lakes more than twelve miles out that were devoid of fish life. Scientists theorized that the survivor fish had been protected by a blanket of ice and were at great depths or were in a lake protected by a ridge from the full brunt of the blast.

The vast majority of land nearest the volcano, tens of thousands of acres, still appeared as a monochrome of gray, seemingly without life. Fires smoldered beneath layers of ash in some areas. Scientists estimated it could take at least a hundred years to produce any semblance of the pre-eruption forest.

Yet appearances could be deceiving when it came to the detection of life. Spirit Lake, once a jewel of the Cascade Range, had been blasted back hundreds of millions of years in geologic time. The lake had become overloaded with nutrients in a process that robbed it of dissolved oxygen. The lake gave off noxious fumes like a sewage lagoon. It looked lifeless, with thousands of stripped trees floating in its mucky soup. On closer inspection, scientists discovered that with the oxygen depletion, a specialized organism had appeared munching on wood, manganese, iron, ammonia, and sulfur. It was a bacterium, thought to be like ones that appeared at the beginning of life on Earth.

When Frenzen got to the ash-covered clear-cut near the mountain, he half-ran to the first plot. At first he couldn't believe what he saw. As he looked under each of the wire cones, he felt his stomach sink. There were no seedlings under most of the cones, and only a few under the others. Some of the protective cones had been upturned by animals, but others were still securely held to the ground yet had no seedlings beneath them. Had his seeds failed to germinate, or was something else to blame?

Frenzen ran from plot to plot. It was the same everywhere. He found few seedlings for the long-term study he had planned. He knew that up to 90 percent of seedlings perish in nature. But that was over a period of years, not within a single season. He had to figure out something fast to save his study.

As he exercised, Jim Scymanky tried stretching both arms out from his sides at the same time. But struggle as he might, he couldn't fully extend his arms. He brought his fists back to his chest and tried again. It was no good. He had not gained any more flexibility in the past several weeks. Scymanky had been a good patient. He had religiously done everything the medical people had asked of him. But maybe this was as good as it was going to get.

What was to become of him? he asked himself. How was he going to provide for his family? The state of Washington continually threatened to cut off its financial support. His only hope of becoming financially whole again was through the lawsuits that his lawyer was working on, one in federal court and the other against Cowlitz County. But the lawyer warned that it would be years before those lawsuits came to trial.

To lower his anxiety he started telling himself, "One day at a time . . . One day at a time . . . " He had come far with that mantra.

"Do you have a pass to get in?" Donna Parker asked the logger. The man had deep blue eyes that captivated her, and he displayed one of the biggest smiles she had ever seen. "See this," he said as he put on his hard hat, "and that *crummy* out there," pointing to the logging truck that had been customized to ride several extra feet off the ground. "That's all the pass I need, Donna. Let's go."

The man, Keith Ross, was a contract logger who had lost two men in the May 18 eruption, along with logging equipment. The employees, Tom Gadwa and Wally Bowers, had worked for another contract logger who in turn worked for Ross. The eruption had left the landscape so altered that even people who had worked there for years could no longer find once-familiar locations. There had been no word on Gadwa and Bowers or on Ross's equipment until Parker and her party discovered the truck parts and demolished equipment. Parker and Steve Gianopoulos were going to lead him back to their find.

At Elk Prairie, near their destination, Parker peered out the window and had to look twice. There in the middle of the flat, gray plain of ash stood a huge mound of snow. No one could figure out why it hadn't melted. "Let's climb it," Parker said. She took a step onto the side of the mound and fell into the snow up to her armpits. The ash beneath the surface of the mound had washed away so that it was like a shell. "This is not a good idea," she said, embarrassed, as the men pulled her out.

Gianopoulos had little trouble guiding Ross to what was left of his truck and equipment. "Sure, that's my stuff," Ross said. He pointed to the twisted log loader. "Monday we were to be up here logging. And we had it all planned. When we saw that volcano start spewing ash in the sky, we were going to dive under that."

Gadwa and Bowers had been at the site cutting wood for extra pay on the Sunday of the eruption. Gadwa just liked being out there. But Bowers' wife was in the hospital, dying of cancer, and he had a lot of bills.

They found an axle that had been ripped from Ross's dump truck. They speculated on the path the truck must have taken when the blast sent it rolling over and tore it apart. "We can still use that axle," Ross said. "Will you help me put that on the back of the truck?"

They found small pieces of the truck strewn all the way back to the road that they had come in on. They looked into the upside-down pickup buried partially in the ash. But they found nothing of the missing men.

At 8:21 a.m. on Friday, April 10, St. Helens exploded, shooting steam and small amounts of ash ten miles into the sky. Clouds obscured the crater and the dome. More activity was expected.

The Forest Service closed the Red and Blue Zones, which now extended twenty miles around the volcano. The Weather Service, fearing the possibility of pyroclastic flows that could melt snow on the mountain and create massive mudflows, issued a flash-flood warning for an area within fifteen miles of the volcano. One minute before the eruption, the Army Corps of Engineers evacuated sixty-four workers from the debris dam on the North Fork of the Toutle River.

The Cowlitz County Humane Society told Stepankowsky there had been reports of small animals acting more aggressively the previous day and a half. Bev Nugent told him she became worried when their dogs started growling early that morning and wouldn't stop. She and her husband, Jim, lived twenty miles from the volcano in an area that would be isolated if floods cut off the Spirit Lake Highway. She hadn't noticed any of the earthquakes that preceded the blast. "My nerves are so bad right now that I wouldn't know if it was me shaking or a quake."

That night Swanson flew in a spotter plane over the crater but couldn't see in because of cloud cover. When he got into the crater the next day he was surprised to see how much pumice had exploded out of the dome. The pumice had helped produce small mudflows from melted snow still in the crater. Swanson and other scientists had thought there wasn't enough gas left in the system to power such an explosion.

Lava poured out of the dome. The prediction based on Swanson's theory of crack widening and other forms of deformation had proven correct. He knew it would have to prove itself repeatedly if it was to be accepted as a reliable tool for predicting eruptions. But he was now confident in its accuracy.

Newspapers called the eruption prediction "unprecedented." The *Daily News* in a headline typical of other publications proclaimed, "Eruption forecast hits mark." Nowhere did the stories mention Swanson's name as the scientist responsible for this "tremendous" achievement.

It was as if the USGS wanted to downplay this scientific first. The man responsible in large part for relaying prediction information to the press was Chris Newhall, the head of hazards assessment. He stressed that deformation wasn't a totally new tool in eruption prediction, a statement that failed to grasp the significance of Swanson's discovery. Deformation had never been used before to provide more than short-range predictions and those had not always been accurate. In fact nothing had proven entirely reliable for either short-range or long-range predictions before Swanson's groundbreaking work.

"The advisories still aren't very specific," Newhall told the news media. "The eruption could have come any time within a few days and even a little later than we said." While this had been Swanson's first successful "public" prediction, the scientist had achieved accuracy in predicting previous eruptions. This was a fact that Newhall chose to ignore. "Even though it has helped us here, we can't be assured that we will see warning signs two weeks in advance of every eruption," Newhall said.

The media left the impression that Newhall had been the one responsible for the science on which the prediction was based. "Chris Newhall's 'extended outlook advisory' . . . came true this morning," the *Daily News* said on April 10. "In an unprecedented move . . . the 32-year-old geologist in charge of Mount St. Helens hazards assessment advised that the volcano would likely erupt in two weeks. It did this morning at 8:21."

What was ironic was that Newhall had led the scientists who had opposed Swanson in his drive to get into the crater as soon as possible after the May 18 eruption. Later, Newhall had argued that the worth of the information that Swanson was bringing back from his forays into the crater did not justify the risk. Now he seemed to be getting credit for the very work he had criticized.

Swanson didn't particularly care. For the most part he was oblivious to the media. In late March of 1980, right after Mount St. Helens first showed signs of life, he got his first taste of working with reporters on this volcano. At Forest Service headquarters in Vancouver, he accidentally walked into an auditorium filled with reporters and television crews. They were in the middle of a news conference. When they found out he was a USGS scientist who had just observed the volcano from a spotter plane, they inundated him with questions. Swanson was caught unaware. In the glare of the bright lights and the frenzy of the reporters, he had made some flubs. The whole experience had been unnerving.

After that it was not that he exactly avoided the press so much as he buried himself in his work. He had no interest in being a media star. And he did not yet have a full appreciation of the role of the media in supplying information to a wary public and bolstering the image of the USGS, an agency that needed to compete for funding. Swanson was more interested in sharing the news of his discoveries in scholarly papers in scientific journals than in taking public bows.

It was shortly after noon. Publisher Ted Natt sat shuffling papers in his office, a glass-enclosed cubicle that gave him a sweeping view of the newsroom and gave the reporters a good view of him. The staff had learned the previous month that the Longview *Daily News* had been nominated for a Pulitzer Prize, the highest award in American journalism, for its coverage of Mount St. Helens. But the time of day that they were to be notified if they won had come and gone. The disappointed journalists tried to get back to work. Still, some reporters in the glum newsroom kept their eyes on Natt, monitoring his body language with every phone call he got. Stepankowsky was one of them.

The reporter watched Natt drop his papers and pick up the telephone. A smile crept across the publisher's face. That was all Stepankowsky needed to spring out of his chair and race toward Natt's office along with a few other reporters who had caught on. "So we won for general local reporting," he heard Natt say.

The publisher, still talking on the phone, looked out over the newsroom. He turned a thumb up. The newsroom broke into instant pandemonium. Stepankowsky grabbed reporter Fran Kaiser and started doing a jig. Staff members were hugging, cheering, jumping up and down, or unabashedly crying.

Natt walked into the room, hugging and patting his reporters on the back. He wrote a check for three hundred dollars and told Stepankowsky and reporter Marlon Villa, "I want you to go out and find the best champagne money can buy. I don't want any of that cheap stuff." Natt did a double take when he saw what his two employees came back with: three bottles of Bollinger Champagne, each in its own wooden carrying case with padlock and key.

As they toasted the Pulitzer, Natt told his staff in a voice choked with emotion that his only regret was that his grandfather John McClelland, founder of the newspaper, was not there to share in the award. Popping champagne corks soon competed in attention with congratulatory telephone calls from the nation's journalism elite and television crews flocking into the newsroom for interviews.

"The Pulitzer affirms that what we did in little old Cowlitz County last year was among the best journalism produced in the entire country," managing editor Gaston told an interviewer. "We are only sorry that many of our readers can't share the joy we feel in winning the Pulitzer. The volcano cost them dearly, and that's why we covered the story so thoroughly—the volcano affected all our readers."

Somehow in the joyful confusion the staff was able to produce the day's newspaper, putting the Pulitzer news at the top. Natt kept sending Stepankowsky out to buy more champagne. The reporter decided to buy more in quantity than in quality, but by now, no one noticed.

For Stepankowsky, the news was starting to sink in. "How in the world are we ever going to top this?" he thought. "How am I ever going to win another Pulitzer in my lifetime? Here I am just turned twenty-five and I've already had the biggest story in my life hit me in the face."

For the moment, though, Stepankowsky felt only joy and pride. The *Daily News* had won an award usually reserved for the likes of the *Los Angeles Times*, the *Washington Post*, or the *New York Times*. The *Daily News* was one of the smallest newspapers to ever win a Pulitzer for news reporting. The volcano coverage by the *Daily News* was "far superior" to that of much larger newspapers, according to the chairman of the Pulitzer jury.

In the first weeks after the May 18 eruption, the reporters and editors at the *Daily News* produced more than four hundred volcano stories. For the period from late March until the end of December, they published more than two thousand news articles. But it was the quality of the work that earned the prize. The dramatic photograph taken by Roger Werth on May 18 when he and Stepankowsky flew over the erupting volcano was given special notice along with Werth's other photos. The jury also judged a core of ten stories submitted by Gaston as the best of the newspaper's journalism on the volcano. Either alone or with another reporter, three of those stories carried the byline of Andre Stepankowsky.

CHAPTER 16

May

"THEY WERE BIG, century-old firs—the Douglas firs that many believe sprang from the ashes of Mount St. Helens' last eruptions in the 1800s. Five months ago, the first of them was shorn to stubble by the Weyerhaeuser Co. in one of its most controversial timber cuts to date."

Stepankowsky read the lead to the front-page story, intently. He was familiar with the area and knew the trees were even older than what the lead sentence said. It wasn't his story. Staff reporter Laurie Smith had written it. But he wished he had. The mountain and what was happening to it had reinforced his already strong environmental interests. The story also touched upon his area of expertise involving erosion of riverbanks.

An area of approximately forty acres in the restricted zone had not been hurt by the May 18 blast. About eight acres of that parcel stretched out over a mile on the Toutle River in Kid Valley. Weyerhaeuser logged the entire forty acres, including the eight acres for which a shoreline permit was required but not obtained. When Cowlitz County had gotten wind of the clear-cut under way, a cease and desist order was issued. Weyerhaeuser ignored the order and proceeded until the logging was finished, leaving Kid Valley looking "as though the whole world had been logged off," according to a company official who asked to remain anonymous. The same official said "the operation was mishandled from the beginning."

The Weyerhaeuser officials who commented for the record called the clear-cut an "honest mistake." They said they thought the company was exempt from needing a shoreline permit in doing what they called their "salvage logging."

Critics charged that the company had taken advantage of the turmoil following the May eruption "to help itself to undamaged, expensive, old-growth timber it otherwise couldn't have touched in years," the story said.

Residents and scientists were outraged. The lush forest along the river had been one of the most scenic spots in the region, shading both waterways and Sunday picnickers. "My heavens, they took everything," said Peggy Floyd, who had owned a home in the area for twenty-two years.

Botanist Bill Arensmeyer said Weyerhaeuser acted with "an arrogant disregard of the law." He pointed out that erosion could eat away the river's banks, further degrading a river already devastated by the volcano. Arensmeyer had worked as a professional environmental witness for the state Department of Ecology. He had resigned a few weeks before to protest the state's lack of action against Weyerhaeuser for what he considered to be serious environmental violations similar to the Kid Valley transgression.

A minority of state legislators termed the Kid Valley cut "a clear criminal act." Bob Jensen, an assistant attorney general for the state, called it "one of the biggest and most blatant violations in the history of the Shoreline Management Act."

In a testament to Weyerhaeuser's political grip on state and local government, officials soon grew silent. They decided Weyerhaeuser had committed only a technical violation of the law resulting from an innocent misunderstanding. They were sympathetic to the company's losses from the St. Helens eruptions, in the region it had always counted on to produce most of its revenues and profits. The Kid Valley cut now melded into the sea of gray that surrounded it.

Again Donna Parker got word that the road to her brother's truck had been cleared. Her teenage son, Jim, went with the group this time. But once more they found the road blocked by eruption debris. They parked the vehicles and walked in for another look at the truck. On the hike back they spotted a man and a woman near Fawn Lake, digging in the ash with shovels. The man was Ralph Killian, the Weyerhaeuser worker they had met earlier. Killian introduced his wife, Jeanette, and they all talked for a while. Then Parker's group left the couple to their task of searching for the body of their son.

At the vehicles Parker looked back. For once, fog or rain did not obscure her line of sight. "Oh, my god," she said. There in the distance, on the ash-covered peak, sat her brother's pickup, small against the sky, just like in the nightmare she had experienced time after time.

Ash pottery, eruption postcards, bottles of ash, volcano T-shirts, souvenir books, sightseeing maps, "inferno dogs," and "crater burgers" were now a part of Stepankowsky's beat. He had added tourism to his roster of reporting duties related to the volcano. With summer approaching and chances for major flooding greatly diminished, he had time to work on other stories.

Experts told him that hordes of St. Helens visitors were expected that summer. Estimates ran as high as three million. Traffic jams were expected from Castle Rock, just off the freeway, to the area beyond Toutle where the Blue Zone began and roads deteriorated. "We're going to have major accidents up there," one county official told Stepankowsky. Sheriff Les Nelson predicted an increase in crime. "Law enforcement is anticipating one tough summer," he said.

Stepankowsky found a new thirty-unit motel and restaurant complex under construction at the foot of Spirit Lake Highway at Castle Rock. Several stores and restaurants were getting facelifts. At Silver Lake a gift shop had been set up in a trailer along with a food stand, fluttering pennants, picnic tables, and a large green tent. Though the location was more than thirty miles from the volcano, it offered the area's best view of St. Helens outside of the Blue Zone that was barred to tourists. "We've been called a carnival and opportunist," said Patti Andrews, who ran the gift shop. "But we're doing a service for the tourists so they can watch the volcano."

At Maple Flats, on Spirit Lake Highway near the Toutle River, Stepankowsky found an orange and brown A-frame house buried halfway up its sides in dried mud. The owner, Blair Barner, had given free tours to visitors the previous summer. He was now going to charge fifty cents a person. "I'd be foolish not to charge. It's hard to find a home that is only halfway destroyed." Before the massive mudflows, fourteen homes stood across the street from Barner. Now there were none.

Barner showed the reporter the mud-caked kitchen counters and the cabinets still crammed with pre-eruption canned and packaged food. Mud was piled deep in the bedrooms and jammed into the bathroom. A clock had stopped at 8:32, the time of the eruption. Barner would continue to serve as tour guide. "Tourists last year enjoyed talking to someone that knew what had gone on. I'm a real volcano victim," he said, smiling.

Back down Spirit Lake Highway, Clara Ottosen was converting an old dance hall into a volcano museum. A blue pickup, its smashed body pitted with rust, stood in front of the rough-hewn building that was surrounded by pines. The pickup had been salvaged from the blast zone. Painters worked at putting finishing touches on a twenty-five-foot mural of the volcano.

Inside, old logging equipment sat in one corner. Photographs of St. Helens before, during, and after the eruption adorned the walls. An exhibit contained a motorcycle wedged between mounted trees and a deer that was now stuffed, all caught in the eruption mudflows.

Ottosen had purchased the inlaid mahogany bar once used by eighty-four-year-old Harry Truman, a public emblem of defiance of the mountain before the eruption killed him. She planned to place a mannequin of Truman across from an eight-foot dummy Bigfoot, or Sasquatch. People were bringing in other artifacts for the museum, which would charge $2.50 for admission.

Stepankowsky wanted a good quote to end his story. He decided to visit Stan Lee, the curmudgeon at the Kid Valley Store. Lee didn't disappoint him. The bespectacled merchant was direct about his feelings on tourists. "I want three coming into this store when one is leaving. We want them fast and thick—with money."

They were going out more with friends they had partied with since high school—just like it was before the eruption, Roald thought. Now they were together at a restaurant. Venus suddenly became overly conscious of the clattering dishes, the loud talk, the laughter. It was like she wasn't really a participant in the scene, but more of an observer watching herself, her friends, and Roald acting out their parts in this performance of fun and games. They were all young, and wasn't this what young people did?

She and Roald had been riding a whirlwind of activity, as if trying to make up for all the time they had lost. At first she felt liberated and rebellious. She wanted to prove to herself that she could be just as young, vital, and even as wild as before with the drinking and partying. But now she felt a deep exhaustion settle over her. She realized she could no longer keep up.

The man on the phone was from the Cowlitz County government. "The road is clear now. You need to get your brother's pickup out of there right away," he told Donna Parker.

"I don't think I can get my crew together this weekend," she said.

"Well, I'm telling you it's in the way. If you don't get it out of there, the workers are just going to tip it over the edge of the cliff." Parker knew the man was lying. Loggers and workers near the site always assured her the truck was not in danger and they promised to call her if there was ever a problem.

Parker grew silent. "Well, are you going to get it out or not?" the man finally asked.

"Okay, fine." That's what she always said to brush off people she didn't like.

Soon after, an Associated Press reporter contacted her. He had heard about her attempts to bring the truck out. "If I knew nobody would make money off the truck, I could be talked into leaving it there," she told him. "But I just couldn't stand the thought of it being part of a roadside attraction." She had heard that some victims' vehicles had mysteriously disappeared from the blast zone only to end up in so-called museums.

"My brother had a very unique sense of humor and I know for a fact it would please him to see the truck cleaned up and on the road again," Parker said. "It's got to stay with people who loved him. . . . It's all very personal. . . . Anybody who lost somebody would understand."

Relatives of eight volcano victims had filed suit against the state, charging that former Governor Dixy Lee Ray, in trying to sustain Weyerhaeuser logging operations, was responsible for the deaths by ignoring scientific warnings. Now Stepankowsky watched as Ray tried to defend her actions during an appearance on ABC's *Good Morning America*.

One of the show's other guests was Seattle filmmaker Otto Sieber. "On the northwest side there was no Blue Zone," he said, referring to one of the mountain's restricted-area designations. "Had the Blue Zone trajectory been carried out all around the volcano, most of the people [who died] would be alive today," he told host David Hartman.

Ray disputed Sieber, who was believed to be writing a book that championed the cause of the victims. Ray said the restricted zones were drawn with the best scientific evidence available. She admitted the state had to consider the area's timber economy, which depended on large tracts of Weyerhaeuser land northwest of the volcano. "There were many theories and speculation about potential eruptions," she said. "We took everything into consideration. Given the same information we would do it the same way again."

Another guest, Donal Mullineaux, head of the USGS hazards assessment team at the time of the May 18 eruption, said no one could have anticipated the force of the blast. But Sieber countered by pointing out that Mullineaux himself had drawn boundaries in a 1978 study indicating that past eruptions far exceeded the boundaries set by the governor.

Stepankowsky headed to the office to write a story about the lawsuit. The suit basically accused the state and Weyerhaeuser of gerrymandering the restricted zones so there was virtually no zone separating the northwest side of the mountain from the area of permissible logging, allowing the company to keep on cutting. This was the area where scientists had found evidence that the volcano directed its blasts in the past. The mountain had grown a bulge more than four hundred feet out on its northwest side before the eruption. Yet the state did nothing to expand restricted zones on that side. All but three of the victims whose bodies had so far been found or who were officially presumed dead were killed on Weyerhaeuser land on the northwest side of the mountain, land outside the restricted zones.

The suit charged that Dixy Lee Ray protected private timber holdings at the expense of human lives. It accused her and other officials of trying to cover up the state's negligence by coldly blaming victims for their own deaths. State officials, the suit said, "publicly and without sympathy proclaimed that people who had died had foolishly ignored warnings and had illegally penetrated the state's hazard zones."

At the office, Stepankowsky started calling the victims' relatives who had lodged the suit. He spoke with Bette Gadwa, the thirty-four-year-old widow of logger Tom Gadwa. "My husband would be alive today if the state of Washington had done its job," she said. "My husband would not have gone near the place if he had been given the proper warning."

Barbara Karr, a teacher in Renton, Washington, lost her husband, Day, and sons Andy and Mike in the blast. Karr said she wanted to "clear the family name." She had heard neighbors call her husband a fool for being in the danger zone, when he and the children had been miles from any restricted zone.

As the first anniversary of the May 18 eruption drew near, Venus Dergan began thinking more about her ordeal on that day, the very thing she had worked so hard to push from her mind. Now she thought about it every day. And she thought about the victims who had not made it out alive. She had been told numerous times after relating her story, "You should not have survived that."

Whenever she was asked what she had done to survive, she never had a good answer. Why had she survived and the others had not? Was there some purpose to her life for the extra time she had been allotted that would some day be revealed to her?

The national and international media—the television networks, the *New York Times, National Geographic, Wall Street Journal,* and many more— were running anniversary stories and many sent reporters. Nearly fifty helicopters and airplanes were reserved by news organizations for eruption anniversary flights.

In the same edition of the *Daily News* that carried his story on the lawsuit, Stepankowsky noticed a headline that quickly grew to infuriate him: "Volcano celebrations to erupt this weekend." Towns across the area, the story said, were planning anniversary festivities. On the schedule were parades featuring volcano floats and events ranging from motorized bar-stool races to mud wrestling and ash-castle building. Entertainers would sing *Legacy of Harry Truman* and *Mr. Weyerhaeuser, I Don't Want To Go.*

Stepankowsky had seen too much destruction and death to believe there was anything to celebrate. He could see having memorial services but he thought the so-called celebrations were sacrilegious and he vowed to stay away. Associate Editor Dick Pollock, however, thought the festivities were appropriate. "Life must go on," he wrote in his column, "and long faces must eventually give way to all the other human emotions."

"What are you going to use it for?" the attendant asked Steve Gianopoulos, who had gone to the U-Haul agency to rent a flatbed trailer for carrying Billy's pickup. "We're going up to St. Helens to get out a truck," Gianopoulos replied. The man looked at Gianopoulos and smiled. "Sure you are," he said as he checked out the paperwork.

When Parker's group met in Portland early the next morning, George Gianopoulos had hooked up the trailer to his International Carryall. It was a small convoy that passed the St. Helens checkpoint on its way to the truck: George and Steve Gianopoulos along with Donna Parker and her son, Jim, in the International, a Bronco carrying Roger Harris and his son, and another Bronco loaded with more of Billy's friends.

The closer they got to Billy's pickup, the more difficult travel became. A great deal of ash remained on the road. The International with the trailer was difficult to handle. The Broncos climbed up the peak to Billy's pickup but the International couldn't make it. Using the winches on the Broncos, the men pulled the International and the trailer to the pickup. Parker thought that if it was this hard just to get to Billy's truck, what would happen after they loaded it onto the trailer and tried to drive back down. A feeling of dread swept over her. What if there was a horrible accident and someone was killed? She would never be able to live with that.

Everyone took a breather as they stared at the pickup buried in ash up to its frame. Then the men started arguing over the best way to get the truck onto the trailer. Parker shook her head and looked on in amusement. "If Billy was here I'm sure he would have an even different way," Parker said. The comment ended their arguing.

They started by digging out the ash around the truck and shoveling a track from each front tire to the trailer. They leaned against the back of the truck and pushed. It wouldn't move. They hooked the pickup to a winch on one of the Broncos and tried again, but the pickup still wouldn't budge. Even with the second Bronco and its winch, plus the International pushing from behind, nothing happened. It was as if the truck was welded to the ground.

Then it began to snow. As Parker watched the flakes flutter down from the sky, she wondered what else could possibly be in store for them.

The men piled out of their vehicles and started arguing again. One of them jabbed at the dried ash beneath the frame. It was like concrete holding the pickup to the earth. Now they brought out picks, axes, shovels, and other tools. Like archaeologists trying to free an ancient treasure, they picked and chopped at the bond that trapped the truck. Finally one of the Broncos tried pushing the pickup. With a creak, the pickup started moving slowly over the ground and up the ramp onto the trailer.

"Wait a minute," she yelled to the men. She walked over to the stump and stared down at the rock that had killed Billy. She didn't know exactly why, but she had to take it with her. It was part of the story. One of the men helped her with the rock.

From the hill they would have to drop down a steep road and negotiate a hard right turn. Beyond the road, a sheer cliff dropped to Coldwater Lake far below. Parker rode in one of the Broncos. She tried not to look over the cliff as the Bronco hugged the road. The dread hit her again. This was all going to end in disaster.

They parked the Broncos at the bottom of the worst part and watched as George prepared to follow with the International and its heavily loaded trailer. "I can't watch," Parker said. "If something happens, don't make a big noise, just come over and tell me." She walked down the road a little more and turned her back to the International poised for its descent.

As George started down the road, the trailer began whipsawing from side to side. Parker heard a never-ending screech of brakes, but she didn't turn around. George was gaining speed even with the brakes engaged on both the International and the trailer. He couldn't stop the rig as it skidded toward the hard right turn. A mound of ash was piled by the side of the road. George managed to turn the International toward the mound. The rig parted the ash, rose slightly up, and hung at the edge of the cliff.

The International had no room to move forward, and George couldn't back it up. They lined up the Broncos, one behind the other one, and combined their winches to pull on the International. Eventually they got it away from the edge and out of the ash mound. Parker then heard more screeching and knew the International was on the move again. Finally the screeching stopped. She felt a tap on her shoulder. "You can turn around now," one of the men said.

Hours later their little convoy pulled into Toutle, coincidentally right on the heels of the town's Volcano Daze parade with its volcano floats, fire trucks, marching band, clowns, and little ballerinas.

"We could have been in it," Steve said.

Parker could see a little humor in the situation, but not much. "Yeah, like I want to be in a parade," Parker said. Then she noticed several people pointing at them and laughing. She wondered if they actually thought the trucks were just a late part of the parade.

While Parker and the others were inside a local restaurant, a crowd gathered around Billy's pickup. As she and Steve walked back to the trucks, they heard a couple of loggers speculating about what happened to the people in the damaged pickup. They heard laughing, and then silence as the crowd parted for them. The clown with the red-bubble nose and floppy shoes did not smile. Even the little girls in their ballet tights and volcano hats grew silent. A teenager in a band uniform suddenly yelled out, "Happy anniversary."

Parker's chest grew tight as she fought to breathe. "You handle it," she said to Steve as she half ran to the cab of the International. She shut the window and turned the radio up loud as she tried to block the scene from her mind. She wouldn't even let herself look outside. By the time Steve got in, she had calmed down. "It doesn't really matter," she said. "I just have to keep telling myself that what matters is we're bringing Billy's truck home."

Swanson and the scientists on his team moved cautiously down the dome, measuring the cracks. Swanson believed he and his crew would have ample warning of a major eruption, but there were other dangers in that hot, stinking environment. Vents along the spidery cracks could spew out heated steam. A misstep into the cracks that were up to ten feet wide could mean a horrible death from burns. The volcano could explode without warning with a blast of scalding water. Even worse were the pocket eruptions that could come from any place on the dome at any time, sending rocks of all sizes flying into the air in the midst of a ball of steam and ash. But for now, all seemed quiet.

With so many planes and helicopters expected over the volcano on the first anniversary, Swanson had flown into the crater the day before to take measurements. The cracks were not widening. Measurements with the geodimeter indicated no movement of the rampart. It was safe, he thought, to say that on anniversary day, there would be little activity on Mount St. Helens.

Swanson had been oblivious to the hoopla. There was one event he wished he had time to attend, though. The night before, the University of Idaho had awarded its first posthumous degree to Jim Fitzgerald, the geology student Swanson had befriended who had died in the eruption. The doctoral degree was presented to Fitzgerald's brother, who marched in the parade of graduates at commencement. The presentation capped a year-long effort by Fitzgerald's friends, colleagues, and family to complete the unfinished work required for the degree.

"Let's compare him to a policeman who goes into a dangerous situation," Tom Johnston said of his son. He was talking to Stepankowsky. David Johnston's parents told the reporter they had tried to warn their son of the dangers when they had their last talk with him, on the Mother's Day before the eruption. But they understood that he was just doing his job on the day the mountain killed him.

The Johnstons were in the Northwest to visit their son's friends and scientists at the USGS. They had hoped to fly to Coldwater II where he perished. But first the threat of heavy air traffic and then poor weather dashed their hopes.

Their son, they said, had a keen interest in volcanoes ever since the 1976 eruption of the Augustine volcano in Alaska, from which the young scientist barely escaped with his life. At St. Helens he had lowered himself several times into the growing crater in order to obtain rock samples before the eruption. "David was a geochemist and he felt he needed the rock samples badly," Tom Johnston said. "He knew the hazards, but he also knew it was an opportunity that would not come back again."

The Johnstons, who lived in Illinois, were staying in the Vancouver hotel where their son lived before his death. "It's frustrating, but we feel closer to David now than we have for a year," Tom Johnston said. He refused to comment on the forthcoming movie *St. Helens* about their son and Harry Truman. The couple and USGS scientists had charged that the movie would present a distorted view of David Johnston's life. The film was due to open soon, with Art Carney as Harry Truman, and David Huffman as Johnston. The Johnstons were negotiating with the producers to make some changes before the film was released.

Washington Governor John Spellman had declared the anniversary of the eruption to be *Daily News* Day across the state. In his proclamation the governor recognized the Longview newspaper for both its Pulitzer Prize and the national award it had won from the Society of Professional Journalists, Sigma Delta Chi, for its coverage of St. Helens. The governor said the awards "effectively designate the *Daily News* as the best newspaper in America for the year of 1980."

A panel of public officials in a tie-in to the anniversary mulled over the media's performance since the eruption. Many participants, while critical of the media elsewhere, tiptoed around criticism of the now hallowed *Daily News*, whose coverage could directly affect the performance of their jobs. "Sensationalism sells papers when you get away from the local area," the Cowlitz County emergency services coordinator, Ben Bena, told participants. "The Seattle media is more dramatic in describing the volcano," said Bill Brown of the Federal Emergency Management Agency.

Stepankowsky often wondered what role the press, the *Daily News* included, had played in the death of Harry Truman. Truman had been built into a media star depicted as a fearless old man who shook his fist at the mountain while remaining stalwart in his refusal to leave his home. Stepankowsky had learned that away from the glare of television cameras and the reporters who Truman supplied with defiant quips, the man was an emotional wreck, terrified by earthquakes that jolted his lodge and confused by conflicting eruption predictions.

Stepankowsky also wondered about his own reporting. He had been accused of beating the drum too often in his effort to alert the region's residents to the flood danger. Two University of Virginia researchers, James Pennebaker and Darren Newtson, suspected that Longview residents were repressing their fears and anxieties because they faced a seemingly unending threat, the anticipated floods.

Of the seven geographical areas the psychologists studied that had suffered from the volcano, only Longview residents showed a high resistance to answering their questions. Only 12 per cent of the people surveyed in towns such as Yakima, Moses Lake, and Centralia refused to participate. Now that the danger was past, "people by and large, were eager to talk with us about their experiences," the psychologists wrote in their study. However, in Longview, 44 per cent refused to participate, many adamantly. "Most of the respondents who refused slammed the phone down without comment."

Almost all who refused lived in the floodplain. The psychologists believed these people hid their feelings because they continued to be threatened by the volcano. The danger was that this suppressed behavior could manifest itself in delayed stress syndrome, much like Vietnam vets had experienced anywhere from a year to ten years after their service. It was feared that area residents, like the veterans, would not seek help until symptoms like broken marriages, depression, and excessive drinking drove them to do so.

Stepankowsky only knew how to do his job to the best of his abilities. He believed that people were better off being prepared for disasters than being caught unaware. While he was disgusted with stories that sensationalized the news, he also believed the newspaper's readers were mature adults who did not need coddling.

Helen helped him take off his Jobst stockings for what he hoped would be the last time. The tight, long, elastic socks that reached to his groin were designed to prevent clotting and promote circulation. They had become part of Scymanky's body. He wore them all day and all night, except when he was in the shower. Now he had been told they were no longer needed. The couple looked at each other. They had learned to share in his victories, and no milepost on the road to recovery was too small to celebrate, even if it was with just a glance and a smile.

Although Scymanky was not speaking to reporters, they were still writing about him. An advance story for the anniversary of the May 18 eruption detailed just how perilous his survival had been.

"It is here . . . that one can best appreciate the destructive force which flattened the forested foothills of Mount St. Helens," the Associated Press story said.

"Once a pristine high Cascade stream, Hoffstadt Creek now mostly runs gray. Along its banks, young trees were tumbled in geometric patterns.

"It was along this creek that four loggers who had been thinning trees the morning of May 18 began a desperate struggle to safety. Three of them died, severely burned, lungs full of ash.

"Two months after the eruption the remains of a herd of elk was found nearby. They had been broiled instantly by 800 degree heat."

The three of them stood there, making small talk, sharing moments of silence that spoke volumes. After all, Swanson thought, what more was there to say? One of his best friends, David Johnston, had died and now his grieving parents were standing before him, not so much to seek answers—they had been told everything about the circumstances surrounding their son's death—but rather to make a connection to his past life by meeting his friends and colleagues and trying to absorb the environment in which he had worked. Swanson felt a great sadness. He and the Johnstons smiled at each other and continued to speak briefly before he had to leave to monitor the mountain that had killed their son.

For the first time in his many trips to the mountain, Rogers planned to let somebody know when he was going and when he could be expected back. He was living in a tiny two-story shack in southeast Portland that a friend had lent him. Now he gave his roommate, Al Garren, an itinerary of his plans for an anniversary-day foray into the crater. He also left his roommate a map of his planned route, a description of his Simca, and his license plate number. Rogers wasn't sure why he was doing this. But his friends worried about him lately. "We don't want you to go up there and die," they told him. "I'm not going to die," he said. "Then, at least tell us where you're going," they said.

"I'm leaving early in the morning," he told Garren. "I'll be back the day after tomorrow. It'll be a day up and a day back."

When he left Portland it was pouring rain, with limited visibility. "Just right for trespassing weather," he thought. "There won't be any airplanes or choppers." This was going to be a new adventure for Rogers. He had never accessed the mountain by the west side. He had always gone through Cougar on the southwest, driving his motorcycle as close to the volcano as possible, and then hiked the rest of the way. But if he took the western approach he could drive his Simca within a few miles of the crater on recently cleared logging roads.

Before reaching the roadblock on Spirit Lake Highway, he turned off onto a logging road. He followed a road that wound along with the North Fork of the Toutle River. The road led him through and around a series of large hills. Eventually he arrived at a point only about three miles from the crater, where a huge boulder was leaning against a big stump. Rogers parked his little Simca beneath the boulder. Now if the clouds cleared, the car couldn't be sighted from the air.

Rogers walked across the plain leading to the breach in the northwest wall of the crater—the great gap created by the immense lateral blast of May 18. He made his way around enormous rocks scattered like giant, disfigured marbles. His feet sank into ash and pumice past his lower calves. Rain poured down. Fog and low-lying clouds wafted around him. Conditions were perfect, he thought. They weren't going to catch him. He didn't have to worry about hiding. Rogers decided he was going to climb the lava dome that dominated the crater floor.

He made his way around steam vents scattered on the ramp past the breach. The vents rumbled like river rapids, chugging out vaporous clouds like a movie fog machine. The wind shifted and he was drenched with a hot spray of steam. He stood on the crater floor, its walls forming an amphitheater two and a quarter miles long by a mile and a quarter wide. A constant cracking sound of rockfall echoed off the walls. Up ahead, rising through the clouds, sat the 425-foot-high crater dome, a massive, steaming rock pile elongated into a form almost a mile long and a half-mile wide. He could taste the sulfur in the air.

He climbed the dome, rain pouring down on his hard hat. He looked down into cracks seeping steam but couldn't see much. He got close enough to the volcano's throat to feel fair in saying he had climbed the dome. He took some photographs. Then he went down.

Though he had been to the crater several times, its geology never ceased to fascinate him. Several chunks of rock as big as small houses nestled against the base of the dome. They had been blasted out from one of the so-called nonexplosive eruptions of the volcano. The crater floor had nooks and crannies to explore. Waterfalls cascaded down the crater walls, eroding away the earth beneath them into new formations.

Rogers started collecting rock samples. He was so intent on studying each one that time got away from him. He realized it was starting to get dark, but he wanted to find one more sample. He found a rock he liked, bent over to pick it up, and threw out his back. Rogers had popped a joint in his back in an auto accident seven years before and now it had gone out on him again.

"What am I going to do now?" he asked himself. He knew he wouldn't be able to handle the three miles of rough terrain that lay between him and the car. But then he thought that if he could climb the west wall of the crater to the rim, he could slide down the snow on the other side, avoid the debris field, and find a more direct route to the Simca. It would not be easy, because the crater wall loomed up more than thirteen hundred feet.

Rogers found a waterfall running off the northwest rim near the breach. It provided a route of sorts up the side of the crater. Partway up he held onto some rocks and looked at the top of the waterfall. Just above it, the slope to the rim was more gradual. But first he would have to heave his body upward to catch hold of a ledge several feet above him. Chilled water ran down his sleeves. He studied the ledge for a moment as his arms began to ache. He looked down. If he missed the ledge or couldn't hold on, he would fall hundreds of feet. He decided not to chance it. Rogers descended back to the crater floor in the dark. He knew he was definitely going to be overdue back at his house. He had a vision of the police waiting by the Simca, ready to haul him away. "Why the hell did I let my friends talk me into making that map?" he asked himself.

He skirted out farther onto the slope covered with rock debris. The slope here was more gradual. Rogers found a hardened lava flow that ran up the slope, covered with about six inches of snow. He tried walking up the slope but kept slipping. He began crawling up the flow, his back in excruciating pain. But he knew if he stopped for long, it would tighten up even more. His body was wet and chilled even as he sweated beneath long underwear and layers of wool clothing. He bumped into something in the inky darkness. He tested it with the instep of his shoe. It was a snow cornice. He knew he was near the top.

He climbed the cornice, expecting it to break off with him at any moment. Finally he lunged and did a belly flop onto the snow. Rogers was at the rim. When he stood, strong winds pummeled him. A two-yard-wide patch of bare gravel-like material ran, like a pathway, along the top of the rim. He took off his backpack and pulled out his poncho. He was out of water, so he ate some snow. Wrapping himself into the poncho, he fell asleep, exhausted, on the gravel.

During the night, he woke. What he saw in the light of a full moon dazzled him. On the outside of the volcano, the thirty-degree slopes were smooth with snow. Inside the crater, the walls were much steeper, with snow cornices at the top and avalanche debris below. But the wind kept snow off the gravel that followed the contour of the horseshoe-shaped rim. Like a broken fluted vase, the rim rolled up and down in horizontal and vertical undulations.

Rogers looked into the volcano. It was shrouded in fog, leaving only the top third of the crater walls showing. Like a slightly raised lid, another fog bank rose just above the rim. The full moon shown through a deep notch on the southeast crater wall, illuminating the fog bank above and showering beams down onto the gravel path, lighting it like a yellow brick road. A column of steam rose from deep in the crater and broke through the fog, bending over the moon with rainbow-like colors as the light hit it and melding into the fog bank above the rim. It was the most beautiful scene Rogers had ever experienced.

The next morning his back felt better. He slid down the outside slope of the mountain and ended up in a blowdown area, with huge trees knocked down every which way. It was now full daylight. The sky was overcast, so he was not worried about being spotted from the air. He just wanted to get back to Portland before his friends turned him in to authorities as being lost.

Then he really did get lost. He climbed over and under fallen logs and around huge boulders. The fallen trees were like a maze. He would come to a ravine filled with crisscrossing logs and have to backtrack. His hopes leaped when he found what he thought were his tracks of yesterday coming from the car. But after a long while, he discovered he was following two sets of tracks. He had made a circle. He had been following his own tracks from that morning.

Rogers gave up. He had never experienced such tiring terrain. He lay down on a log. "Damn, they're going to catch me," he told himself. "I'm dead meat." Then he looked up to see a dead standing tree in front of him, twisted into a cloverleaf-like deformity. "Wait a minute, I know that tree," he thought. He walked to the other side of the tree. There were his tracks from the day before. He looked up at the sky. "God, it's about time I got a break."

When he reached the car, it was getting dark. He had spent the whole day finding his way through about three miles of blowdown. He took off on the logging road. But this was an unfamiliar road network for him, and he made some wrong turns. He was lost again. Traveling down one road, he slammed on the brakes at a sudden washout. A portion of the road had collapsed down into a twenty-foot-deep ravine. Rogers was pretty sure the road led to a road system he knew. Somehow he had to get the Simca to the other side of the fifty-foot-wide ravine. A very narrow part of the road was still intact, but there was no margin for error. Rogers maneuvered the Simca halfway across the ravine before part of the remaining strip of road gave way beneath one of the tires. The car balanced precariously on the edge. He managed to get out of the car. He figured he could jack up the car on the side where the edge had given way and just push the vehicle back onto the road. The jack lifted the car a short way before the device broke and the car leaned back to the side again. Rogers looked up at the sky. "You're watching, aren't you?"

Immediately below him, two alder trees about six inches wide at the butt and fifteen feet long lay in the ravine, still attached to their root systems. Rogers got a folding pruning saw out of the car. He limbed one of the trees and then cut it above the roots. He dragged the heavy alder out of the ravine.

With some sturdy limbs he built a tripod for a fulcrum, placed the limbed tree trunk across it, and wedged the tree tip under the rear of the Simca. He managed to raise the car a short distance before it started wobbling. Gradually, with several tries, he inched the car back onto the road. He climbed back into the Simca, started the engine, and gingerly drove the car the rest of the way across.

Safely on the other side, he gunned the tiny car as it shot down the logging roads. Finally, under a gray sky just before dawn, he could see down to a Weyerhaeuser camp where logging trucks were parked in rows, ready to start a new day of work. The road flattened out as he approached the camp and became wider and smoother. He saw a long line of pickups, filled with loggers on their way to work, filing into the camp past a guard shack. Rogers was on a road owned by Weyerhaeuser. But if he could get past the guard shack and onto the county road on the other side, he didn't think company guards would have the right to stop him.

Rogers pushed the accelerator to the floor. It seemed the little Simca would never make it across the field where the logging trucks were parked. Through the windows of the guard shack, he could see two men inside. He drove the car between the log scale and the guard shack, watching the startled expressions on the faces of the security men as he waved, passing within a few feet of their window.

On the county road he didn't slow down. He raced to get onto I-5. He still had an hour of driving to get home. He had been gone two nights. He was overdue, and he knew his friends had reported him missing by now. He had just passed Longview when smoke began pouring from beneath the hood. He pulled to the side of the road. Some electrical wiring was on fire and Rogers smothered it with his jacket. But the Simca wouldn't start. "Oh, damn," he thought. "Al Garren has got my map. He's got a description of where I was going. He's probably given it to the cops by now. I'm dead meat. I'm dead meat."

He rigged the starter wire to the battery and got the car going. When he reached Portland, it was morning rush hour. Rogers wove his car in and out of traffic. In front of his house, he jumped out of the car, unlocked the front door, and ran up the stairs to Garren's bedroom. He burst into the room, jumped on the bed, and shook his roommate by the shoulders. "I'm back, Al, I'm back. Did you turn me in?"

Garren wiped the sleep from his eyes. "Nah, I decided to give you another day."

—✳—

Shortly after Rogers' anniversary trek into the crater, authorities launched a crackdown on trespassers in the Red Zone. Several people were arrested and either pleaded guilty or were found guilty and received fines ranging up to a thousand dollars and jail sentences up to a year. "What we're doing is giving notice to the fact that we are doing an effective enforcement program," said Skamania County Sheriff Bill Closner. Rogers was not fazed by his difficult trip or news of the arrests. He was already planning his next venture into the crater.

CHAPTER 17

June and July

HE COULD HEAR the building-up of the muffled roar, seeming to come from deep within the bowels of the volcano. Swanson was on the dome, looking down its side away from the breach of the crater. He froze as he listened to the rumble grow. There was nowhere to go. For the first time while in the crater, he allowed himself to feel fear. How could he have been so wrong? There had been no seismic activity. There had been no widening cracks. No warning at all. The scientist took a deep breath and waited as the roar grew deafening and the ground vibrated.

A military jet swooped up over the crest of the crater and thundered above him.

When Frenzen reported to his professor about the destruction of most of the seedlings at the St. Helens plots, the two men came up with a new plan. What had probably happened, he and Jerry Franklin thought, was that they had underestimated the tenacity of hungry animals that lost so much of their feeding ground.

They decided Frenzen needed to stratify a new group of seeds. Before seeds sprout in nature, they go through a period of cold and wet as typically found in winter. This process is called stratification and can be replicated in the laboratory, which Frenzen did. The surviving seeds still planted at the plots would form the basis for one set of experiments, while the new seeds would be used for an allied experiment. When Frenzen returned to the plots with the new pre-germinated seeds, he had planted them under wire-mesh screen cones and he had also buried a mesh cylinder six inches beneath the plants to keep burrowing animals out.

Now he hurriedly dropped off his things at the lodge at Cispus, where he would live for the next four months, and drove to his plots. Once again he was surprised. Some of the mesh cones over his pre-germinated seeds had been knocked over, even though they had been securely spiked to the ground. The young scientist was baffled. Animals large enough to knock over the cones were not typically interested in seeds for food. He found some elk tracks near a few of the overturned cones and figured the animals had inadvertently knocked them over. But it was the overturned cones with no visible tracks that intrigued him.

As he entered Otto Sieber's top-floor Seattle apartment with its view of Puget Sound, Stepankowsky had his guard up. The filmmaker, now turned aspiring author, had an agenda that Stepankowsky the skeptical journalist had to take into consideration. Sieber had gone into the blast zone with a film crew a few days after the May 18 eruption. They became lost. A sheriff's helicopter found them, and Sieber was issued a citation for being in the Red Zone, but the sheriff's deputy refused to help them get out. The men thought they were going to die. A few days later Sieber and his crew were rescued.

Sieber had been in the same general area filming only a few weeks before the eruption and he knew it was not in the Red Zone. He successfully fought his citation, sued Skamania County for failing to rescue him, and started the research that revealed to him that almost all the victims had been killed outside of any restricted zone.

He contacted survivors and relatives of the victims. "I found that these people were voiceless, that they didn't know how to take up these issues," Sieber told Stepankowsky. He introduced these "voiceless" people to Gerald Parks, a friend who happened to be an attorney for one of the state's largest law firms, Graham & Dunn. Sieber said a sense of indignation moved him to help the survivors and relatives. He said he would not make money directly from any resulting lawsuit. As for his planned book, Sieber agreed that the publicity would help sales, but he said he was not writing it for the money. It was a "labor of love."

"I will make money," Sieber said, "but I won't apologize for that. This country was built by hustlers. I am part hustler, half artist, and part filmmaker."

Stepankowsky sensed Sieber's anger and bitterness toward the authorities. After 150 interviews and numerous searches through public records, Sieber had become convinced the small restricted zones were the result of a "concord of indifference. . . . A group of bureaucrats did not care what happened to people as long as they got their checks and put up a good front."

Victims, he said, had been lulled into a false sense of security by the Red Zone borders. Scientific evidence had warranted larger restricted zones. When Stepankowsky pushed him on what scientific evidence he was talking about, Sieber could cite little beyond studies published years before, predicting a large eruption. As the reporter probed deeper, Sieber kept saying over and over with a slight foreign accent, "You are very good. You are very good."

Still, Stepankowsky had to give the man the benefit of the doubt. He already knew that Sieber's allegations were to a large extent correct. The restricted zones on the northwest side of the mountain had been essentially nonexistent before the eruption.

Back at the office Stepankowsky reviewed more documents and made more phone calls. He tried to suppress his anger. Before the eruption he and other journalists had been led to believe there was a substantial restricted zone around the mountain. He felt state officials had deceived the media and the public.

Stepankowsky found that a week before the state's restricted-zone boundaries were drawn on April 24, the USGS had publicly confirmed that St. Helens' north face had been growing a bulge at the rate of five feet a day since March. A massive landslide could occur if the bulge continued to grow, the USGS warned.

Stepankowsky also discovered that the USGS had failed to disclose that Barry Voight, an independent contractor specializing in landslides, had told the agency that a huge landslide was likely—and that such a landslide could release pressurized super-hot groundwater and magma in a catastrophic eruption. While he did not include the word "lateral" in his study, Voight compared what could happen at St. Helens to lateral eruptions on mountains in the past. Only a few days after the restricted zones were established by the governor, USGS scientists estimated that 80 million cubic feet of explosive magma had now entered the chamber inside the volcano, providing credence for Voight's prediction.

Scientists were not in agreement with each other. Many of them believed that a landslide would be accompanied by mudflows and flows of superheated rock, gas, and ash, but they expected any actual eruptive blast to be vertical—not lateral. Those scientists said there should be some seismic warning of an eruption. Others said it could come without warning.

Even the two USGS scientists charged with keeping public officials informed could not agree. Rocky Crandell warned federal officials in late April that the volcano's north face was "sitting on marbles" and could slide off without warning. On May 10 Donal Mullineaux told a public meeting in Toutle that the bulge would probably slide down a little at a time and "a large event is unlikely." Authorities and landowners cited this confusion as the basis for keeping the restricted zones small on the northwest side and for the lack of stronger warnings.

All this left a lot of questions in the reporter's mind. The restricted zones had not included the area northwest of the volcano that Mullineaux and Crandell in a 1978 report had declared to be hazardous based on past eruptions. The Forest Service had no authority to establish restricted areas for private land. But in making its own off-limit zones the agency had recommended to the state that at least a small area in the upper Toutle Valley mentioned in the 1978 report be included in state restricted zones. On April 30 the governor signed an executive order that not only failed to heed the Forest Service advice but also left all private lands outside of any restricted zone. Weyerhaeuser land was untouched.

The order did little more than replicate the Forest Service boundaries, with one exception. The state closed Spirit Lake Highway six miles out from the Forest Service boundary on the northwest side. This had the effect, Stepankowsky pointed out in his story, of actually bringing the state's roadblock of that highway twelve miles closer to the mountain than it had been before the governor signed the law. This allowed unwary tourists and visitors to the mountain to take any one of dozens of logging roads to get closer to St. Helens while thinking they were safe because they were not in any restricted zone.

Donna Parker couldn't get her immediate boss at Tektronix to understand what she was going through. She told him up front after Billy's death that there would be times she was going to be late, or wouldn't come in at all, but that she would always let him know ahead of time. There would be times, she told him, that she would need to be on the phone at work. But he could not understand. He kept saying to her, "Donna, you got to have control. You got to focus your mind."

One day when he started the same refrain, she asked him, "Did you ever lose anybody?"

"I lost my dad," he replied.

"How old was he?"

"Eighty-seven."

"That's not the same as losing somebody suddenly," she said. "My mind has nothing to do with it. My stomach does it. Then my mind reacts to my stomach and I have no control over my stomach."

Every time she picked up a newspaper or magazine and read about the volcano and its victims, Parker's anger grew and her depression with it. The publication would inevitably portray the victims as foolhardy adventurers who illegally went around roadblocks to meet a death that they deserved for being so stupid. Even some friends and relatives inferred that somehow Billy's death had been his own fault. "Everybody knew it was going to blow up," they would say. "If everybody knew it was going to blow," she would retort, "then why in the hell did they let George Weyerhaeuser have his men working right next to the mountain?"

"Donna, you have to get over this," she was told many times. People said to her, "Time heals all wounds," but she didn't believe it. Parker started saying to them, "I have a right. I have a right to cry. I have a right to yell. I have a right to be mad." She would look them straight in the eye. "I have a right. You did not lose anybody. You have not gone through it." That usually shut them up.

They could not know how the nature of her brother's death affected her. They could not know how the image of Billy's burnt and blackened body haunted her. Even the lawsuit brought against the state by some of the victims' relatives bolstered her spirits for only a short while. Wrongful death suits in Washington could only be filed by spouses and children or by economically dependent parents and siblings. Neither she nor her mother could be parties to the suit. But that didn't matter to her. She only wanted everyone to know the truth—that Billy and the others died because they were misled by the state into thinking they were in safe areas.

One day she felt she had had enough. Parker decided to take a leave of absence and spend part of the summer at a relative's cabin outside of Fairbanks, Alaska. Her brother had spoken fondly of his trips to the cabin. His plans for another visit ended with his death. Now she would take the trip for him.

"We think we have a job up here in Washington for you," the man from the state's workmen's compensation agency told Scymanky.

"Doing what?"

"As an assistant to an electrician."

"Whereabouts?"

"It's in a little town in Eastern Washington."

"You mean out in the sticks?"

"You could call it that," the man said. "It doesn't pay much, but the cost of living isn't that high there."

"Yeah, sure I'll take it," Scymanky said. "Are you going to pay for us to move out there?"

"No."

"Well, listen, you know I don't have that kind of money."

"I'm sorry. That's the offer."

"Well, then I won't do it."

"Suit yourself. I'm going to put you down as uncooperative and we're going to cut you off again."

Scymanky slammed down the phone. It was a game with these state officials. They would come up with ridiculous offers for jobs in fields where he had no experience, paying very little, and in isolated parts of the state where he could not receive the medical care he needed. If he refused the job or if they could find some technicality, they would halt his benefits.

It would often take over a month with letters from doctors or threats of a lawsuit before his benefits were reinstated. In the meantime Scymanky and his wife would have to scramble to come up with enough money to feed their family. It was nerve-racking and depressing.

With a blast from St. Helens on June 18, Don Swanson now had predicted four eruptions, counting the December eruption whose prediction he had kept to himself and his team. He had called every eruption since October—and he had not issued any false alarms. But Swanson was not working alone. Other discoveries on St. Helens and the refinement of existing techniques were adding to the array of prediction methods.

Seismic monitoring had become a powerful tool for providing short-term predictions, usually a few hours to a day in advance. Scientists found that certain earthquakes increased in number and size just before eruptions. Telemetry from electronic tiltmeters in the crater also helped. Tilting of the crater floor started several weeks before an eruption, accelerated rapidly over several days, and then suddenly changed direction, anywhere from minutes to days before an eruption. Measurements of carbon dioxide and sulfur dioxide emissions also were linked with the timing of eruptions.

None of the methods had proven as successful as Swanson's in predicting an eruption far enough in advance to give people time to prepare for it. But Swanson maintained that the future of eruption predictions lay in the interrelated use of all these methods, and perhaps others not yet discovered

The race was on between the logging companies and the people who wanted to see the area around Mount St. Helens preserved as a

national monument. State officials claimed Weyerhaeuser and Champion-International were building roads and cutting timber on land designated for scientific and interpretive reserves once the necessary legislation passed Congress. A spokesman for the State Department of Natural Resources termed the timber companies' actions regrettable. "Congress has got to act before more options are foreclosed," he said.

Pockets of living trees that were the last hope of much of the remaining wildlife on the mountain were being clear-cut. "It's business as usual around Mount St. Helens, said Susan Saul, co-chair of the Mount St. Helens Protective Association. "Land management decisions are being made with the bulldozer and chain saw."

Stepankowsky found a more immediate threat to the region. June rains had swollen Coldwater Lake. Water had risen two more feet behind the mud pile of debris damming the lake six miles northwest of St. Helens. This was on top of the 125 feet the lake had risen since October. With the lake's only monitoring gauge known to malfunction, plans to warn the public of a breach were "woefully lacking," according to Cowlitz County Sheriff Les Nelson. The Army Corps of Engineers hired a contractor to start work on a drainage channel to stabilize the lake. But even with that work, scientists said a violent earthquake could breach the dam.

By now the Corps had dredged 56 million cubic yards of debris out of the Cowlitz River—more than half of the 97 million cubic yards the French dug in building the Suez Canal over a ten-year period. The only problem was that experts expected much more than that to seep into the river from the seemingly endless sea of ash and rock the volcano had deposited.

They sat together on the couch in her house watching television, but they might as well have been a thousand miles apart. Every once in a while one of them would try to start a conversation, but it didn't last long. In the past this wouldn't have worried Venus. She and Roald had grown to have an easy familiarity with each other where talk wasn't always necessary. But now they sometimes felt they had to fill the void with words, straining to bring back something that was slipping from their grasp.

Roald had always known Venus as a determined, focused woman. Now she was pushing for a stronger personal commitment from him. But the harder she pushed, the more he found himself drawing away. He couldn't help it. He was finding it hard to understand why the relationship couldn't go back to being like it was before the eruption.

Venus knew the eruption had changed them and their relationship. Perhaps it had given them a dose of adulthood before they were ready, she thought. The fame, the interviews, the seemingly endless appetite of the public to hear their story, also had taken a toll. The mountain was driving them apart, even as it had bound their lives closer together.

It had grown late. Roald decided he needed to go home to sleep at his parents' house that night.

It was morning and there was a hint of the heat to come as the rising sun began to warm the chilled air. Parker looked up at the mountain from the porch and took a deep breath. Up here in the little cabin tucked into the Alaska wilderness she slowly felt herself regaining control of her own soul. She no longer had to contend with people calling her brother a lawbreaker or with employers, friends, and relatives who did not understand. She could breathe freely again.

Parker tried to do things she knew her brother would do, like bathe in one of the nearby hot springs, walk for endless hours with her dog out onto the tundra, or hike one of the trails disappearing into the dense, surrounding forest. It was then that she felt close to Billy, that she could feel his presence.

As she looked at the mountain, her eyes locked onto the small plane that a bush pilot had crashed onto the slope years ago. She remembered a gift Billy had brought to her one year. "Here, Donna, look at what I found for you." It was the antenna from the plane she now looked at, and he had presented it to her like a coveted trophy.

Parker decided that on this day, her task would be to climb to the wreckage. She followed a path that zigzagged up the mountain. When she reached the plane, she tentatively looked through the windows and wondered if the pilot had escaped safely. She walked out onto a nearby ridge as she imagined Billy had once done. The wind blew back her hair. The view was magnificent. She could see forever.

She now knew that if she were to survive, she would have to channel her anger and despair into something that would bring purpose back to her life and at the same time help redress the wrongs done to Billy and the other victims.

The critical letters and phone calls had begun to come in right after the newspaper won the Pulitzer. They caught Stepankowsky by surprise.

They were all pretty much in the same vein—some reader upset because his or her pet project had not received coverage in the *Daily News*. "Now that you've won the Pulitzer Prize, do you think you're too good to cover community news?" In fact the newspaper had not changed its policy on what events it would cover and it had freed some reporters from volcano coverage to concentrate on other local stories.

Stepankowsky did not yet fully appreciate the role a newspaper often plays in a small town. If the newspaper does a good job over the years, its readers start to feel ownership. They come to accept the newspaper as part of the family. And as in any family, there are bound to be disagreements and hurt feelings. Stepankowsky was not sure the people of the Longview-Kelso area appreciated the work that he and the rest of the staff had done during a trying time to keep them informed.

All that changed the night he and his fellow employees walked into the banquet room of Kelso's Thunderbird Inn. There had to be a thousand people there, he thought. There was the governor of the state, the executive editor of the *Seattle Times*, and a host of other dignitaries. But what was more important to Stepankowsky were the hundreds of people from the community.

This banquet to honor the *Daily News* and its Pulitzer Prize had been the idea of the citizens of the area. These people had dug into their own pockets to help pay for the event. The keynote speaker was Osborn Elliott, former *Newsweek* reporter and editor and now dean of Columbia University's Graduate School of Journalism. He was a member of the Pulitzer board that had chosen the *Daily News* for the prize based on its coverage of the volcano.

"The *Daily News* was ahead from the very beginning," Elliott said. "The paper stayed on top of the story week after week. . . . You cannot teach insight or compassion or the kind of fire in the belly that drives. That's what kept the *Daily News* staff writing and kept photographer Roger Werth shooting and editing thirty hours straight." The crowd rose in standing ovation. When they introduced the staff and called out his name as thunderous applause swept over the hall, Stepankowsky knew he was now a member of the family.

Chapter 18

August

THE TWO GEOLOGISTS walked over the gray landscape as one of them pointed out the different types of rock debris left by the May 18 eruption. Don Swanson listened closely as Harry Glicken presented an autopsy-like report on avalanche material and what it meant. The two men shared a connection to David Johnston's death. Glicken had been Johnston's field assistant. He was to have been the one Swanson relieved at the Coldwater II observation post before Swanson persuaded Johnston to take his turn at the duty. But Swanson usually did not think of Glicken in terms of grief and death.

When Swanson thought of Harry Glicken, it was hard for him not to smile. The scientist was a first-class eccentric. Glicken decided when fish he was cooking was done by the sound of the smoke alarm. Swanson had heard the story about Glicken getting off a bus in his stocking feet and then realizing his boots were continuing down the road in the bus. He tried not to travel in a car with Glicken when he was at the wheel. It scared him more than going into an active volcano. Glicken would drive at full speed, all the while engrossed in geological discussion, not even noticing the four-way stops he sped through.

The wiry young man bubbled with ideas. But to those who didn't know him, Glicken seemed disorganized and his ideas disjointed. Swanson knew otherwise. Glicken had the ability to pull together disparate pieces of information and observations that other scientists might have missed and organize them to determine cause and effect. Glicken had been a doctoral student when the mountain blew. That summer he tried to get the USGS to hire him as part of the survey team. But the powers-that-be did not think the quirky Glicken would fit in. He hired on with an independent contractor, however, that worked for the USGS. He was determined to map out the eruption's debris avalanche, a huge and tedious undertaking. The work, he let it be known, was to be dedicated to the fallen scientist.

The eruption in its avalanche had thrown out pieces of the mountain ranging from tiny fragments to intact sections up to hundreds of yards wide. The avalanche had created a bumpy landscape with whole chains of miniature mountains, called hummocks, as high as two hundred feet extending for miles into the Toutle Valley. The top layer of the mountain's slopes, consisting of a mixture of fine sand, weathered rocks, and plant life, had avalanched first. Then came huge boulders and glacial ice blasting out of the mountain during the eruption. The topping was a layer of fine silt, sand, and ash, in some places many feet deep, that drifted down during and after the eruption.

As they hiked over the debris field, Swanson appreciated the quality of the work accomplished by Glicken and the other two members of his team. They had mapped out in meticulous detail the location of the debris left by the volcano. Then they reconstructed where the sections of deposits had originated and how they had gotten to their present locations.

Listening to Glicken, Swanson recognized this was the most detailed and important work ever done about a debris avalanche. It had yielded findings that could save thousands of lives. For years scientists had been trying to figure out where volcano hummocks originated. Some thought they were mudflow remnants, explosive deposits, or debris left from retreating glaciers. Glicken proved they were the result of volcano avalanches that could travel for miles—and a new lethal threat was added to the known list of volcano-related killers such as mudflows and pyroclastic flows.

The discovery caused a reassessment of volcanoes around the world. Previously undetermined avalanche fields were found around scores of steep-sided volcanoes. Now that scientists had a good fix on the nature of avalanche flows, they would be better at determining the hazards a volcano might pose in future eruptions. This was especially important for slumbering volcanoes, such as St. Helens, that suddenly spring to life after decades of inactivity.

However, the discovery had not changed the minds of USGS bosses, who still refused to hire Glicken. As he realized that nothing he did would make a difference with the USGS, he fell into a depression. Still, Glicken pressed on like a man possessed, working excessive hours in what many in the USGS considered to be a guilt-driven state over Johnston's death.

For a while after Johnston was killed, Swanson tried to console the grieving Glicken. If Glicken had remained on the mountain with Johnston the day of the eruption, it would have not made any difference, except there would now be two dead scientists. But Glicken still carried on in his grief. Swanson gave up. And there was little counsel he could offer Glicken on the subject of overworking. Swanson, too, was a driven man.

"They're cutting off our workmen's comp payments because Jim won't move us and take some ridiculous job," Helen Scymanky told the psychologist.

The couple watched as the psychologist grew visibly upset. "It's not right," he said. "I've written them. They know Jim's condition. They're just playing games. I'll give them a call." He made himself a note.

"How are you doing otherwise?"

"My arms," Jim said. "I'm worried. I don't seem to be making any progress lately and I still can't extend them out."

"Sometimes we hit plateaus, and it takes time to get beyond that," the psychologist said. "How's your sleeping."

"Better. Much better. I can almost get through the night."

Helen told the psychologist how weary she felt in dealing with Jim, the children, the finances, while still trying to do the household chores. Yet she often felt guilty for not doing enough.

"Quit beating up on yourself," he told her. "Look how far you've come. You're both doing great. Learn to take some time-outs."

Helen smiled. Time-outs? She was afraid that if she ever took a break, she might never find the strength to get up and go on again.

Oregon State University researchers warned public health officials for the first time that they had detected "highly infectious" Legionella bacteria in the lakes around St. Helens. They asked the Army Corps of Engineers to limit the amount of water that drained out of Coldwater Lake and into the Toutle and Cowlitz Rivers, a prime source of the area's drinking water. But Stepankowsky found that as usual in stories dealing with the volcano, there were conflicting opinions and theories.

The bacteria believed to cause Legionnaire's disease had been detected in St. Helens lakes the previous summer, but the public had never been told. The researchers said they were now issuing a public warning because temperatures were rising during a hot spell. The warmer the water, the more virulent the bacteria. Scientists had already discovered concentrations of the bacteria at levels ten thousand times greater than they had found elsewhere under normal conditions. Not much was known about the disease that had been discovered only a few years before, when it killed more than two dozen people at an American Legion convention in Philadelphia.

The Corps of Engineers, sponsors of the study that led to the researchers' findings, now branded the warning as "ridiculous" and based on "wild speculation." Local and state health officials told Stepankowsky they saw no reason to take any special measures to protect the public. "Studies throughout the United States have not been able to associate any illness with the Legionnaire's organism," said Bill Harper, director of environmental health for the local health district. Other officials said they believed the organism to be the cause of the disease, but only in airborne form. In any case, water treatment plants would kill off the organisms if they were waterborne.

Stepankowsky contacted microbiologist Dave Tison, a recognized authority on the disease. Tison said any one of the six known strains of the bacteria could be infectious under one set of conditions but not under others. Scientists had not been able to establish any pattern. The type of bacteria found in the St. Helens lakes had resulted in harm to laboratory animals, the OSU researchers pointed out. Tison agreed that this was probably a strong caution sign that should be heeded.

More than a dozen scientists working near the lakes had come down with flu-like symptoms believed at first to be associated with Legionnaire's disease. While health workers were not able to specifically identify the disease in the scientists, they had not been able to rule it out either. They had found no other plausible explanation for what was being called the Red Zone flu.

As he worked, his back ached and sweat dripped off his body. Frenzen was on his hands and knees in the ash, meticulously counting the seedlings in one of his 225-square-meter subplots. With each seedling, he measured its height and recorded its condition. The sun reflected off ash that it heated at times to 160 degrees Fahrenheit. The ash irritated his eyes and it filled his ears. It gave the sandwich he packed for lunch a crunchy feel as he bit down on the grit.

The endless counting and recording could be mind numbing, but he had trained himself to do it hour after hour without stopping, sometimes without even moving his position. He carefully spread the nascent branches of a seedling apart in his fingers, noting the yellowish condition that indicated the plant might not make it.

He rose up and was startled. Surrounding him was a herd of elk lying on the ground where they had chosen to rest. The elk, some only a few feet away, were as surprised to see him as he was to see them. Frenzen surmised that in the stillness doing his count, the elk had not recognized him. Now one after another and then in unison, they rose to their feet and leaped into the air around him. In watching their graceful prancing, he was so awed by their beauty and the surprise of them being there that he had little fear of what their sharp hooves could do to him.

When they saw the sign advertising Reid Blackburn's car on display at the museum near Silver Lake, Parker didn't hesitate. "We have to go see that," she told the woman sitting next to her in the car. Jeanette Killian was the wife of Weyerhaeuser worker Ralph Killian, who had helped Parker find her brother's truck. The Killians had lost their son and daughter-in-law in the eruption. Besides their grief, the two women shared a determination to bring to light the fact that almost all those killed by the mountain had been innocent of wrongdoing. Both women believed their relatives' deaths had been caused by negligence on the part of authorities.

After Parker returned from Alaska, she had mapped out her strategy. She decided to concentrate her energies on a new organization that she and others were forming. It was called SHAFT—St. Helens Alliance of Families for Truth. The organization might be able to help set the record straight. Her long-term hopes lay with the upcoming trial of the lawsuit filed against the state by victims' relatives. She knew there was an attempt being made to include Weyerhaeuser as a defendant. She thought that ultimately the trial would bring justice. Until then she planned to keep busy doing what she could.

They stopped in front of what looked to Parker like an old barn. It was actually an abandoned dance hall that had been turned into a privately operated volcano museum. Inside they found the car in which photographer Reid Blackburn had been suffocated to death by ash from the erupting volcano. The car sat in a pile of ash brought into the building.

A mannequin of the late Spirit Lake innkeeper Harry Truman stood nearby. Other displays ranged from animals stuffed after having been supposedly caught in the blast to damaged logging equipment, before-and-after photos of St. Helens, and models of the volcano. A battered pickup truck caught Parker's eye. The two women walked over to it. This could have been her brother's truck showing up on public display, Parker thought, if she hadn't stepped in to rescue it. They confronted the owner of the museum, who told them she had permission from the families to display the vehicles. For Parker and Killian, the whole museum was sacrilegious, making money from the dead.

The museum display didn't identify the owner of the truck. But Parker knew whose it was from the description of where it was found. The truck belonged to Arlene Edwards. Her body had been found more than a month after the eruption, lodged in a hemlock tree six hundred feet below where she had been standing on Elk Rock with her daughter. The body of her nineteen-year-old daughter, Jolene, was discovered near the truck. Parker believed that Arlene Edwards had another daughter, a young woman who had not been with them on the mountain. Perhaps Parker could find the surviving daughter.

"See, the alder's coming back. Does that make you feel better?" The blond thirty-six-year-old scientist was talking to four research assistants and Stepankowsky as he pointed to a two-foot-tall alder seedling on the banks of Clearwater Creek. The scientist, Art McKee, gently touched the upper leaves of the seedling. For miles there appeared to be no other vegetation. The stream gurgled as it made its way around and through heaps of gnarled logs and rock debris at the site nine miles northeast of the volcano.

"We're still a long ways from being normal," McKee told Stepankowsky. The May 18 eruption had "zapped to zero" the vegetation along the stream, he said. As the crew moved down the creek, recording types of plants and number of fallen trees, McKee explained the importance of plant life in water environments.

Twigs and dead leaves normally fall into streams from overhanging vegetation, he said. This attracts flies that are in turn eaten by fish. But when that overhanging vegetation is stripped from the land, a food source is eliminated. This forces the stream to manufacture its own food in the form of algae and fungus. As a result, different types of insects and fish are attracted. The lack of overhanging vegetation also causes the water to warm, killing off some fish. Without adjacent vegetation, stream banks become more likely to erode. Turbidity of the water increases, further changing the habitat for fish.

McKee said his work would help in understanding how man-made actions such as clear-cutting affect the environment. The scientist rubbed the fuzz on his unshaven chin. "This is an opportunity to work on an extreme case," he said. "In the past, we worked on streams disturbed by logging or landslides. We have never had a whole watershed wiped clean before. The St. Helens eruption provided a grand scale experiment."

The USGS scientists at St. Helens were having such success in predicting eruptions that they now had a luxury never afforded in previous volcano studies. They had codified what a successful prediction should entail, an area that greatly interested Swanson.

Time, place, type, and magnitude of eruption, they decided, were all key elements in making not only successful but *useful* predictions. A prediction should not simply say an eruption was going to take place on a volcano, but should specify *where* on the volcano. Scientists should designate a reasonably short period in which the eruption was to occur. As the eruption neared and more information was obtained, scientists should narrow the predictive window. In the case of St. Helens they were now able to predict an eruption two to three weeks ahead and gradually refine and pinpoint its time to a few hours before the event. Predictions also should be able to specify to some degree whether an eruption was likely to be explosive or nonexplosive—that is, whether it would contain both gas and superheated ash or rock (explosive), or ash or rock without the gas (nonexplosive). The prediction also should state whether the eruption would be large or small.

The scientists decided it was not a good idea to use a percentage of probability in issuing predictions. In the past in many areas of science, use of percentages had only confused the public and diminished the impact of predictions. They also wanted a clear separation between a forecast and a prediction. A forecast was a general statement about future activity. It could be made based on the geological past of a volcano. A forecast would not be nearly as precise as a prediction.

Swanson believed the predictive methods used on Mount St. Helens could be applied to similar volcanoes around the world if personnel got the right training and equipment. He loved the simplicity and low-tech nature of his crack-measuring method. While tiltmeter measurements, monitoring with a geodimeter, and use of seismographs all required highly trained specialists and expensive technology, his method did not. Even poor countries could afford a tape measure.

Venus Dergan could tell something was wrong when she let Roald into the house. As he sat down he didn't slump back as he usually did, but sat on the edge of the chair and stared at the floor for a moment. Then he looked up. "Venus, I don't think it's going to work out between us." He thought they were too young to get any more serious about each other than they already were. It would be best if they dated other people to see how it went.

Venus knew they had been drifting slowly apart. Still, she never thought it would come to this. How could he leave her, this man who had saved her life? She blamed the mountain. She knew that if they had not been caught in the eruption and the flood, this would not be happening. It had changed everything. She could not think of anything to do but let him go, though she knew it was going to be terribly hard. After all, they had always been there for each other. He had been her first and only real love.

Lately it seemed to Stepankowsky that all he was writing was bad news where the mountain was concerned. First came the reports about the tourist season. It had been a bust. The state had projected three to four million tourists visiting the St. Helens area. But tourists in these numbers were nowhere to be found. In fact tourism was off more than a third from what it had been the previous year when many people had stayed away under the constant threat of eruptions.

The people who did show up didn't seem to have deep pockets. Stacks of volcano T-shirts remained on store shelves along with Mount St. Helens playing cards, key chains, ashtrays, games, necklaces, and other trinkets. And local businesses no longer had the steady stream of hunters, fishermen, and campers that had come in the past for the exquisite mountain's beauty and resources. This tourist season was going down as one of the worst ever.

As if that wasn't bad enough, the Army Corps of Engineers now estimated it would take an additional fifteen years and $1 billion to continue its flood-control efforts, a dramatic increase in cost. "The increase results from new studies showing that awesome amounts of sediment will continue to flow down the Toutle—amounts that far exceed last winter's calculations," Stepankowsky wrote in the front-page lead story with a banner headline proclaiming, "Corps to curtail dredging, concentrate on Toutle dams."

The Corps had concluded that it needed to dredge some 372 million cubic yards of volcanic sand from the Toutle River, more than six times the material already removed from the Cowlitz River. Without the work, the Corps said, the chances for major flooding would increase fourfold in Kelso and Longview and many times that in Castle Rock and Lexington.

Unfortunately for the region, the Reagan administration had picked this time to rein in the often high-spending Corps. When $35 million was cut out of its budget for St. Helens work, it decided to concentrate its efforts on continued operation of two sediment dams on the Toutle along with some dredging on that waterway. But within a few weeks dredging work was stopped even at the sediment dams. Future work remained doubtful.

The Reagan administration came up with another surprise. A top official now wanted the Forest Service to reconsider its final plan for land use around the mountain. John Crowell Jr., assistant secretary of agriculture in charge of the Forest Service, wanted to change the plan to permit more salvaging of downed timber and to reduce the area the Forest Service had recommended as off-limits to logging.

The plan had been a compromise pleasing no one. It called for about 116,000 acres for interpretive areas, prime animal habitat, and scientific study. Logging would be prohibited on this acreage. This was closer to the 61,000 acres the logging industry had agreed to support than to the 216,000 acres that environmental interests had said needed to be left untouched. Now the Reagan administration wanted to reduce the protected area even more.

Stepankowsky pointed out that Crowell was former chief counsel for the Louisiana-Pacific Corporation, "one of the nation's largest buyers of federal timber. In that role, Crowell became widely known for big attacks on the Forest Service." Crowell especially questioned the necessity to preserve land in the upper Clearwater Creek basin. If that land was opened for logging, many scientific studies would be destroyed.

Peter Frenzen listened intently to what his fellow scientist was saying about plant growth on St. Helens. The man's research team worked in many of the same areas as Frenzen did. Almost as an afterthought the scientist told Frenzen about some birds he had seen, the sparrowlike birds known as juncos that could grow over six inches high. "By the way, some of my crew has seen the juncos go after your seed," he said. "They come down like dive bombers and repeatedly hit your cone screens until the screens come loose and tip over."

CHAPTER 19

September and October

ROBERT ROGERS was so engrossed in examining a rock that the helicopter seemed to come from nowhere. He had just enough time to look up and see it fly over the volcano's rim as its roar filled the crater. There was nowhere to run. He stood near the top of the dome. He could only watch helplessly as the helicopter banked and returned to hover over him and then drop to the surface.

Don Swanson and another geologist jumped out of the helicopter. Rogers hiked up the slope slowly to the two men. "Do you arrest trespassers?" he asked Swanson.

Swanson scrutinized him for a moment and then smiled. "No. We're scientists, not cops." Swanson knew that if he himself was not a geologist who already had access to the mountain, he would be right there with this man and the other trespassers. With his natural curiosity, his need to know, and the pull of the volcano, nothing could have kept him away.

The USGS issued an advisory that another eruption was expected. Swanson made use not only of his visual observations of cracks and faults, but also of his geodetic monitoring network that now included five stations inside the crater trained on almost twenty targets on the dome. Increased emissions of sulfur dioxide also indicated an eruption might be near. Seismic activity was not on the rise, but it was expected to increase a day or two before the event. Swanson believed it would be a nonexplosive dome-building eruption.

Scymanky threw the soccer ball downfield to his children on the school playground. The ball sailed through the air and he felt good. It had become important to him that it go a little farther each day. If he couldn't get back the full use of his arms, he would make the most of the function he did have. He found he could do quite well and the accomplishment began to make him feel whole again.

This time of his life had not been wasted, he kept telling himself. He was alive. He had been given a second chance. And he had grown closer to his children.

The boys, all four of them, now raced toward him as the oldest carried the ball above his head like a trophy. Scymanky opened his arms as wide as he could and gathered the boys close to him.

When the volcano erupted, Don Swanson was in Tokyo for an international symposium on volcanology. The conference was taking place in a country that consisted of almost four thousand islands, many the result of volcanic activity. Japan had sixty active volcanoes including the still steaming Fujiyama, a symbol of the country's strength and beauty. It was the mountain that St. Helens was most often compared to before the May 18 eruption. Mount St. Helens—its eruptions, and the scientific discoveries it had generated—was the topic of the hour.

When the latest eruption, a non-explosive one, was announced, it was Don Swanson that the Japanese media flocked to hear. Only a small article about the eruption had been transmitted around the world from the United States. As in the case of previous eruptions, it said nothing about Swanson. But the Japanese media had been primed by scientists who knew of Swanson's work. He was overwhelmed when Japanese television crews and news reporters descended upon him in a rush. Swanson had little inkling that a public beyond the scientific community was interested in his work. He was dumbfounded.

Throughout the day they pestered him for interviews. The Japanese journalists were entranced with Swanson and his discoveries. They knew they were onto a great story that had eluded even the press of Swanson's homeland. This volcanologist was the first man in history to provide an accurate long-range prediction of a volcanic eruption, and he had done it not just once, but five times.

Swanson was the first to admit that his formula was not perfect. He did not think it would work with all volcanoes in all situations. After all, even in the case of St. Helens, he didn't think it would have accurately predicted the May 18 blast because that eruption had been initiated by an earthquake. The earthquake released a landslide, in essence uncorking the pressure building in the volcano. But Swanson's work was at least a beginning in the science of long-range eruption prediction.

As Swanson's face appeared on millions of Japanese televisions that day, the face of a fictionalized David Johnston flickered bigger than life on the cinema screen a half world away in Longview, Washington, where the movie *St. Helens* had just been released.

———✳———

Rogers and a friend from the San Francisco Suicide Club, Mark Northcross, stopped by his parents' house in Vancouver on their way to the volcano. Rogers didn't know exactly what possessed him to stop. He hardly ever saw his parents.

The visit did not go well, as usual. His parents knew about his trips to the volcano and did not understand, let alone approve. The fact that Rogers placed himself in danger and enjoyed it was just one more strange aspect to a son they considered to be weird. Rogers castigated himself for having made the stop, then realized what he had been doing. He wanted Northcross to somehow understand, to know why he was the way he was.

"I can see now why you latch onto groups and do crazy things all the time to get recognition," Northcross told him as they sped away in the Simca. "You get no support from those people."

This wasn't the whole answer, though. Rogers did a lot of crazy things simply for the sheer thrill of it. The adrenaline rush heightened the sensitivity of every atom in his body and blocked out everything extraneous. It allowed his spirit to soar. There was nothing that compared to playing with death while fighting off the fear of being caught, being a rebel defying the powers-that-be, telling the world to go screw itself.

They hiked into the crater at night. It was dark and overcast, perfect trespassing weather. Rogers hadn't expected the patches of snow in the crater this early in the season. A thirty-foot-high rock sat a couple of hundred yards in front of the dome. They set up camp there.

Then they went exploring. Rogers had built a device—a pyrometer—to measure temperatures inside the volcano. He was going to shove the pyrometer on its ten-foot telescoping wand into cracks in the crater floor and take readings. Before he could take his first reading he walked over a patch of snow that gave way beneath him. He fell into a fumarole, a deep hole that was releasing hot gases from the volcano. He landed upside down on a slight ledge while the pyrometer continued its fall into the fumarole, bouncing off walls as it disappeared.

Rogers caught his breath, righted himself, and grabbed his hard hat. After climbing out of the fumarole, he didn't say anything for a moment as he brushed off his clothes. Then he looked up at Northcross. "A guy could get hurt up here," he said without smiling.

They bedded down in their ponchos by the big rock for the rest of the night. After breakfast they went exploring again. Rogers showed Northcross the geological highlights of the crater, as if he were a tour guide who did this all the time. The two men split up as they walked around the dome. Rogers was taking photographs when a steam eruption blasted a spout of water to the sky. Northcross came running back from the other side of the dome. "It's okay," Rogers told him.

Rogers posed Northcross in front of the steaming dome. He gave him a sign to hold that he had made up for this occasion. Beneath a headline proclaiming in large letters "Red Zone Rambler," Rogers had written: "Volcano Lecture Tonight. Don't bring the briquettes, just the hamburger. We'll do the roasting for you on the volcano."

Then Rogers took a kite kit from his backpack and put it together. A kite store owner had asked him to take a photograph with the promotional kite flying over the dome. He had already taken a photograph of a ruler standing on end in the ash with the dome in the background. The ruler displayed the name and phone number of his girlfriend's answering service business. For the friend who had supplied him with motorcycle parts, he took a hat with the name of the man's business embroidered on it, set it on a round rock that looked like a head, and snapped away with the billowing volcano in the background. There had been other promotional photographs for which Rogers was not averse to being paid.

He let out the kite as the wind swept it up into the air. In the windy crater he had little trouble keeping the kite aloft. He directed Northcross where to stand to align Rogers, the kite, and the steaming dome above it into the frame of the photograph. Northcross took several photos. For much of the rest of the afternoon, Rogers flew the kite in and out of the steam of the volcano, playing tag with it.

Stepankowsky no longer had what could be called separate beats for the mountain. His beat *was* Mount St. Helens and almost anything to do with it. He had grown possessive. He was jealous of anyone else receiving assignments having to do with the volcano or was afraid others might not do an accurate job. The mountain had become more to him than just a story. As he was getting off the airplane after a recent trip, the volcano erupted. He went around saying, "She waited for me to get back home before she blew."

Recently he had spent days on the mountain interviewing scientists as Roger Werth took photographs for a series of stories. Stepankowsky had learned that hundreds of scientists from forty-six universities, and a multitude of organizations, industries, and government agencies, were conducting at least 250 studies in the huge laboratory that St. Helens presented to them.

Each day the mountain spewed from four hundred to five hundred tons of sulfur dioxide and several thousand tons of carbon dioxide into the air. These emissions were much lower than immediately after the May 18 blast, but were considered a strong indication of a much-alive volcano that still had the power to erupt with deadly force, Stepankowsky learned.

One theory behind the string of recent, less powerful eruptions was that magma was now degassing before it reached the surface, a view shared by Don Swanson. Many scientists in the USGS thought Mount St. Helens had started a long, slow wind-down to dormancy. "The eruptions are now working from a magma body that is pooping out of gas," USGS hazards chief Chris Newhall told Stepankowsky. Seismic testing had not been able to detect a magma chamber believed to be a shallow, spherical reservoir beneath the crater that held the volcano's supply of molten rock. Some scientists thought this indicated the pool of super-hot material was too small to detect. More eruptions were expected, but not on the scale of the May 18 blast.

Some of the agency's elite scientists disagreed. Dwight "Rocky" Crandell, who had retired and now worked part-time for the USGS, urged caution in dealing with the volcano: "There's still a chance you could get another mass of magma moving up that could renew explosive activity smaller, as large as, or larger than May 18."

All the scientists readily admitted, however, that they didn't know precisely what was going on beneath St. Helens. It was possible a large magma chamber existed deep beneath the ground with several tubes of magma rising from it to narrow into a smaller conduit before reaching the shallow crust above, said Steve Malone, who headed the University of Washington seismology laboratory. Such a configuration could result in an event equal to the eruption of Mount Mazama in southern Oregon in 6000 BC. Mount Mazama had exploded with a force forty times that of the May 18 St. Helens blast.

The reporter and photographer also detailed the latest findings of other scientists. They found thousands of Douglas fir seedlings a half-inch tall springing from a thick carpet of dead needles beneath towering old-growth trees that were dying as a result of the eruption. But they were told the seedlings might die the following summer, once the nutrients that the dead needles had injected into the volcanic ash had been exhausted. The ash, contrary to popular belief, would be virtually useless in the support of plants until it broke down into digestive chemicals and minerals a hundred years hence.

Another scientist was excited to find a hundred salmon spawning nests along a tributary of the Toutle River. But the story was the same here. With the presence of the contaminating ash and without plants to eat and provide stream cover, the salmon would die.

The recovery story was one of yin and yang, with seeds of hope containing both positive and negative aspects. While erosion had filled the rivers and streams with fish-killing sediment, in some areas it had worn away the ash, exposing the fertile original soil beneath. While gophers were helping bring back plant life by burrowing and kicking up nutritious original soil, there was some question about how long the gophers themselves would survive once the dead roots they fed on were exhausted. Birds like the juncos attacking Frenzen's seed plantings had been spotted in small numbers in the devastated areas but had not yet nested in any of the standing dead trees. At the same time, for some unknown reason, hummingbirds were found everywhere, even in the volcano's crater.

Algaes had been discovered that put nitrogen, a key plant nutrient, into the soil. Lupine plants that took nitrogen from the air to the soil were starting to thrive. But few plants other than those planted by scientists had germinated from seed. Nearly all the plants now to be found on the mountain were from roots that survived the eruption.

"There hasn't been a spectacular recovery," state biologist Bob Everett told Stepankowsky. "It will take a long time for some areas to recover. But it's not a desert anymore. The area is sort of alive. It's coming back to life. Nature is more resilient than we give it credit for."

Twice she drove past the house located off Terwilliger Boulevard in Portland. But Donna Parker couldn't bring herself to stop and knock on the door unannounced to see Roxanne Edwards, a high school student who had lost her mother and sister in the eruption. Parker didn't know how to approach her. She didn't know what to say. But the fact that the mother's truck was on hideous display at the volcano museum kept nagging at her. Finally she called the girl on the phone.

"Hello, Roxanne, you don't know me, but my name is Donna Parker and I know you lost someone and I lost someone, too." Parker told the girl how much she admired her for all Edwards had been through. Edwards was a young girl alone, fighting to keep the small inheritance she had been left. Authorities had made a strict interpretation of the law regarding people who were killed on the mountain but whose bodies had not been found. First, photographic proof and affidavits were required to prove the deceased had been on the mountain at the time of the eruption. Then it had been decided that presumptive death certificates would not be issued until the state legislature passed a special law covering the St. Helens situation. Relatives of the dead were left to navigate the legal morass. Without a death certificate, companies refused to pay off on life insurance policies, and most other forms of assistance were denied.

The body of Roxanne's mother, Arlene Edwards, was found more than a month after the eruption. But until then Roxanne suffered the indignities of the system and had to endure the resulting financial pressures. The authorities had found the bodies of her sister, Jolene, her sister's dog, her mother's dog, and the personal belongings of her mother, but they said there couldn't be any ruling of death for her mother because the body had not been found. Only nine months before losing her mother, her father had died.

The girl wouldn't let Parker hang up. It seemed like she would talk forever. Finally they agreed to get together. Then Parker called the girl right back. She hadn't been able to get up enough courage to tell her about the truck.

"Roxanne, I have something to tell you. And I don't want you to start crying and get upset. But I have to tell you, because I know you are going to want to know."

"What is it?

"I saw your truck. I saw your mother's truck."

"Where?"

"They've got it in a museum before you get to the mountain."

"Oh, god. I can't believe it."

"Then you did not give your permission?"

"Absolutely not."

In the weeks that followed, Parker and the girl grew close. Roxanne had put up with many of the same trials that Parker had endured, with people suggesting that her mother and sister had been foolhardy in being so close to the mountain. The older woman found herself taking on a motherly role. Even though Parker had few assets of her own, she bought some needed tires for Roxanne's car. When Roxanne couldn't make mortgage payments, Parker was there with her checkbook. Parker thought she made her greatest contribution by giving the girl someone to talk to, someone who understood.

They hoped to visit the museum and see the truck. Roxanne remembered that her mother had often stashed things under the front seat.

From the ridge, Swanson looked down upon Spirit Lake, still thick with muck and thousands of floating logs. The Corps of Engineers had begun draining Coldwater Lake to reduce the danger that its debris dam would give way. But not many people realized that at Spirit Lake, there was a threat potentially many times more deadly than the one posed by a breach of Coldwater Lake.

If anything, Spirit Lake was in worse shape and more dangerous than when Swanson had seen it the day after the May 18 eruption. He was awed at that time by the sight of the massive debris dam holding back billions of gallons of water and mud. Since then the lake had risen almost an additional sixty feet, doubling the amount of liquid material that would be released in a breach.

Swanson had become increasingly alarmed. New studies of the dam's materials proved what he had suspected. The dam, made of highly erodable material, was structurally a disaster waiting to happen. The Corps of Engineers had been able to blast and bulldoze outlets to drain smaller lakes, but the precarious makeup of the debris dam seemed to make this impossible at Spirit Lake.

Hydrologists provided a scenario for what would happen if the rubble holding back the lake broke. They determined that water would jet down the valley of the North Fork of the Toutle at a rate of two million cubic feet per second. Mudflows ten times greater than those that followed the May 18 eruption would demolish every city between the volcano and the Columbia River, including Longview. But only Swanson and a few other scientists and government officials knew of this threat facing the population that lived within sixty miles of Mount St. Helens. It had been determined that for now the public would not be told. Meanwhile the experts would be debating how to start draining the lake without disturbing the unstable dam.

More than a hundred and twenty volunteer searchers stood waiting for instructions on the grounds of the destroyed Weyerhaeuser Camp Baker logging facility. It was a cold and wet Saturday morning. Winter would soon make searches impossible.

Some of the volunteers drank steaming coffee as they stood in front of the armada of vehicles—Jeeps, four-wheel-drive Blazers, buses, vans, and horse trailers housing the steeds that would be used in the search. The squawk and crackle of walkie-talkies filled the air.

These people included members of search and rescue teams, reserve deputies, and amateur radio operators. Their mission was to enter the restricted zone to look for the remains of the more than two dozen people "officially" declared dead from the May 18 eruption whose bodies had never been found. Some of the volunteers thought that the number of missing dead on the mountain was actually much higher than that. "This is our last-ditch effort to find these people," Cowlitz County Sheriff's Deputy Mike Nichols told the throng. The last body had been found a couple of months before near the Green Mountain Mill. It was his hope that erosion from recent rains had uncovered more bodies.

They broke into fifteen teams. The largest team, a ten-member group led by tracker Maurice Saxon, drove vehicles up tortuous logging roads toward the mountain. A thick fog enveloped them. Suddenly the lead vehicle braked to a stop. Towering over the convoy, the arm of a monstrous logging crane rose through the clouds of mist. The jawlike grippers at the end of the extended arm were open, as if trying to clutch the sky. The crane blocked the road, so searchers piled out of their vehicles. They hiked the rest of the way to their destination, the steep, V-shaped ravine of Disappointment Creek, littered with fallen logs. Here, it was believed, Paul Schmidt of Oregon had met his death while on a hiking trip. "We want Mr. Schmidt," Saxon yelled to his crew like a football coach admonishing his team to go win. "Let's find him."

Slowly, very slowly, the volunteers moved down the hillside, their eyes locked to the ground, leaving not even an inch of the ash mantle unexamined as they navigated between fallen logs, splintered tree trunks, and boulders. The hillside was soon festooned in bright orange and pink tape tied to tree branches every hundred yards or so, marking the areas that had been searched. At the bottom of the ravine, the searchers trod across the damp, volcanic mud and up the other side.

By the end of the day the team had found only a shredded piece of faded blue denim. Saxon remained undaunted. "It was successful in one respect—we know where he's not," he said. The other teams had not been much luckier. It was decided to try again the next day. Each area would be combed again.

On Sunday, Sgt. Bob Dieter of the Cowlitz County Sheriff's Reserve led another ten-person team. First they searched an area near Schultz Creek at Fawn Lake, where it was believed John Killian had vanished and where the body of his wife had been found. The team had no success.

They moved on to the Green River about twelve miles northwest of the volcano. They were looking for any clues to lead them to the bodies of Tom Gadwa and Wally Bowers. The crew found a stream with its milky water sending off a foul smell. Slowly they made their way alongside the stream, stumbling over rocks and searching through piles of blown-down timber. The searchers grew excited with the finds of pieces of fiberglass, paneling, cloth, and rusted metal thought to belong to a red pickup. On the other side of the stream they found the tattered front seat of the pickup. That was the end of their discoveries.

Back at the base camp, the search teams gathered their finds together. It amounted mainly to a collection of clothing scraps and pieces of metal. The biggest find was a torn blue jacket with a bone in its sleeve. Deputy Nichols doubted it was human, though he had no evidence to the contrary. The last official, large-scale search for the missing on Mount St. Helens had ended.

On October 20, 1981, five days after the Forest Service submitted its final land use plan, Mount St. Helens was declared the nation's first officially designated volcanic area. This move was seen as a precursor to naming the mountain a national monument.

The Forest Service plan provided for a total area of 109,000 acres. For environmentalists and many scientists who originally recommended that over 200,000 acres be preserved, the decision was devastating. It was seen by environmentalists and others as a cave-in to Reagan administration pressure and special interests.

As Stepankowsky wrote, it reduced by more than 3,000 acres the amount of land the Forest Service originally deemed necessary for a protected wildlife habitat. A multitude of wildlife sites were eliminated. A 6,520-acre research area on Clearwater Creek was slashed from the final plan, leaving a host of scientific studies to perish. The Miners Creek area along the Green River, greatly coveted by scientists and environmentalists, was eliminated from the plan because it still had some old-growth trees that timber interests wanted to cut. Of thirty-nine geological points of interest that scientists deemed vital in telling future generations the story of the volcano, eleven were lost, now to be used at the discretion of private interests.

More than 820 million board feet of timber, almost 50 million more than originally in the plan, would be allowed to be salvaged or logged on publicly owned lands. This represented over 75 per cent of the marketable timber in the area. The plan also allowed for 89 per cent of salvageable timber to be taken off state and private lands that would ultimately be in the monument area. The area's largest private landowner by far was Weyerhaeuser.

When Stepankowsky sat down at his desk a few days later, he was not sure that the eruption prediction story he was assigned to write would even make the front page. There was little threat to life or property, and these prediction stories had almost become routine. The USGS now had so much confidence in its ability to predict eruptions on Mount St. Helens that some scientists were calling for the public to be allowed into the Red Zone.

"Mount St. Helens will erupt non-violently and add to its stadium-sized lava dome within two weeks—perhaps in the next few days—scientists said this morning," he wrote. "The U.S. Geological Survey issued an extended outlook advisory late this morning after scientists noted an increased rate of ground deformation in the volcano's crater."

He quickly finished typing the story. He had other, more important stories to write about St. Helens and its impact on the area. And on October 30, the mountain erupted once again, adding nearly five million cubic yards of material to its dome.

PART 3
1984-1986

The Trial

The eruption was an act of God.
But the deaths were an act of man.

—Ron Franklin

CHAPTER 20

1984–1985

THE TRAIL NARROWED. The trees closed in around her, and the sky descended like a gray lid. Her legs grew wobbly on the path through the deepening snow. Venus Dergan thought the mountain might be shaking. Her heart pounded. She couldn't breathe enough of the cold, thin air at this elevation. Then she was sinking again, down into the thick muddy water, her hand reaching through the surface above as though she could somehow pull down the air her body craved. Now she was being brought out of the water, but she was lost, so lost. She needed to get out of here.

Slowly Venus realized where she was. She had been having a flashback to the flood that almost killed her the day Mount St. Helens exploded. But that didn't stop her panic attack. She still gasped for air.

Why did she think she could ever set foot on a volcanic mountain without paying the price? Friends had invited her on this hike up Mount Baker, a dormant volcano in the same Cascade Range as St. Helens. But several miles up the trail and into the snow, she realized that just like that day on St. Helens, this was uncharted territory. She had never been here before. They had hiked farther than she anticipated. What if they got lost? She had no idea what might lie ahead, and her imagination vaulted her into foreboding, and from there into the morass of her memories of that terrible day.

As the anxiety swept her up once more, it didn't seem like she was merely remembering. She was there again. Roald kept grabbing her and fighting to pull her from the water.

"I've got to get out of here," she yelled to her friends. She turned and made her way down the mountain.

---

*

Suddenly, the plume of steam and ash rocketed to an altitude of more than twenty thousand feet. It was the third time in less than a month that such blasts had exploded on Mount St. Helens. They produced the greatest plume clouds and resulting excitement since 1980. There had been many eruptions since then but most of those had been events where lava—that is, magma that had reached the surface—was extruded quietly without giant plumes. But scientists didn't consider these three new bursts to be eruptions. There had been no magma involved. In the past such bursts could prove to be precursors to eruptions, but scientists didn't think so this time. It was believed that earthquakes on the dome had released pressurized, superheated ground water.

Twelve days later scientists were stunned. Don Swanson's team landed in the crater without Swanson, who was in New Zealand on assignment. Molten lava poured out of the dome in a new eruption. It was the first time new lava had broken through the surface of the dome without warning. There had been only slight increases in earthquake activity and swelling on the sides of the lava dome that stood more than seven hundred feet high. No eruption prediction had been made.

The headline in the *Seattle Post-Intelligencer* blared out at Andre Stepankowsky. "Ray: State agency was to blame for Red Zone," it said. The P-I had been leaked a copy of former governor Dixy Lee Ray's deposition leading up to trial of the wrongful-death claim brought by the injured, such as Jim Scymanky who had now joined the suit, and relatives of victims. By this time the Weyerhaeuser corporation had been added as a defendant, along with the State of Washington. Now Stepankowsky had just a few hours to provide coverage of the deposition story for the *Daily News*. He started making phone calls.

In a previously taped interview played during the deposition, the governor blasted her own Department of Emergency Services for its poor performance during the days surrounding the May 18, 1980, eruption. The department was "totally incompetent" and was to blame for the odd shape of the restricted zone that in its contortions had spared Weyerhaeuser land.

The most important thing to come out was information about Ray's sworn testimony in an earlier affidavit, referenced during the deposition, that she had spoken several times with George Weyerhaeuser about the volcano. Previously she had maintained she had never talked to Weyerhaeuser.

"Before May 18, I spoke with many people and representatives of many organizations, both public and private. I spoke several times with George Weyerhaeuser," Ray testified. "Weyerhaeuser Company was closely following the developments on the mountain because of its ongoing logging activities in the area. Our discussions included measures taken by the company to protect its employees and others present on Weyerhaeuser land in the event of an eruption. In my judgment, these measures as well as others being taken by other public and private organizations were sufficient to protect human life from foreseeable risks.

"This judgment resulted in my decision as to the establishment of boundaries and which lands would be included within the restricted zones."

The affidavit had been made several months before the deposition. Under questioning from plaintiffs' attorney Ron Franklin, Ray now asserted that when she had said "our discussions" in her affidavit, she was referring to herself and her head of emergency services—and not to George Weyerhaeuser. The state's attorneys maintained that Ray's reference to George Weyerhaeuser was to "general discussions" about the volcano and not about the restricted zone.

In a telephone interview with Franklin for his story, Stepankowsky learned that Ray had met with Weyerhaeuser attorneys two to three weeks before she gave her deposition with what Franklin called a "reinterpretation" of her affidavit. "I think it's strange that Weyerhaeuser officials are explaining to a former governor what she meant," Franklin said.

While the former governor scrambled to recast what she had said in her affidavit, she gave the Weyerhaeuser lawyers a new bombshell they hadn't expected. Ray said she had met at least once, face-to-face, with George Weyerhaeuser while the restricted zone was being drawn before the May 18 eruption. Previously she had claimed such a meeting never took place. George Weyerhaeuser in his deposition continued to emphatically deny any such meeting. But Weyerhaeuser was a member of Ray's Economic Council, had been one of her strongest supporters in getting her elected, and headed a company that was one of the state's largest employers. It was not beyond reason that he would meet with the governor.

As he made his way up to the ridge, he couldn't figure out what he was hearing. Strains of distant music wafted through the night air above a rising din. Once on top of the hill Peter Frenzen realized it was country music. He was startled by the scene below. Amid barracks and construction shacks, he saw earthmovers, dump trucks, cranes, and workers bustling about. An ethereal haze of light lit up what looked like a surreal children's play-set plunked down in the gray, barren earth of the valley below the volcano. Then he remembered reading about the tunnel being built to drain Spirit Lake to help prevent collapse of its debris dam and the massive flooding that would result.

Only a few years before, Frenzen had looked down into this same valley from a ridge. Then in the distance the volcano had steamed in the sunset, with shadows darkening the vast, prehistoric-looking terrain. He almost expected to see a herd of dinosaurs lumbering across the land. Now as he listened to the groans and squeaks of the heavy equipment, blending with the country music, and watched the ash clouds swirl from the ground and linger as dust in the light, he felt a profound sense of loss.

As the rock-chewing monster burrowed ever deeper into the bowels of the earth, Stepankowsky thought this must be how Jonah felt being inside the whale. He sat behind the operator of the machine that was digging the Spirit Lake tunnel. Nicknamed "the mole," the machine actually looked more like a hundred-yard-long snake, stretched to its full length. This snake had a head of rotating cutters that ate into the earth. A conveyor belt inside its eleven-foot-wide body carried the diggings to waiting muck cars. The machine hissed and growled in the dank, dark air on its journey to cut the 1.6-mile-long tunnel to lower the water behind the dam.

"It's not really a hard job to learn," operator Bobby Lee yelled at Stepankowsky above the crackle, grind, and rumble. "It's mostly watching dials to make sure everything is all right." The wet rock dust caked on Lee's face made him look like a clay model of a wrinkled old man. A thin, red laser beam pierced the darkness to keep the mole on target.

"I go wherever the machine goes," Lee told Stepankowsky. "I don't mind jumping around. I get to see a lot of the country that way."

Huge hydraulic arms spread from the machine, pressing against the tunnel wall to hold the mole in place while the cutters were thrust against the rock. Knee-deep water surrounded the mole from water sprayed on the rock for the cutters. This was a wet, dirty, noisy, smelly place to work, Stepankowsky thought. He didn't know how the workers did it. The four electric motors driving the twenty-seven disc-shaped cutters strained as the mole shuddered. The sickening, burning smell of overworked motors filled his nostrils. The machine had encountered mud.

The mole did much better with hard rock. Pockets of softer rock, like that in this hill near the lake, slowed the process. Big chunks broke off and clogged the conveyor system that carried debris to the muck cars that hauled the debris from the tunnel. Sometimes, like now, the mole encountered a stratum of mud that could only be dug by hand. It was not a good sign.

All the letters were written on stationery headed with a drawing of the volcano and the organization name of SHAFT, standing for St. Helens Alliance of Families for Truth. Donna Parker and others had formed the group to correct the misconception that the volcano victims had ignored danger zones.

To *Ripley's Believe It or Not*, she wrote, "Your presentation on Mount St. Helens . . . stated that 'people ignored warnings.' I am writing to inform you that the above statement is in error and should be corrected."

There were other letters, to the *Oregonian*, *Time-Life*, *Holiday Films*, and many other media outlets. Often, as she wrote a letter, she would find herself growing so angry she would have to take a break.

Parker saved her most acidic writing for the *Daily News* and publisher Ted Natt in what had become a running battle. She believed that because the newspaper had been right there on the scene, day after day, that if any publication had an obligation to set the record straight, it was the *Daily News*.

"You don't really understand," she said in one letter to Natt. "Is it because you are not familiar with the area, Spirit Lake Highway, and 3500 Road before the eruption? That can be your only excuse. Maybe my case will be clear to you this time. Number one, no one snuck around roadblocks. Number two, no fool died there. Unless you have information that no one knows . . . there was only one roadblock."

One day she phoned Natt. The *Daily News* had brought out a book in the early months after the May 18 eruption. The book, filled with reprints from the newspaper, had become a surprise best seller. Parker wanted Natt to issue a recall of the book, just like an auto manufacturer would recall a defective vehicle.

"Lady, I think you need some help," Natt told her.

"Yes, Mr. Natt. I *know* I need some help."

The attorneys at Graham & Dunn had contracted with a Texas law firm to represent the injured and the relatives of victims in the trial against the State of Washington and Weyerhaeuser. The Scymankys and the other plaintiffs had been told that the lead attorney, Ron Franklin, was one of the best wrongful-death and injury attorneys in the country.

When Franklin and his associate knocked on the Scymankys' door, Helen excitedly jumped up to welcome them. The attorneys, dressed in jeans and sweatshirts, with their Texas drawls, had a down-home manner that put the Scymankys at ease. They probed Jim for hours on every little detail of events leading up to the eruption and on the tragic day and afterward. He was more than ready for them. He had suffered injuries that cost over a half-million dollars so far in medical treatment and rehabilitation, plus the pain and suffering that entailed.

———✳———

Don Swanson drove past the construction site where the mechanical mole had almost finished punching through Harrys Ridge for the tunnel to help drain Spirit Lake to a safe level. In front of him stood ridges scarred with new roads built for the tunnel project that had begun only a month after Congress had declared the area part of a national monument, "allowing geologic forces and ecological succession to continue unimpeded."

Members of the construction crews had demonstrated how easily man could upend the fragile environment of a recovering area. To provide eggs for breakfast they brought in chickens, and to feed the chickens they brought in grain that sprouted into a grass species foreign to the area. The new grass brought mice, so workers introduced cats into the mix. This ecological chain reaction soon skewed the data of biological researchers studying natural succession in the blast zone.

What worried Swanson as a geologist was what the mole had encountered in its digging. The machine found sections of mud that its blades could not cut through and that had to be dug by hand to get to solid rock again. The mud meant to Swanson that the mole had found geological faults running vertically through Coldwater Ridge. This was evidence of the St. Helens seismic zone lying beneath the ridge and the mountain itself.

Swanson's fellow scientists agreed. They determined, based on information from the February 13, 1981, earthquake of 5.5 magnitude on the mountain, that an earthquake of 7.0, many times more powerful, was a definite possibility. If an earthquake occurred in conjunction with another eruption, it was possible the tunnel could be blocked and the dam holding back Spirit Lake could break open.

A new government study, not yet released to the public, outlined the devastating effects that could be expected of a Spirit Lake breakout. The new study was different from past ones in that it was based not on a worst-case scenario but instead on what was called "more probable" conditions.

Longview, Kelso, and surrounding towns would be destroyed by flooding, the study said. The impact of the water and volcanic debris hitting the Columbia would be so great that the river, one of America's largest, would change course and actually flow backward, bringing huge floods upriver. The river would be nearly choked off near the Cowlitz River mouth with volcanic debris, according to the report. A giant sandbar would act as a dam, causing the Columbia to rise steadily over a two-week period. Dikes on the river would collapse and all the land southward, to and including the Portland Airport, over fifty miles from Mount St. Helens, would be under water. The projected loss of life was in the tens of thousands.

Venus Dergan walked up to the front door of Roald's house with trepidation. She hadn't seen him for a long time. Then unexpectedly he had called to say a newspaper reporter wanted to interview them. During the interview in his home, Roald kept turning and smiling at her. They joked easily. But other times she found him almost studying her, and she wondered what he was thinking.

The Toutle River raged not far below the brink of the cliff where the isolated house sat as symbol of a longtime marriage torn asunder. The initial mudflows had missed the upscale house, but the rains that came afterward engorged the silt-filled river. The waters ate away the front yard and a huge chunk of road, leaving the house cut off from any neighbors. The insurance company would not pay off on a flood policy on the now virtually worthless house because the structure had not yet actually been damaged by a flood.

Owners Jim and Jerri Hastings, married for twenty-five years, sat hostage in their dream home with their three teenage sons, a color television, a kitchen table, a set of chairs, and little else. They had moved most of their possessions out because they believed it was only a matter of time before the river finished the job and the house toppled into the water. Many times the family felt the house shake as huge stumps rammed the riverbank, now just feet away from them. Stepankowsky had followed the family over the years in several stories for the *Daily News*.

"After the mountain blew, we didn't do anything," Jerri Hastings told the reporter as she twisted a straw around her finger at a local restaurant where he interviewed her. They had experienced a happy marriage until that point, she said. "Then all the laughing stopped. I really knew for the first time in my life what it was like to be scared." They figured out ways to get to and from the house, but the daily fear stayed with them.

The house never did fall into the river. "The people who lost homes and came out of that mess were luckier," she said. "It happened and they readjusted and started over. We sat there and looked at it every day. It was a constant reminder."

Jim Hastings, an electrician for Weyerhaeuser, tried to scrape together enough money to build a house on another lot. He got the foundation poured but Jerri realized it would take a long time before they had the money and time to finish it. Four years after the May 18 eruption, she divorced Jim. "I just wanted a normal life," she said.

Jim, who still lived in the house with one of his sons, was not bitter about his wife leaving him. "Sometimes I wonder, 'Why me?' But why is it anyone?" he asked Stepankowsky in a separate interview. "Things are going to work out. I'm optimistic. There's bound to be better and happier things in the future."

There was a chance the Army Corps of Engineers was going to start dredging the Toutle River in front of the house and dump dredged material onto the eroded area, returning some value to Jim Hastings' home. It was ironic, though, he said. He blamed an earlier dredging project by the Corps for causing the erosion in the first place.

The call from Roald surprised her. He wanted to get together to talk. When they met, she was taken aback even more. "I'm still interested in making it work with us," he told Venus. "I would like to give it another try." It dawned on her that their recent meeting for the newspaper interview had somehow renewed his interest in her.

She felt confused—angry and stunned at the same time. She thought of reasons it wouldn't work. Their lives still seemed topsy-turvy, five years after the eruption. She had survived by putting emotional distance between herself and Roald. Her romantic attachment to him had gradually worn off over the years, she told herself. She no longer loved him in that way, did she?

She didn't want to chance the hurt again, if once more it didn't work out. She wasn't willing to take the risk. "I'm in a steady relationship," she told him.

The scientist stood on the dome for the photographer, hands on his hips and elbows extended, like he was king of the mountain. He was now known by his colleagues and other scientists as Mr. Crater. He had continued to do what no one else had ever done—accurately provide long-term predictions of eruptions. His string of successes now stood at twelve for Mount St. Helens.

Yet Don Swanson knew as well as anyone that you never really conquer a volcano. The failure to predict the June 1984 eruption had proved that fact in the eyes of some, even though that failure was not nearly as detrimental to Swanson's prediction theories as originally thought. Swanson was in New Zealand at the time, but after his return he took a close look at the data. They indicated that some lava had poured out of a crack on the dome, but that pressurized magma had not forced its way through rock to the surface. Thus the event should not be categorized as an eruption. The USGS continued to call it a eruption, though that determination was debatable among the agency's scientists.

Swanson realized there had been no failure of his prediction methods. However, he continued to modify his techniques to rely more on forms of ground deformation other than crack widening.

He had been digging out instruments in the snow when the newspaper photographer wandered by. The photographer asked him to pose for an article the *Seattle Post-Intelligencer* was doing for the fifth anniversary of the big blast. Then he went back to work. What instruments were showing, he believed, was that the volcano was running out of gas. This had been the longest period without an eruption since the mountain came alive almost five years before—more than two hundred days of quiet. The last eruption, on September 14, 1984, had sent limited ash to the sky and caused minor mudflows. Still, he thought, it was a little early for the mountain to fall back into dormant slumber.

At the Monticello Hotel in Longview a red carpet had been rolled out. A giant cake in the shape of the digging machine known as the mole was ready to be cut. Souvenir pieces of Harrys Ridge, bored from the tunnel designed to lower Spirit Lake by fifteen feet, were waiting. Just as four dignitaries opened a gate to allow water to start running through the tunnel from the lake, five volleys of 75-millimeter blanks boomed over the town from the National Guard Armory, one for each year since the tragic May 18 eruption.

Despite opening of the new tunnel, flooding problems were not over. "Our main worry is that people will jump from the opening of the Spirit Lake tunnel to thinking there are no flood problems in Cowlitz County," a spokesman for the Federal Emergency Management Agency told Stepankowsky. "We've only crossed one of them out." A local citizen enjoying the festive mood, Ethel Mayclin, tended to agree. She told Stepankowsky, "Let's worry about the others tomorrow."

Donna Parker nailed the fourteen large crosses together and spray-painted them white. Then she carefully wrote the names on them with black paint. The crosses were for the volcano victims whose bodies had never been found. Relatives of the rest of the twenty-one people missing planned their own means of remembrance. On the fifth anniversary of the eruption, Parker intended to plant the crosses at the locations where it was believed these people had met their end. She now knew many of the relatives and friends of the victims. She hoped that by having a place to focus on—the site of a wooden cross—each person might find some peace.

It had been years since Rogers had seen his father, and now here he was in the lobby of the dentist's office. The men stared at each other in surprise.

His father had wanted to be an engineer but never had the money to go to school. He could never understand why his son dropped out of Portland State University. Even more so he couldn't fathom why Rogers risked life and limb scrambling up bridges or climbing a live volcano. When Rogers wanted to borrow money to buy and fix up rundown houses, his father refused to loan him the funds. Yet he gave money to a neighbor.

At the dentist's office each man tried to mumble something to the other. Soon Rogers fled out the door as he felt his chest tightening and he struggled to breathe.

"Since the mountain blew, the Army Corps of Engineers has spent millions of dollars to keep the Cowlitz River out of our basements," Stepankowsky wrote in an overview article. "But the Corps had made mistakes."

When the Corps dredged the Cowlitz in 1980–81, it dug below the pre-eruption riverbed, piercing the underlying groundwater table and draining the wells of residents along the river, leaving them dry. Erosion resulting from dredging sent tons of earth from people's property tumbling into the rivers, severely damaging many homes. "The Corps admitted it was at fault," Stepankowsky wrote, "but wouldn't pay to repair the damage, citing a policy not to pay for damage caused by emergency work."

Insurance companies had not treated victims and their belongings any better, Stepankowsky discovered. "Most insurance companies refused to pay claims for mudflow damage, saying earth movements are excluded from their coverage." A state court in 1983 ruled the companies had to consider the damage a result of an explosion, which most policies covered. But it was too late for the majority of home and property owners. Most policies allowed only a year to appeal rejected claims and many people had not bothered to appeal because they thought their chances were nil.

The federal government also shortchanged residents. In 1980 Congress approved $945 million for volcano relief and repairs to be used by residents of the area. Investigators found, however, that federal agencies had submitted false and inflated damage claims to grab more than their share. The money had then been spent on pet projects and contractors in other places. In the end, residents of the area saw only half of the relief funds.

The bugling of elk rose up from the valley in the blowdown zone, echoed off the walls of canyons, and grew into a chorus. Peter Frenzen could not see them, but the sound sent a thrill up his spine. Often in the past he had felt guilty that he was so absorbed in his work of documenting the now-thriving seedlings in his plots that he couldn't see the forest for the trees, so to speak. But now after five years he could not miss it.

About fifteen hundred elk had been lost in the May 18 eruption. State game scientists had predicted that elk wouldn't repopulate the area for ten to fifteen years, when enough trees had grown up to provide cover. Now elk could be found even on the slopes of the volcano itself. Since he had been coming here in 1980 he had seen no hawks. But now that vegetation was starting to thrive in spots, mice had returned and so had the hawks.

As he hiked over the blast zone, Frenzen noticed that where he had found scattered seedlings years before had now become thickets of alders. Scientists in surveying the devastation had declared it would take at least ten years before shrubs and small trees would take root again. But nature in ways scientists never anticipated had provided for the healing of the land without man's help. Animals such as gophers that had been deep in their burrows during the eruption had survived to mix fertile soil with the ash above in their burrowing, allowing seeds to take root. Erosion had washed away the ash along waterways, exposing the soil beneath that now supported stands of cottonwoods and willows. Many dormant plants had come alive to push up through the ash after the blast.

Now strong, weedy species such as pearly everlasting, fireweed, tansy, and thistles splashed color across the gray landscape each spring and summer. These patches of weeds helped trap more seeds that sprouted into more vegetation attracting the elk and other animals. Conifer seedlings were beginning to appear again naturally with seeds blown in from patches of trees on the edge of the blast zone, Frenzen told one of the reporters who now sought him out. He was becoming recognized as something of an authority on the biological changes on the mountain.

Frenzen cautioned the reporter that rebirth had been spotty near the volcano. But he also told him that looks could be deceiving. In the gray, barren ugliness on the north side of the volcano there appeared to be no life. Even here, though, nature was at work in just one of its many acts of self-healing. Heat trapped in the ground helped bring warm water and steam to the surface. Water condensed out of the steam, allowing halos of moss and algae to develop. The moss and algae in turn helped trap airborne seeds that were now developing into patches of willows, fireweeds, and thistles.

Small trees that had been protected by snow during the blast were also now thriving on some of the desolate north-facing slopes of the mountain. "That's been the really striking thing," Frenzen told the reporter. "How really resilient nature is. Pretty much, nature has been taking care of itself."

Five years after the eruption and resulting flood that nearly took her life, Venus Dergan talked with Stepankowsky for a story on the aftermath of that day. She told him that she had become more loving toward her family and friends and felt "closer to God." She considered the relationship she had with Roald to be between "just friends."

"Now I can look forward to doing things, and accomplishing things I've not been able to do," she told Stepankowsky. "I was young when that happened and never had a chance to get my feet on the ground. . . .

"I turned out to be a normal person after it all happened. I'm leading a normal life. They'll always be memories in the back of my head. But now it's like . . . I don't want to say a dream, but it gets foggier and foggier year after year. It's like it's laying itself to rest."

CHAPTER 21
1985: Volcano Watch

SWANSON WASN'T TOO SURE about the idea, but his bosses thought it would be a great public relations coup for the fifth anniversary of the May 18 eruption. Swanson would fly into the crater with a television crew to give a live TV interview from inside an active stratovolcano. More than a dozen technicians with the help of a helicopter from a Portland television station struggled with high winds to place microwave transmitters on surrounding ridges. The station touted the upcoming broadcast as a major engineering and broadcast feat.

The volcano had stirred with minor earthquakes all week. On Friday, May 17, Swanson took measurements confirming that the now nearly 750-foot-high lava dome was swelling, but at a slow rate. The USGS determined the quakes were still at what was called background levels while approaching low levels, the next step of activity. Swanson was fairly sure an eruption would not happen the next day, but he couldn't entirely rule it out.

Next morning they landed in the crater and the crew set up for broadcasting. When they started the interview, the wind rushed around them and Swanson found himself almost shouting into the microphone. Then the ground began to shake. The dumbfounded crew tried to keep on their feet as the shaking grew in intensity. A shocked look spread across the face of the reporter interviewing him. It was like a scene from a B-grade disaster movie, Swanson thought. He knew the level of activity for earthquakes had just jolted past background, and perhaps even past low. Swanson realized the earthquake that was now being watched on TV screens across the Northwest could be a precursor to another eruption. He just didn't know when it would come.

—✳—

After hearing about the increased seismic activity on the mountain, Donna Parker decided she wasn't going to chance going into a danger zone to place her crosses. She stored them for a safer day.

While reporters clambered to see Jim Scymanky on this anniversary day of the eruption, he went fishing. He refused to give any interviews.

The newspaper had interviewed Venus and Roald so long ago that Venus had forgotten about it. Now she called Roald to tell him the news. "Have you seen this morning's paper?" she asked. "We're all over the front page."

When Stepankowsky looked at the *Daily News* Monday afternoon, he went straight for Bob Gaston's column, "News Notes." Under the headline, "County's economic hard times shrink newspaper's size, staff," the managing editor detailed the problems that were resulting in cutbacks at the *Daily News*. "Economic ailments besetting Cowlitz County are seriously affecting this newspaper," Stepankowsky read.

The newspaper's total page count in April had been 762, which was 110 fewer than for the same month the previous year. Since the May 18, 1980, eruption, the newspaper had lost nearly 35 percent of its advertising lineage. The salvage logging that had propped up the local economy for a while had ended. The effects of a national recession were lingering.

Newsroom personnel retiring or leaving would not be replaced. Reporters would have to adjust to the diminished news hole resulting from the decline in advertising. "We've addressed our space problem by setting length limits on local stories and trimming news service stories," Stepankowsky read.

That night shallow earthquakes increased in the crater. The next morning Swanson returned to measure for any increase in swelling. In the past few years rockfalls had made it more difficult to measure the cracks because they could no longer be easily seen. For the most part he had retired his tape measure in favor of an electronic distance meter and tiltmeters that radioed data back to the observatory constantly, even during severe storms. But in using his tape measure less, Swanson had developed techniques based on his original theories, using types of dome deformation other than cracks. Now he used the electronic meter to measure distances from points on the crater floor to points on the dome. He found the dome was definitely expanding at a more rapid rate than before.

The USGS issued an eruption alert, predicting an occurrence within the next two weeks. It would be the first eruption in eight months. Somehow Northwest television stations got it garbled and took to the airways predicting an eruption comparable to the original blowout. The story spread across the country. Phone lines into the USGS were jammed, but finally Stepankowsky was able to get through. USGS volcanologist

Chris Newhall told him the TV reports were "absolutely" an exaggeration. There had been no large swelling or deep earthquakes to indicate the volcano was about to explode in an eruption to equal the lethal May 18, 1980, blast.

It was just after 4 a.m. when the airplane broke through the clouds, revealing the volcano below for Swanson and his team. It had been less than a week since the eruption alert was issued. For the previous three days, tiltmeters had radioed information back to the observatory in Vancouver that indicated the dome had sharply increased its swelling. Hundreds of earthquakes in growing magnitude were shaking the crater. The scientists in the airplane looked for a new lobe of lava, but found none. However, Swanson could tell from the location and intensity of the glow inside the dome that molten rock had moved high into its confines. A dome-building eruption had started and the dome was growing internally with it.

The plane banked and the scientists on board became excited. Down below, a U-shaped trench, known as a graben, had developed on the south side of the dome. The graben was several hundred feet long, 150 feet wide, and anywhere from 15 to 45 feet deep. It was formed, Swanson believed, when molten rock pumping into the dome pushed the south side out laterally. The stretching caused the top of the dome to sink, forming the graben. The graben and the swelling of the south side told Swanson that the volcano's internal plumbing had changed. Any possible explosion would now occur on the south side rather than the north.

The next day a USGS hydrologist aboard a helicopter spotted a new lobe of lava pouring out of the southeast side of the dome where Swanson had discovered the graben. But in the following few days Swanson and the other scientists became baffled. The newly formed lava lobe did not seem to be growing. There had been no explosions. Yet unlike with any previous eruption, earthquakes continued at an extremely high level of activity. The situation was a reminder to Swanson that as much as he knew about St. Helens, there remained a great deal of mystery, and so much more to learn.

The earthquakes gradually died back. There was no explosion. On the first clear day, he went back into the crater. He was not disappointed. The dome-building eruption was one of the largest on the mountain in five years. Lava had pushed the southern third of the dome out by three hundred feet. More than 250 million cubic feet of material had been added to the dome. Swanson's techniques had again succeeded in predicting an eruption.

There was a lot to report in Stepankowsky's regular newspaper column, "Volcano Watch." One item told of a study that determined there had been no lasting harm to Washington state's agriculture from St. Helens ashfall. The cost of initial damage, however, was pegged at $55.5 million. In another item Stepankowsky reported that the actor who played David Johnston in the movie *St. Helens* had been found stabbed to death. David Huffman had tried to prevent a car robbery. His body was later discovered in the woods of San Diego's Balboa Park.

Stepankowsky added that the movie "was roundly criticized for making mincemeat of the true story line."

"'Volcano Watch' Dormant" read the subheading over the copy ending his column. "This will be the last Volcano Watch column to appear in the Daily News for three months," Stepankowsky wrote. "I am taking time off from my reporting duties to start writing a book about the volcano." He also knew the years of covering the volcano beat had taken its toll. He was exhausted.

The lawyers sat around a long table in the conference room at the Marriott Hotel in Portland. Jim Scymanky had been waiting a long time to have his say. He had never been involved in legal proceedings, but he found himself surprisingly calm, even welcoming this deposition and the trial beyond. He wore casual slacks and a short-sleeve shirt. "Let the attorneys dress up and wear the suits," he thought as he dressed that morning. "I'm not going to a ball or anything."

Scymanky was comfortable with what he was going to say. He knew what had happened to himself and the others. He had his story down pat, and nobody was going to change it. Over the years he had relived it hundreds of times.

The questions from Weyerhaeuser's lead attorney, Mark Clark, were straightforward. The lawyer was knowledgeable and made his points in a nonabrasive manner. He had a self-assured confidence and sincerity about him, with none of the arrogance that could drive Northwesterners mad. By the end of the deposition, Scymanky found himself wishing Clark was *his* lawyer rather than being the attorney for the opposition. He thought this lawyer from Seattle would play well with local audiences, including the jury, who were often suspicious of outsiders, such as Californians and Texans.

—✳—

At first, for several days, he stared at the blank page before him. Then Stepankowsky typed a few words and stared some more. Each day he promised himself to do better tomorrow. Few people had his overall knowledge of the volcano, and this was part of the problem. He was overwhelmed with information and memories. The right words wouldn't come. His book on St. Helens was going nowhere.

Gradually he found himself spending less time in front of the typewriter as he increasingly sought solace in the music he made with his beloved grand piano. His father, a European immigrant, had wanted him to be a concert pianist. At the age of eleven, Stepankowsky began dedicating his time to the instrument. While other boys were out playing baseball, he sat at his piano, endlessly practicing or taking lessons. Somewhat small for his age, he became the target of bullies who tagged him as a sissy. He held his natural aggressiveness in check, until one day he discovered hockey. He not only made a team, but was good enough to go on to play at the college level. Through it all his father always made it clear that the piano came first.

His father hired faculty members from the famed Juilliard School as piano teachers for his son. Unlike many young music students who rebel against demanding parents, Stepankowsky did not grow to hate the piano. He enjoyed making music. But he discovered his love of literature and writing was stronger. In his senior year of high school, he announced to his father he was no longer going to pursue a career as a concert pianist. The senior Stepankowsky was devastated.

There had been other writers in the family. Andre Stepankowsky's grandfather had been a writer and a Ukrainian revolutionary who once shared jail time with Leon Trotsky, one of the founders of the Soviet Union. When the grandfather found himself out of favor with his comrades, he left the country for other European destinations. He went on to become a respected journalist who founded a news service, the Ukrainian Press Bureau in Lausanne, Switzerland.

A few months before, when he couldn't sleep one night, Stepankowsky had read some lines in a sonnet by the seventeenth-century English poet John Milton. They got him thinking again about his talent for the piano.

When I consider how my light is spent,
Ere half my days, in this dark world and wide,
And that one Talent which is death to hide,

> *Lodg'd with me useless, though my Soul is more bent*
> *To serve therewith my Maker, and present*
> *My true account, least he returning chide.*

The lines and the thoughts they inspired grew increasingly to haunt him. He had told himself, after first reading the poem, that he would give a piano recital on his upcoming thirtieth birthday. Now the recital began to obsess him, the first time in a long while that he wasn't consumed by the mountain.

He decided he would concentrate on music that emphasized his Polish and Russian roots. Day after day and into many nights, the sounds of Chopin, Tchaikovsky, and Khachaturian resonated from the apartment he shared with Paula and through their working-class neighborhood. He would give the recital to raise funds for charity, he thought. And he would invite his father.

It was a major blow to Scymanky and the other plaintiffs. King County Superior Court Judge James McCutcheon ruled that the state had acted properly in establishing restricted hazard zones around Mount St. Helens and that the state was not responsible for injuries or death caused by the eruption. In ruling favorably on the state's pretrial motion for dismissal, the judge said he found no evidence of conspiracy or other improper action by the state.

The plaintiffs' legal brief had argued otherwise. "On March 4, 1981," the brief presented to the court said, "Governor Ray was asked during a recorded interview whether there was any 'special precaution taken' before the decision was made to make the zone 'on this side' [the northwest side of Mount St. Helens where Weyerhaeuser owned land] smaller than on the other side. Her response was 'oh, yes, of course. And we didn't do it until the Weyerhaeuser Company agreed that they would take full responsibility for both instructing their people, . . . setting up a system of rapid warning and evacuation.' She later confirmed that 'Mr. Weyerhaeuser told [her] that he would accept certain responsibility for the workers.'" The brief quoted Ray as saying, "There was an understanding."

George Weyerhaeuser had hedged in his deposition, according to the plaintiffs, on whether he had discussions with Ray about drawing up the restricted zones. "Although Mr. Weyerhaeuser 'cannot now recall' whether such discussions took place before or after May 18, 1980, he does recall expressing to the Governor his 'concerns about the State's procedures for granting and restricting access to the restricted zones around Mt. St.

Helens, and the impact of these procedures on Weyerhaeuser's logging operations.'"

McCutcheon ruled that then-Governor Ray's executive order of April 30, 1980, establishing restricted areas was a discretionary governmental decision. Ray, as the highest officer of the state, had every right to make such a decision when it benefited public safety. The state argued successfully that it should not be held liable, because there was no way to predict what the volcano would do.

However, the judge ruled that the trial of the civil suit against Weyerhaeuser could proceed because there was some evidence the company had indicated the areas around the mountain were safe. But whether this was true or not was up to a jury, the judge said.

Five years after he started his study, Frenzen's paper on his experiments with seed planting in volcanic ash, or tephra, was published. The study was titled "Establishment of Conifers from Seed on Tephra Deposited by the 1980 Eruption of Mount St. Helens, Washington." His professor, Jerry Franklin, also was given credit, but Frenzen was listed as the lead author.

It had been a prodigious effort. In just one of several experiments, the young scientist had followed the progress of more than thirty-five thousand planted seeds over several years. A seed planting study on this scale had never before been done on a volcanic landscape.

Frenzen found that under the right conditions, most conifers would grow in the ash, though to varying degrees. Douglas fir seemed to do the best. But what was fascinating to some colleagues was Frenzen's minute descriptions of the specific conditions, such as shading, under which certain species did best. The study had obvious implications for aiding natural recovery after disasters in the future.

At the end of his leave of absence, Stepankowsky was anxious to get back to work as a reporter. He never returned to writing his book. The Mount St. Helens trial was the biggest story on the horizon. His pretrial article in the *Daily News* presented a sharp rundown on the principal issues and players. Though the article didn't mention it, eruption survivor Jim Scymanky was among the plaintiffs.

Did scientists suspect that Mount St. Helens would erupt
as violently as it did on May 18, 1980, when it killed 57 people?

If so, did they tell state and Weyerhaeuser Co. officials of their concerns? Did Weyerhaeuser suppress or ignore warnings?

Those basic questions underlie one of the Northwest's most publicized lawsuits, brought by relatives of 10 people who died in the eruption.

Trial of the suit begins next week in the King County Superior Court before Judge James McCutcheon. . . .

Attorneys for the relatives will try to prove that Weyerhaeuser misled employees about the safety of camping and working near the volcano. All 10 victims were Weyerhaeuser employees, their relatives, or contract loggers who were working for Weyerhaeuser.

Only three of 57 volcano victims were believed to have been in the restricted zone established by the state. Most victims were on Weyerhaeuser land north and west of the mountain.

On Sept. 23, McCutcheon dismissed the state as a defendant, but the case against Weyerhaeuser will proceed.

Attorneys and McCutcheon say the trial will last from three to eight weeks and as many as 40 witnesses might be called. Former Gov. Dixy Lee Ray and George Weyerhaeuser, president of the timber company, probably will testify, said Ron Franklin, the Houston, Texas, attorney for the plaintiffs.

"It's my responsibility to bring to 12 jurors the events surrounding one of the most significant occurrences in the Northwest in recent times. It's going to take a lot of evidence," Franklin said.

If court documents are any indication, Weyerhaeuser's attorneys will call much attention to the deposition of U.S Geological Survey geologist Dwight Crandell. A deposition is a pre-trial interview in which attorneys interview potential witnesses.

Crandell advised public officials about the volcano before it erupted catastrophically on May 18, 1980.

Crandell, according to a transcript of his deposition, said, "It was my belief that the likelihood of a lateral blast was very low until a [lava] dome" appeared.

A lava dome appeared at the mountain after the May 18 eruption, not before. For that reason, Crandell said, he did not include a lateral blast on a pre-May 18 map outlining potential hazards for public officials.

A lateral blast from the volcano—a sideways explosion of hot rock, steam, ash and gas—is what killed most, if not all, of the 57 victims of the eruption.

In court papers, Weyerhaeuser attorneys have also pointed out that public officials who participated in hazard meetings claim they never heard the USGS speak of a potential lateral blast.

Attorneys for the victims' families have highlighted the deposition of Dick Nesbit, Weyerhaeuser Camp Baker superintendent, who said company officials guaranteed an advance warning of an eruption. Attorneys point out that scientists had said an eruption could occur without any warning.

The jet swooped down out of the sky. It had been a short ride for Jim and Helen Scymanky from the Portland Airport. Attorney Ron Franklin's secretary, a vivacious woman with a Texas drawl, picked them up at Sea-Tac Airport. They checked into a hotel in Seattle's Pioneer Square district. That evening they were chauffeured to the skyscraper offices of Graham & Dunn.

In a conference room high above the city, they met the other nine plaintiffs, including relatives of the members of Scymanky's work crew who had perished. When the trial judge dropped the State of Washington from the suit and left Weyerhaeuser as the sole defendant, the action had eliminated others vested in the case. Only Weyerhaeuser employees, their relatives, and contract loggers working for Weyerhaeuser were allowed to proceed as plaintiffs. A team of lawyers was present, but Franklin did most of the talking.

The Scymankys had never met any of the other plaintiffs. As Franklin opened the discussion for comments, they paid close attention to other plaintiffs' attitudes in asking questions or making statements. Some people were adamant about what they wanted to see brought out at trial. Many of them emphasized they were not as interested in the money as they were in establishing the innocence of their loved ones in meeting death on the mountain.

The Scymankys were enthralled with it all. Jim didn't ask any questions. He thought he had a good handle on things. The next day at the hotel, Jim Scymanky met Ralph Killian. Scymanky listened attentively to Killian's story of how he had spent much of the past five years digging in the volcanic wasteland for his son's body.

---※---

"What was Weyerhaeuser's response?" Ron Franklin asked the jury in his opening statement. He was referring to company reaction in the spring of 1980 to the danger that Mount St. Helens could erupt. He leaned his lanky frame forward and stared into the eyes of the eight women and four men as he answered his own question. "It was business as usual," he said. "Weyerhaeuser did not vary the work location of one living person as a result of the volcano's activity."

They would hear testimony, Franklin said, showing Weyerhaeuser officials were more interested in maintaining logging schedules than confronting the dangers of an active volcano. All morning Franklin outlined for the jury the course his case would take in trying to show that the world's largest wood products company was to blame in the deaths and injuries of its own workers.

"The eruption was an act of God," Franklin said. "But the deaths were an act of man, when man ignored weeks and weeks of God's warnings."

Weyerhaeuser had ignored pre-1980 USGS warnings of an eruption and even the briefings that agency provided after the volcano sprang back to life, the attorney said. Franklin said he would prove Weyerhaeuser had misled employees in downplaying the dangers of working and camping on company lands. He said Weyerhaeuser had "guaranteed" balking workers that there would be advance warning of any eruption. "We will prove to you that the concept of a guaranteed warning is a joke," he said.

Franklin emphasized to the jurors the historical gravity of the case. They would be asked to answer questions never before considered by a jury, he said. "What's the value of timber compared with human life? What is the obligation of a corporation to describe to workers what the true hazards were? What is the role of science—its capabilities and limits?"

"The twelve of you are the conscience of this community and the conscience of this state," he said.

In the afternoon Weyerhaeuser attorney Mark Clark told the jury that the company should not be held responsible for events no one could foresee. Even the USGS scientist Don Swanson had no inkling of what was to happen, Clark said. Swanson had been scheduled to change places with scientist David Johnston when the mountain erupted and took Johnston's life. "Dr. Swanson did not believe there was a life-threatening danger on that ridge," Clark said—adding that Swanson would testify to that fact.

Clark maintained that evidence would show the only apparent danger from the bulge that scientists considered at that point was from a possible mudslide or landslide. Weyerhaeuser would also be calling several other scientists as witnesses. USGS volcanologist Dwight Crandell would testify as to why he didn't include a possible lateral blast on the volcanic hazards map he prepared for officials.

The question, Clark said, is simply, did Weyerhaeuser officials respond to the threat as reasonable men? He would convince the jury that the answer should be yes, he said.

The next day, a Saturday, the jury would fly around the mountain to gain a better grasp of locations they would be hearing about in the coming weeks. The trial would resume on Monday, but Stepankowsky, who sat through the opening statements, wouldn't be there. The *Daily News* couldn't afford to staff the trial on a regular basis. The reporter would have to pick and choose what days would produce the best copy. Few other media organizations chose to staff the trial. A cult leader and his following, based on a ranch in central Oregon, were now at center stage. The Bhagwan, leader of the cult, and related death plots now dominated the Northwest's airwaves and news pages. But Donna Parker, who also sat through the opening statements, would be there on Monday.

CHAPTER 22

1985: In the Courtroom

WHEN JUDGE JAMES McCutcheon resumed trial on Monday he cautioned jurors that although their "delightful" ride above St. Helens over the weekend had been in two Weyerhaeuser helicopters, they should draw no conclusions from that fact. Both sides had paid equally for the trip. Nothing the jurors had seen could be considered as evidence. The trip was simply a way to familiarize them with landmarks that would be mentioned in the trial.

What was being called the jurors' "volcano education" continued with the calling by lawyer Ron Franklin of Stephen L. Harris, author of *Fire and Ice: The Cascade Volcanoes*. Harris explained volcanic terms such as bulge, lateral blast, and pyroclastic flows. Other scientists and experts followed.

By the end of the day, Donna Parker had heard enough. At this pace it would take forever to get to the heart of the trial. She thought all this scientific testimony could only serve to bore and confuse the jury, as it had done to her. She might as well go back to work, she told herself. Parker decided she would follow the trial as best she could through Jeanette Killian, who had taken up the task of sitting alongside Franklin each day in court as the representative plaintiff.

The logic of some of Judge McCutcheon's rulings perplexed the plaintiffs. Franklin was allowed to use the testimony of a volcanologist but was not permitted to tell the jury that the man was part of a Weyerhaeuser hazards assessment team prior to the May 18 eruption. With testimony of the man, Clive Kienle, Franklin hoped to establish that Weyerhaeuser erred when it "guaranteed" its employees advance notice of an eruption.

Kienle said he agreed with a 1978 USGS assessment that mudflows, pyroclastic flows of superheated rock and gas, and lateral blasts were all potential hazards in any St. Helens eruption. Almost all the dead were the result of the volcano's lateral blast. "Did the USGS ever advise anyone in your presence [before May 18] to discount anything in the [1978] assessment?" Franklin asked the scientist.

"No, not that I am aware of," Kienle said.

The private volcanologist also said the USGS never guaranteed prior to May 18 that it could provide advance notice of a major eruption. "But the feeling was general that any major event would have at least a couple of hours of warning."

University of Washington seismologist Steve Malone testified that no guarantee of a warning would have been possible. But then Weyerhaeuser attorney Mark Clark sought to turn Malone's testimony to his side's advantage by emphasizing that scientists saw no clues of the big blast to come.

"In hindsight, was there anything in the week of May 12 to 18 that would have enabled you to foresee May 18?" Clark asked.

"No," Malone said.

Don Swanson was taken aback when he heard that Clark had said the scientist would testify in support of Weyerhaeuser's position. Clark had said that Swanson saw no life-threatening danger on the mountain on May 18, or else he wouldn't have been getting ready that eruption morning to replace David Johnston. If that was what the Weyerhaeuser attorney believed the scientist had meant, then he was mistaken, Swanson thought. There was always a chance for danger on a volcano, and both he and Johnston had accepted that as part of the job.

Swanson had been asked by the plaintiffs to testify as a "neutral" expert witness. But in giving his deposition before trial, he had made comments relating to dangers on the mountain that defense attorneys wanted to use. As he now sat down in the witness chair to testify in the plaintiffs' case, he had another worry. When he gave his deposition, it had come out that he had returned to observation of the volcano only two days before it erupted. For the previous three weeks he had been doing scheduled contract work in Eastern Washington. Swanson had asked Franklin to make sure and bring this out in his testimony before Clark had a chance to ask the question.

When Franklin finished questioning him without asking the question, Swanson was flabbergasted.

The Weyerhaeuser attorney, Clark, lost little time in homing in on his key questions for Swanson. Did Swanson think there was an "impending imminent threat to life" on the morning of May 18?

"No more so than there had been several days before," Swanson answered. But "there was some finite threat."

The courtroom was silent for a moment. Clark picked up a file that was Swanson's deposition. Didn't Swanson answer the same question during his deposition, Clark asked, by answering "No, I did not because I had agreed to go out on May 18."

"I stand by both statements," Swanson replied. "I think the discussion revolved around what was meant by 'imminent.'"

Clark grilled Swanson on into the next day. Then he asked his last question. Where had Swanson been the three weeks before the eruption? The question was asked as if the scientist had been trying to hide something all along.

When Swanson replied that he had been working elsewhere in the state and had returned to St. Helens only two days before the eruption, a gasp went up from the jury. It was like everything he had explained in his testimony about the dangers of the volcano had lost its credibility. He still couldn't fathom why Franklin had not brought out the fact of his absence before the eruption. It had to have damaged the plaintiffs' case.

The most damning evidence to date in demonstrating negligence on the part of Weyerhaeuser and the state came in the deposition of USGS hazards expert Dwight "Rocky" Crandell, read at the trial. He had no direct hand, he said, in drawing the restricted area boundaries announced by the state. In fact it was as if officials had ignored the hazards map he had prepared for use by the USGS, the state, Weyerhaeuser, the Forest Service, and others.

The contorted Red Zone boundary on the northwest side came within a few miles of the volcano's peak and followed Weyerhaeuser property lines, leaving the timber giant's lands completely outside the restricted area. Yet this was the very area Crandell's hazards map had shown to be the most dangerous. And it was the area where almost all the victims of the May 18 blast met their deaths.

Crandell had plotted the dangers of flows of superheated rock and gas, as well as mudflows and flooding, on his hazards map. He said he used the record of mudflows and other hazards from previous eruptions and then "added 25 percent" to get his estimates. Using this historical record, by far the area that posed the greatest risk was the area the state had chosen to give its least protection, the Weyerhaeuser lands.

In several meetings with state and Forest Service officials before the restricted zone boundaries were drawn, Crandell had stated that a large avalanche off the unstable north slope could "uncork" a volcanic eruption. A week before the boundaries were announced, Crandell put his fears into writing. In a letter to the supervisor of Gifford Pinchot National

Forest, Crandell officially supplemented his hazards map by warning that part of a growing bulge on the north face could break loose in a massive avalanche. That avalanche could not only result in an eruption, but could also cascade into Spirit Lake and generate huge floods and mudflows.

The trial, entering its third week, shifted focus from the scientists to Weyerhaeuser employees. Weyerhaeuser had sent crews to within two miles of the volcano just days before the giant eruption, according to testimony of the company's woods planning coordinator for southwest Washington. Walt Metzer said Weyerhaeuser could have met its long-range harvest projections even if it had pulled its crews back from the mountain. Instead, the company sent crews even closer as snow melted at higher elevations.

Donald Vernon, a Weyerhaeuser employee for thirty-four years, told of members of his logging crew asking supervisors about volcano dangers. "Generally, they'd say 'Hey, you're going to get twenty-four hours' advance notice. Quit your griping and go to work.'"

Vernon's work site had been near Coldwater Ridge, where scientist David Johnston lost his life. Vernon was not at work the Sunday the mountain exploded. But two men on his crew died in the blast.

Two days before the massive eruption, a Weyerhaeuser land use supervisor ordered warning signs that read "Hazardous Operating Area—Keep Out—Trespassers Will Be Prosecuted." Ross Graham testified that he tried to word the signs carefully because the company was worried they might prompt loggers to demand hazard pay.

Franklin grilled Lora Murphy, a coordinator for the state Department of Emergency Services, about why there was a proposal to expand the state's restricted zone around the volcano. "Isn't it true one reason for the proposed expansion . . . was the bulge on the volcano's north face?" Franklin asked. "That there was a heightened sense of urgency at DES because of these conditions?"

"There was an increased need to protect people . . . [to] make sure they didn't get into the area," Murphy replied. The proposal had languished on the governor's desk through the weekend of the May 18 eruption.

Weyerhaeuser officials had made it clear they wanted to continue logging, Murphy testified. "There was a need expressed by the entire timber industry to protect people and [to] log," she said.

The story poured forth from Jim Scymanky as if it had been bottled up for years. "I heard someone yelling and coming through the brush," he testified, telling how Jose Dias ran down the hill as St. Helens erupted. "He was yelling in Spanish that the volcano was exploding. He didn't even have any boots on and he looked like he was scared to death."

Scymanky recalled how he was momentarily puzzled. Then the force of the eruption hit him and the other three men in his logging crew. "It was like a freight train coming through the woods. There were rocks shooting through the trees. Then it turned black."

He started to tell the jury about being knocked around by the blast, when Judge McCutcheon cut him short. The judge said Scymanky's testimony was encroaching on an agreement to keep a lid on emotional appeals and outbursts that might unfairly sway jurors against Weyerhaeuser. It was another of McCutcheon's hard-to-fathom rulings against the plaintiffs that seemed timed to coincide with some of their strongest testimony. Scymanky's matter-of-fact descriptions of what happened to him had been presented with little emotion.

This was Franklin's last day in putting on the plaintiffs' case. He called Jack Schoening, the company's woods manager for the region. Schoening said he was the official who determined where Weyerhaeuser's employees and contract loggers worked around the mountain each day. On May 7, eleven days before the cataclysmic eruption, the USGS reported new cracks on the mountain's northwest side that could endanger people on Weyerhaeuser land. Schoening said he flew over the peak in a helicopter but could not see any cracks. When the chopper landed to refuel he discovered that local supervisors had ordered the evacuation of "two or three crews." He sent these men back to work.

In resting his case, Franklin stunned the courtroom. He would not be calling Weyerhaeuser President George Weyerhaeuser or former Governor Dixy Lee Ray as witnesses. For months the media had whipped up an anticipation of hearing these two key players in the Mount St. Helens tragedy tell all they knew, under oath in a court of law.

Mark Clark wasted no time in getting to the heart of the Weyerhaeuser defense: that no one could have known the eruption would be as deadly as it was. The defense called Dwight Crandell, the USGS hazards expert who had been so damaging to the Weyerhaeuser case with his deposition testimony.

Clark referred to an article co-authored by Crandell that stated of the May 18 eruption: "The lateral blast extended about three times further . . .

than the largest known previous blast at Mount St. Helens and devastated an area that is probably 10-15 times larger."

"Did you foresee the magnitude of the May 18 eruption," the attorney asked the scientist.

"I did not," Crandell answered.

However, under cross-examination from Franklin, the scientist said his hazards map and warning did anticipate the mudflows, ash flows, and avalanches of superheated rock and gas that devastated Weyerhaeuser land and probably contributed to many deaths. He also said that while he may not have anticipated a lateral blast, he thought an equally deadly vertical eruption had been a "distinct possibility."

"How would the destruction from such a [vertical] event have compared with May 18?" Franklin asked.

"It would have been as destructive or more destructive," Crandell replied. He said St. Helens had experienced at least six massive vertical eruptions in the past 4,500 years of the same type that had destroyed Pompeii. The chances for another one, he thought before the May 18 eruption, were great.

Clark scrambled to bring on another scientist to refute Crandell's testimony. The next day, Cliff Hopson, a University of California–Santa Barbara geology professor, testified for Weyerhaeuser.

Hopson testified that the full force of a vertical eruption is directed up in the air—unlike the May 18 lateral explosion that directed its fury against the earth. While the eruption of Mount Vesuvius did annihilate Pompeii in AD 79, Hopson said, residents had eighteen hours to flee before superheated dust and rock hit the town. He testified that if a vertical eruption matching the largest in St. Helens history had occurred May 18, the chances of life-threatening danger in the area where the victims died would have been "extremely remote."

"There was no concept in my mind that what did happen could happen," testified Weyerhaeuser's John Wilkinson. He was manager of the company's southwest Washington region at the time of the eruption. "Had I had any such inkling . . . we would have done a lot of things differently."

Wilkinson said he had believed that scientists could give "a few days to at least two hours warning." He was now aware, however, that no warning was given and that testimony had revealed that no scientist had ever guaranteed one.

It had now been over a month since the trial started. In its fifth week, Clark called his final witness, Jim Rombach, who had helped develop Weyerhaeuser's contingency plans for the volcano. He said Weyerhaeuser relied on reports from USGS geologists to determine whether work crews should be moved from job sites in the two months before May 18. He testified that the worst damage was expected to result from mudflows and flooding in the valley floor. As a consequence, Weyerhaeuser ordered its workers to keep to higher ground. Major events foreseen by scientists, Rombach said, did not include the massive eruption that occurred.

Government agencies in the region began meeting shortly after March 20, 1980, when the volcano first stirred, Rombach testified. Weyerhaeuser officials attended and left phone numbers with the Forest Service and other agencies. "We wanted to be involved in that information and notification process," he said.

However, under Franklin's cross-examination he admitted the company had no procedure to receive USGS updates. He did not recall seeing the update of May 5 that reported increased activity linked to the bulge on the volcano.

Franklin walked to a map and pointed to to Weyerhaeuser work sites close to the mountain. "Tell me why it was the Weyerhaeuser company had to work in this area," Franklin said.

"We were working in that area against [based on] the USGS assessment," Rombach said. "We were hearing mudflow. We were hearing avalanche."

"Were you hearing the mountain might blow its top?" Franklin asked.

That was considered a possibility, Rombach testified. But warnings indicated major dangers wouldn't extend far from St. Helens. "If we had had an assessment from the USGS not to work in those areas, I'm convinced we wouldn't have worked in those areas."

As Andre Stepankowsky worked at the *Daily News* office, preparing to cover final arguments in the St. Helens trial the next day, volcano news of a far more tragic nature was being made four thousand miles away, in Colombia. The Nevado del Ruiz volcano began shaking and spouting ash. Searing ash and pumice melted the glacier on the mountain's eastern slope. Soon billions of gallons of heated water cascaded down the side of the volcano, racing to speeds of near one hundred miles an hour, ripping up the soil to bedrock and shearing off slabs of ice.

The roaring water picked up entire forests and huge rocks on its journey. Mudflows joined the headwaters of the Azufrado River, already swollen by a steady rain. The churning mass dumped into the Lagunillas River and grew again as it slid down the three vertical miles to the sleeping farm community of Armero.

In the canyon above the town, the waters and mud slowed and swelled behind debris to a height of over 130 feet. Then the water burst through with a thundering roar, sweeping away more than four thousand buildings. In minutes, twenty-one thousand people lost their lives, making this the worst volcanic disaster in over one hundred years.

News of the disaster devastated Swanson. It was like none of the lessons of St. Helens had been learned. The techniques and technology spawned by the Washington eruption had not been used on Nevado del Ruiz.

In the months before the November 13 eruption, as Nevado del Ruiz was stirring, Colombian officials had asked the USGS for help. The USGS replied that with eruptions to deal with in Hawaii and Washington, no one could be spared. A little more than a month before the eruption, the USGS sent an administrator who was also a scientist from its headquarters in Reston, Virginia. The administrator stayed for a week, determined the volcano was little threat to the city of Manizales on the other side of the volcano from Armero, and left.

The only problem was that people in cities all around the volcano thought the assurances applied to their city also. By the time the administrator left, very little hazard mapping or seismic monitoring had been undertaken. There was no radio-linked seismograph on the mountain. Data was collected by hand and then mailed to Bogota for analysis. The USGS later sent six telemetered seismographs and a seismologist to the volcano. But when guerrillas seized the presidential palace in Bogota, the mission was scrapped.

Finally, a little more than a week before the eruption, a preliminary hazards map was issued by Colombia's Bureau of Geology and Mines. The map drew the wrath of business groups concerned that fears of an eruption would bring an economic downturn. A national magazine said the map would result in lower real estate values. A local archbishop labeled the map "volcanic terrorism." National officials in Bogota called the hazards map needlessly alarming. The map and assessment said there was a 100 percent certainty in the event of an eruption that mudflows would reach Armero, but that there would be a two-hour warning for evacuation.

The politics and the confusing science were all too reminiscent of St. Helens, Swanson thought. What was needed was what he and other top scientists had pleaded for since the St. Helens eruption: a rapid response team of USGS scientists who could hop around the globe at short notice to help at volcanic hot spots. Instead of shelling out millions of dollars in relief after a disaster, the United States might be able to take lifesaving steps beforehand.

All that had been learned at St. Helens was useless, Swanson knew, unless it was carried forward by knowledgeable scientists in assessing new situations. There was a glaring example of this fact in regard to St. Helens and Nevado del Ruiz.

By the standards of St. Helens, the 1985 Nevado del Ruiz eruption was puny. It produced only one-sixth of the volcanic matter that St. Helens released in its May 18 eruption, but had horrific results in loss of human life. A 1982 incident on St. Helens had already demonstrated how a very small eruption might result in great flooding.

On March 12 of that year, a small spray of pumice, hot gas, and steam burst from St. Helens. It was a minor event except that it occurred in winter and melted most of the snow in the crater. Within minutes a lake formed. Warm water gushed out of the breach in the northwest crater wall and raced down the slope, cutting a ninety-foot-deep ravine. The water increased in speed and mass, picking up debris from past eruptions until it was a gigantic mudflow. Twenty miles from the crater, the flow breached a retention dam and sent the Toutle River surging within a few yards of its height during the mudflows from the May 18 eruption.

USGS scientists published a paper that showed scientifically how such a small eruption could result in catastrophic floods under the right conditions. The paper could have been a blueprint for what occurred at Nevado del Ruiz. But no one with the proper knowledge and training had been sent to the Colombian volcano.

Stepankowsky sat transfixed, busily scrawling notes as he listened to Ron Franklin's impassioned closing arguments. "There was no reason whatsoever to expose these people to an active volcano," Franklin told the jury. Weyerhaeuser put logging quotas ahead of people, haste ahead of prudence, he said. Scientists had testified that a life-threatening explosive eruption was a distinct possibility, Franklin said. Enough information and warning were available that a prudent person would have taken more precautions.

Weyerhaeuser, he said, did nothing in response to the volcano. Logging plans went unchanged. The company failed to keep posted on the latest reports. "I couldn't believe it. Is that too much to ask, to read what scientists were saying?" Franklin asked.

The attorney cited testimony by Jack Schoening, Weyerhaeuser's woods manager, that he ordered loggers back to work after the USGS expressed concern that they were too close to the volcano. "Jack Schoening flew around the mountain and said, 'Get back to work.' He didn't call the USGS, the Forest Service, or anyone. That's arrogance."

Jurors who had scooted forward in their seats to hear Franklin's often emotional closing now sat back with chins cupped in their hands as they listened to Mark Clark present his summation in soft, measured tones.

"In hindsight," Clark said in his closing arguments for the defense, "things would have been handled differently." But, he said, decisions were based on what was known at the time, not on what was unknown. Clark said that scientists believed the most serious danger was from pyroclastic flows and mudflows that would be confined to valleys. Victims who died were on ridges, he said.

In all of known history, there had never been a lateral blast that devastated so much area at St. Helens. With the May 18 blast, scientists had learned a volcano can erupt more powerfully than at any time in its past. "What happened was unprecedented," the attorney said. Negligence could not be gauged against a "standard of perfection," Clark continued. "It must be measured against reasonableness."

When Scymanky heard the judge's instructions to the jury, he lost all hope. Jurors would have to navigate through twenty-eight separate jury instructions regarding issues in the case. Some instructions covered more than a page each and contained a multitude of subsections. For each of the twenty-eight issues, jurors would have to vote against Weyerhaeuser on every subsection for the plaintiffs to prevail. But the timber company would be off the hook on an issue if jurors found for the company on only a single subsection.

Because this was a civil suit, not a criminal trial, a unanimous verdict was not required. With a twelve-person jury, ten votes for or against was all that was needed for a finding on any issue.

"I'm glad its over," Jeanette Killian told a reporter. She was the only family member among the plaintiffs to sit through every day of trial. "I'm going home to rest. No matter what happens, we had a chance to tell our story. I don't care about the money. If we get some, I'll give it to my grandson who lost his father."

CHAPTER 23

1985–1986

EIGHTEEN DAYS after beginning deliberations, the St. Helens jury reported its decisions. Of the twenty-eight issues the judge handed to the jury, it deadlocked on the majority of claims against Weyerhaeuser.

The jury dismissed claims of two of the victims' relatives on 10–2 votes. The jury ruled that Tom Gadwa was working for a Weyerhaeuser subcontractor when he was killed but otherwise was not employed by the company and should have had access on his own to information on volcano dangers. And the jury concluded that John Killian, while relying on Weyerhaeuser information that the volcano was not a threat, was involved in recreational camping with his wife when they were killed. Under Washington state law any large landowner who opened areas for public recreation could not be held liable for injury or death.

The jury deadlocked, with votes of 9–3 or 8–4 in Weyerhaeuser's favor, on a variety of issues, including assertions of negligent misrepresentation, breach of duty to provide a safe workplace, and breach of duty to warn of the volcano. No issues were resolved in favor of the plaintiffs.

"We got real close to a victory," Weyerhaeuser attorney Mark Clark said. "The biggest thing we had to be concerned with wasn't the facts as such, but the emotion and sympathy by having eight people in there who had lost their husbands, or sons, or daughters."

Juror Debra Trotto, who consistently voted for the plaintiffs, said, "There wasn't any sympathy factor. I didn't feel it at all." She said she believed Weyerhaeuser had been negligent by not moving its operations away from the mountain after having been informed how dangerous it was. But other jurors thought that Weyerhaeuser's contingency plans were little different from those of other companies facing the threat of an eruption. Kathleen Reed, another juror who sided with the victims and relatives, said she thought that jurors who favored Weyerhaeuser believed the plaintiffs were only out for the company's money.

For days the Longview area had been gripped by cold. Streets were slick with ice. The sponsors of Stepankowsky's piano recital asked if he would like to postpone it. He explained that his father was flying cross-country and was already on his way. On the day of the event, the weather warmed and townspeople packed the McClelland Arts Center. Tickets were priced at five dollars, with proceeds to a local organization that helped the poor.

The selections he would be playing, Stepankowsky explained on stage, were reflections of his Polish and Russian roots. First came *The Seasons* by Tchaikovsky, a collection of pieces, one for each month, that is "very, very Russian." As he played the section that evoked the month of June, he thought of his father in the audience. Memories of hot summer days, with the sound of a baseball game wafting through the open windows as he practiced piano, flooded back to Stepankowsky. Later came a flashy piece written by Armenian-born Aram Khachaturian when he was a student in Moscow. Then a quieter, melodic composition by Poland's Frédéric Chopin.

Stepankowsky saved his showiest piece for last—Chopin's Ballade in G minor. As he came to the most difficult part, his fingers flew like they had a life of their own. In an interview with the *Daily News*, Stepankowsky had likened the mounting intensity of this music to the great bulge that had built on the volcano's north side before the eruption. "The piece builds and builds and builds and finally explodes. It overpowers me. It's almost too much to control emotionally."

Now the perspiration dripped down his face and into his eyes, but he did not notice. When he finished playing, the crowd rose to its feet and would not stop clapping. He squinted into the blinding stage lights at the audience beyond. He could not see his father, though he knew he was there.

Venus had known the man now sitting on the couch with her since they had been in junior high school together. He had just returned from five years in the Navy. Venus had practically grown up with him and he had run in the same crowd with her and Roald.

As they talked she gradually brought her scarred hand out from hiding under her leg without even thinking about it. They talked about how they wanted to settle down, how they each wanted kids, and then they talked about getting married to each other. She knew he had been involved with drugs and alcohol in high school, but so had a lot of her friends. Now he seemed so much more mature than she had remembered him being. He knew what he wanted out of life and she liked that.

Venus had heard that Roald had gotten married not too long ago. She told herself that maybe it was all for the best.

Judge James McCutcheon refused to dismiss the remaining claims against Weyerhaeuser, the ones the jury deadlocked on. After reviewing the dismissal motion by the company's attorneys, he said, "That's a jury question, not for me."

A reporter called Jim Scymanky with the news. "I think it's great," Scymanky told him. "I just want to go up against Weyerhaeuser again. I know they're in the wrong." No date was set, but there was talk by Franklin of a second trial in the fall of next year, 1986.

The blue and gray Hughes 500 helicopter rose up into the night sky from Harrys Ridge in front of the volcano. Flakes of snow showered down. The pilot radioed the airport in Vancouver that he was returning to base. On board besides the pilot was a maintenance man and a photographer named Ralph Perry. Perry had been assigned by *National Geographic* to use infrared photography for a nighttime shot of the laser beams Swanson now employed to measure dome deformation.

Since the May 18 eruption, Perry, a thirty-two-year-old freelancer, had been to the mountain dozens of times. He traveled the world for such media giants as *Time, Newsweek, People,* and the Black Star picture agency. But the mountain always lured him back. He had lost one of his best friends, photographer Reid Blackburn, when Blackburn died in the lethal eruption nearly six years before. Even with Blackburn's death, however, Perry never seemed worried about the dangers of the mountain, until this flight. His wife had never seen him so nervous about a trip. It just didn't feel right, he told her.

Now the helicopter banked and flew across the snow-dusted pumice plain in front of the volcano. But for some reason it began to lose altitude. It crumpled to the ground, its rotors spewing ash and snow into the air. The emergency locator beacon on the chopper went dead.

The next morning searchers in planes and helicopters scoured the area between the volcano's crater and the airport in Vancouver. Snow, fog, and low clouds hampered the search. The three missing men, however, were well equipped with clothing and supplies and there was hope they would be found alive.

By the following day the weather had improved somewhat. A television news helicopter spotted the wreckage that afternoon. Inside were the bodies of the men. The volcano, where weather could change at a moment's notice and wind shear seemed to come from nowhere, had claimed three more lives. Since the big eruption thirteen people had died in aviation accidents on the mountain and an additional twenty-three had been injured. They included photographers, tourists, and even some of Don Swanson's fellow scientists.

It had rained hard for more than a day before the Sunday concert by the Southwest Washington Symphony. So Stepankowsky was not surprised to find the Columbia Theater in Longview less than half full when he arrived with Paula to review the performance for the *Daily News*. Stepankowsky tried to concentrate on the orchestra and guest pianist Joel Salsman. But while Stepankowsky had the ear of a musician, he also had a reporter's ear for the rain and what it meant. He worried about the ceaseless roar of the rain outside.

As soon as the concert ended, Stepankowsky and Paula were out the door and on their way to the office for what was developing into a major story. It was decided that Stepankowsky would coordinate flood coverage from the office, directing other reporters, including Paula, in the field. Stepankowsky knew a major flood was on its way. The only question was how damaging it would be.

As the night wore on and the rain kept falling, the area's major rivers—the Cowlitz, the Toutle, the Kalama, even the mighty Columbia—crested and overran their banks. Sirens wailed and television shows were interrupted as authorities ordered hundreds of homes evacuated. Neighbors joined sandbag brigades or frantically dug ditches around each others' homes, trying to protect themselves against the advancing waters.

While there were victories, for the most part people fought a losing battle. In some houses residents looked on helplessly as the water rose slowly through heating ducts. In other homes the flood came crashing through garage or front doors. Mudslides closed roads and slammed into houses. In some areas residents took to boats, stopping to pick up neighbors as they rowed. A Red Cross evacuation center opened, but many people on high ground welcomed friends, family, and even strangers into their homes.

Stepankowsky could hardly believe what he was seeing after Roger Werth developed photos he and other photographers had taken. A man with water up to his shoulders pulled a raft that carried his wife and the few belongings they had managed to rescue. Buildings lay half crumpled by mudflows. Rows of abandoned cars waited in rising floodwaters. Men waded through water with people on their backs.

The most telling of all was Werth's photo of the local bowling alley. In the foreground was the very top of a submerged car. Then a great expanse of water rose toward the "Bowl" sign that adorned the top of the huge multistory building. The photograph had been taken at noon. Now it was late at night. The waters had advanced much farther, Stepankowsky was told.

The rain stopped the next morning. Stepankowsky checked the rainfall figures. In a forty-eight-hour period ending at seven that morning, 6.22 inches of rain had fallen, the most ever recorded in the area for such a period. The rainfall on Sunday, 3.8 inches, was the second-highest single-day total on record.

Since the eruption of the volcano in 1980, Stepankowsky had been recording river depths and flows, estimating silt amounts, and checking government figures and long-range weather predictions. The reporter was still haunted by the vast area of ash that he had seen from the air after the eruption, ash that eroded away to clog the rivers. For six years his figures had indicated that a heavy rainfall would result in massive flooding, despite the dredging done.

Over those years there had been some flooding, but not on the scale he knew was possible. The region had been blessed with some of the driest winters on record. Key officials continued to dismiss the silt threat and flooding at the magnitude suggested by Stepankowsky's reporting. Local representatives had repeatedly taken his articles to Washington, D.C., to lobby for flood-control dollars. But in the nation's capital many agency heads and members of the Reagan administration continued to dismiss the flood threat as "exaggeration."

Now Stepankowsky called sources to find the peak height of river flows during the new flood. What he found indicated the erosion of Mount St. Helens had contributed to the magnitude of the flood. The Cowlitz River had shown the effect of silt in its channel. Its peak at Castle Rock during the flood was as high as that recorded during the last great flood in 1977, though the volume of water flowing through the channel was 25 per cent less than in the 1977 storm.

This time the Federal Emergency Management Agency launched its own investigation. Its team of scientists and experts culled from twenty-three agencies agreed with Stepankowsky. Volcanic sediment deposits had raised the bed of the lower Cowlitz River by as much as seven feet, aggravating flood damage. The filling of the riverbed had raised water tables in the lower Cowlitz Valley by three to four feet, backing up drainage, submerging sewer lines, and causing more than twenty giant sinkholes in the Longview-Kelso area alone. Only two such sinkholes had developed in the eight years before the flood.

The storm "suggests a pattern of relationships that relate back to the Mount St. Helens eruption," the study said. Damage from the flood would run into many millions of dollars to the still depressed area.

The attorney for the families, Ron Franklin, was trying to get them to accept a Weyerhaeuser offer to settle the case. His office had contacted each of his clients in the case individually, outlining the risks of being on the hook for the defendants' legal costs if they didn't settle. When Donna Parker heard about it from Jeanette Killian, she suggested that the plaintiffs meet together to make the decision.

Though she wasn't a plaintiff, Parker attended the meeting at the Bowers family home not far from the volcano. She led the argument against settling and in favor of holding out for a new trial. By the time she was done, her audience, which included most of the plaintiffs, was fired with emotion. They all signed a letter telling their attorney and Weyerhaeuser basically, "Hell, no."

At 259 days the volcano had just set a record for quiet time between eruptions. The previous record was 258 days between September 14, 1984, and May 30, 1985. Scientists did not place any significance on these back-to-back periods of quiet. They did not believe the mountain was quite ready to return to another centuries-long dormancy.

Still, St. Helens' lingering silence was one of the factors reflected in USGS staffing. Tight budgets, fewer eruptions, and the siren call of other volcanoes had drawn many scientists away. More than ten full-time USGS scientists had once worked on the mountain, but now there were usually only two. Swanson now had other projects in addition to his work with St. Helens.

Many of the St. Helens scientists now served overseas in a program the USGS and the State Department had developed to help volcanologists in New Zealand, Indonesia, Colombia, Italy, and the Philippines. Swanson's hope for sharing knowledge from St. Helens had become a reality after the Colombian volcano disaster that claimed twenty-one thousand lives.

Stepankowsky's wife told him about it. As business reporter for the newspaper, Paula routinely wrote about the sale of local businesses. But she was surprised when she was told to write a story about the sale of the *Daily News.*

A large chain, by making an offer, had put the sale of the *Daily News* into play along with the rest of the media operation that family interests had built, including three daily newspapers, a weekly newspaper, a state magazine, a printing company, and a local cable concern. The properties were worth over $61 million.

Several family members wanted to sell but publisher Ted Natt, grandson of the founder John McClelland, managed to raise enough money to save the *Daily News* from the transaction that resulted in the sale of the rest of the properties.

"I'm in the phone book. If you didn't get your paper, you can call up the guy at home and give him what for," Natt told Paula Stepankowsky when she asked for a quote. "You don't have the same level of concern with a group ownership."

Swanson rushed to take measurements in the crater, in an area north of the dome. It was much too dangerous to take measurements on the dome itself, which now periodically spewed rock fragments along with ash, gas, and steam. Over the next several weeks the volcano recorded nearly fifty explosions, some sending ash plumes more than twenty-five thousand feet into the sky and raining rocks down as wide as three feet and weighing hundreds of pounds on the area over a mile away. But there had not yet been any sign of molten rock, signaling an eruption.

Weather for the next several days kept Swanson and his team out of the crater. Then earthquake activity beneath the dome increased from moderate to high within a twenty-four-hour period. Tiltmeters transmitted information that indicated swelling of the north flank of the dome. Swanson returned to the crater, but the top half of the eight-hundred-foot-high dome was obscured by steam. However, he got enough information to feel comfortable in joining with other scientists in having the USGS issue a warning that an eruption could occur within the next week

The mountain erupted molten rock from the dome a few days later to create one of its largest lava lobes ever. But no one was there to witness it. Bad weather kept scientists away. When Swanson and others were able to get back into the crater, more than just the new lava lobe greeted them. An avalanche, the largest ever recorded from the dome, had occurred. It was less than a week before the sixth anniversary of the May 18, 1980, blast.

CHAPTER 24

1986

"THE VOLCANO GETS SO HOT it has to explode because the sun is so hot," kindergartner Nathan Richmond told the reporter. "I'm not afraid of it. I was born in the volcano. My dad told me."

Stepankowsky was interviewing a class for a story to appear on the sixth anniversary. Most of the children were born in 1980, the year Mount St. Helens awakened and unleashed its lethal blast. These were children growing up in the shadow of an active volcano.

"My dad says the mountain burps," offered Kate Gray.

"It's because of rocks that get hot and melt when the sun comes up," chimed in Lillie Campbell.

"It's red stuff," said David Hobbs.

"It's that hot stuff that comes out of the mountain when there's too much water in it," said Brian Masten.

"But it can't reach here," assured Brian Ashley.

No one knew what a lava dome was, and only a couple of students knew the volcano had erupted in the past week. But all the children seemed confident in their descriptions of the size of the volcano. "It's a little bigger than a school," said Mirina Carlson. "I've seen it up close. I haven't touched it."

"When did the volcano have its really big eruption?" Stepankowsky asked.

"When the dinosaurs were alive—about a hundred years ago," answered Michael Jorgensen. His classmates offered other estimates varying from ten days before to fifty years.

It struck the reporter as odd when he asked the children if their parents or older siblings ever talked about the volcano and most of the pupils said no. One girl, Erin Ward, knew that her parents' home on the Toutle River had been destroyed by mudflows, though she was only three months old at the time.

Stepankowsky contacted Erin's mother, who told him Erin was beginning to understand how deeply the volcano had wounded her family. "Every time we pass the road where our house was, she wants to see where we used to live," Janice Ward said. "She asks, 'Nothing could get us now, Mom, could it?' I say, 'No, Erin.'"

As her son Jim held the cross, Donna Parker pounded it into the ash-covered ground. It was the day of the anniversary. A cool, crisp wind blew off the mountain and swirled ash around them. She tasted the ash grit in her teeth and felt it irritating her face. Parker paused for a moment to wipe ash from her watering eyes.

This was the last of the fourteen crosses. They were for the missing and presumed dead—the ones, she said, "who never got home." She had permission from their families to place the crosses, a victim's name on each. Near as she could determine, each cross was planted at the spot where the person died. She hoped the crosses would give the surviving friends and family members a place to center their grief and their memories.

After placing the crosses, Parker and her son stopped at a spot near a former log-loading operation. Ever since reading the investigative report in the *Oregonian*, she felt the state had covered up what really happened on the mountain. She remained skeptical that officials had made a full accounting of the dead and missing. Parker and her son brought out their shovels. This site seemed a likely place to search for more clues.

Her shovel hit a piece of cloth. She carefully pushed the ash away from it with her hand and pulled it out. It was a pair of child's play pants. She found a piece of candy in one of the pockets. No children had been recorded lost from this spot. In fact Parker knew that no children had been reported missing or dead from the entire road that went by this site.

They drove up another road that Parker had never been on. She spotted a gully with several ragged and torn tents in it. She and her son looked up to a peak above the gully. It seemed obvious to her that the volcano's blast on this day six years ago had blown the tents off the peak. But what happened to the campers? There had been no reports of missing or dead.

"Did they settle yet?" Otto Sieber asked Jeanette Killian and Donna Parker. After conducting some business in Seattle, the women had decided to visit Sieber at his nearby apartment. They had considered him a friend ever since he approached them five years earlier about a book he said he was writing.

Sieber was visibly shaken when Killian told him that she and the other plaintiffs were still spurning the Weyerhaeuser settlement offer. They had met that morning with their attorney, Ron Franklin, who argued in favor of the settlement. Franklin barred Parker from the meeting.

"Damn," Sieber yelled as he walked away into another room. He paced back and forth across the room, muttering cuss words to himself. Sieber's wife, who had been sitting with them, looked startled.

"Why is he so upset?" Killian asked.

"What does he have to be upset about?" Parker asked as she looked at Sieber's wife. The woman shrugged her shoulders.

Ron Franklin's office made it clear to Scymanky and Franklin's other clients that if they didn't take the Weyerhaeuser offer, there could be negative consequences. The plaintiffs could be responsible for some legal costs involved with the trial for both the state and Weyerhaeuser. And there was a chance the plaintiffs would not win a retrial.

Almost all the plaintiffs caved in and decided to accept Weyerhaeuser's deal, but Scymanky held out. After months of pressure he relented. Helen Scymanky called Franklin. The appeal of the judge's decision that Washington state could not be sued for negligence was still winding its way through the courts. She extracted what she thought was a promise from Franklin that if Jim agreed to the settlement, the attorney would continue to pursue the case against the state to the highest courts in the land, if necessary.

It was a shocking low-ball settlement. It provided between $19,727 and $21,727 for each of the eleven plaintiffs, including Jeanette Killian, and $15,000 for an additional plaintiff, for a total of $240,000 to be paid by Weyerhaeuser. Out of this total, $125,000 would go to their lawyers. Scymanky netted just over $10,000. The estate of Scymanky's friend Leonty Skorohodoff saw $3,333.33 go to his widow and the same amount to trusts set up for his two children.

Robert Rogers watched as one of the other players in the pickup volleyball game ran out of bounds chasing the ball. For no particular reason, Rogers stepped backward, shifting his weight to one leg. The leg gave way and he collapsed to the ground in pain. He tried standing, only to collapse once more. He couldn't believe it. He had climbed mountains, bridges, and buildings and never been injured. And now he had hurt himself while standing around during a volleyball game. It seemed preposterous.

The doctor told him he had torn ligaments. Once they healed, the leg could give out again, the doctor warned—especially if uneven pressure was put on it, such as from climbing a rough, rocky surface. Rogers sensed that his days of climbing St. Helens were over. He had made over forty trips to the crater.

Parker and Jeanette Killian continued to wonder why Otto Sieber seemed so upset when the plaintiffs initially rejected the settlement deal. They contacted attorneys at Graham & Dunn, the law firm that had hired Ron Franklin. The women wanted to know if Sieber had any financial connection with the firm. One of its attorneys, on condition his name stay out of it, indicated this was the case. Parker searched public records and found a deposition that Sieber had given during proceedings involving the Weyerhaeuser liability trial. In the deposition, Sieber testified that "I was retained and still am acting as an investigator for the plaintiffs."

"When were you retained?" Sieber was asked.

"I was retained in 1981," he replied.

Sieber went on to say he had at first established informal verbal agreements with families as to the sharing of information.

"Was Barbara Karr the first plaintiff that you ever spoke to?"

"I believe it was; I can't be absolutely certain," Sieber replied.

"And when was that?"

"I think sometime in mid-February."

Parker then discovered a state attorney general's brief that said, "Sieber's interviews with over 30 people all occurred in the February to April 27, 1981 period." One section stated that Sieber "presented himself as an author writing a book and did not advise those he interviewed that he was an 'investigator' or that he had entered into contingent fee contracts with certain of the families to obtain and share information as a basis for a lawsuit for 9% of the eventual recoveries." Parker also found a claim of $49,510 submitted by Sieber to Graham & Dunn for services rendered, which the law firm had used as part of its request for payment in any settlement. Parker knew Sieber had finished a draft of a book. She always thought of him as being a concerned author and filmmaker and nothing more.

In a June 8, 1981, story in the *Daily News*, Stepankowsky had reported on his interview with Sieber. The article said that based on the information Sieber provided during the interview, Sieber "doesn't stand to make money directly from the case." Sieber indicated that his motives in helping the survivors were altruistic. The article said Sieber told Stepankowsky that he had introduced relatives of the victims "to Gerald Parks, his friend and an attorney for the Seattle law firm that has taken the case." The law firm was Graham & Dunn.

In questioning attorneys and other sources, they found that Graham & Dunn was not a specialist in personal injury claims. Its specialty was corporate law. As one of the largest and most prestigious law firms in the state, its client roster read like a who's who of Washington's powerful and elite. The women asked themselves why such a firm had taken a case that pitted it against one of the state's largest employers and in an area of law that wasn't its specialty. And they wondered why Graham & Dunn had gone to a Texas law firm to find an attorney to work with, when it was well-known that outsiders did not always do well with Northwest juries.

They also dug up information on the presiding judge at the trial, James McCutcheon. A stalwart of the Democratic Party, he had been appointed Superior Court judge along with other judges in a controversial action by Governor Dixy Lee Ray just one month after the May 18, 1980, eruption. A law had been passed requiring election of judges, but the law was somewhat unclear on how it should be implemented. The governor interpreted it to mean she had additional time to keep making appointments. McCutcheon had also been taken out of the usual judge rotation when the St. Helens case was assigned.

It was all okay now, Venus thought. They were all together at the party, she with her new husband and Roald with his wife. The two couples chatted for a while, laughed, and drank. Then she watched Roald take his wife into his arms and dance a slow number. She turned to her husband. It was all okay now.

Frenzen had known the road work was coming. He had been assured that reconstruction to get rid of a hairpin turn would not damage his tree-planting plots on the nearby hillside. Now he stood in the straightened roadway. The road building had taken out the entire hillside, and his plots with it. The plots constituted one of his most important sites and their destruction meant the loss of years of work.

Other scientists had complained to him about experiments destroyed by construction crews. He was now considered to be the unofficial "resident scientist" at the mountain outside of the USGS geologists, and he was expected to pass these complaints on to the Forest Service. It was just one of the many duties that had gravitated to him, including dealing with the press, because he spent so much time on the mountain. He also found himself helping to prepare information about the mountain for public presentations and planned visitor centers.

There were rumors the Forest Service was about to launch a nationwide search to hire an official monument scientist for what was now known as the Mount St. Helens National Volcanic Monument. He was sure the chosen scientist would take over many of his ad hoc duties. Then he would have more time to get his work done.

He looked one more time at where his plots had been. He kept staring, hoping to spot some shred of his work still lingering in the gray dust. But the bulldozers had left nothing.

Parker was driving near the mountain when she encountered the deputy sheriff. They got into a conversation about what had been called International Camp, named for the many foreigners who had camped there just outside the restricted area before the May 18 blast. Over the years Parker had continued to question whether all the deaths at that camp had been accounted for.

"Every night I would go up there and take down every license plate number in case something happened," the deputy told Parker. "I took down a list of license plates there the night before the eruption." The deputy had handed in the list to county officials. This was just what Parker was looking for. She hoped to use the plate numbers to track down people who were at the camp and find out what happened to them.

Earlier in her search Parker had located a man who had taken photographs showing vehicles parked at the site the night of May 17. But not many license plate numbers were discernible. She also found a Weyerhaeuser photographer, Jessie Wilt, who had published a few photographs of the camp. Parker visited Wilt, who told her she had no more photos of the site. Parker felt that the woman was hiding something. She threatened to sue her unless the photographer produced her series of negatives for inspection. The threat got her nowhere.

Now Parker went to the county courthouse to see if the deputy's list of license numbers had been put on file. No one at the courthouse knew anything about it. And she had been so caught up in what the deputy was telling her that she had forgotten to take down his name.

The years since the 1980 eruption had been one of the driest spans of weather for the region in a century, the National Weather Service hydrologist told the audience. The floods earlier that year could be just a taste of what was to come, Lee Krough said.

"Nature has been good to you people since the eruption," he said. "If you get some storms like you had in the seventies, you'll have a considerable flood." Ten feet of silt that wasn't there before the eruption now coated the bottom of the Cowlitz River, he said.

Stepankowsky checked out the sparse audience. They were almost all officials who had gathered for this public education meeting. The meeting was held the same night as a World Series game and Monday Night Football. Only a handful of local residents came. The event was scheduled to kick off Flood Awareness Week.

He often camped out in the open near his plots without a tent. Frenzen threw down his tarp and settled in for some stargazing. Late that night he fell asleep. He woke when he heard raindrops falling on nearby foliage. Before he could open his eyes he felt the drops falling on his face. But the night had been clear and no rain was forecast. Then he realized he was not getting wet.

He opened his eyes and shook his head, as though trying to shake the sleep away. Ash fell from his hair. Frenzen touched his face. It was covered with ash. He had been caught in one of the dozens of ash blasts the volcano had spewed into the sky since its recent eruption.

The eruption added over thirty feet to the height of the lava dome, raising it to 875 feet, slightly higher than the RCA building at Rockefeller Center in New York City. This was the highest dome ever. Don Swanson thought it wouldn't go much higher. The eruption had been the fifteenth he had predicted correctly. All indications suggested that the mountain was about out of both gas and the molten lava that accompanied it.

She hadn't been invited to the grand opening of the Forest Service's first volcano visitor center, but Donna Parker went anyway. No expense was spared in the $5.3 million building designed to tell the St. Helens story. The Silver Lake center was meant to represent the Forest Service as a progressive steward of the volcano and, not incidentally, to lure a good share of the expected trove of tourist dollars. The building was strategically located just five miles off Interstate 5 where the public could use telescopes to view the volcano thirty miles away.

Parker entered the building that was constructed using trees and rocks that had been torn up by the May 18 eruption. Huge glass panels rose from the entrance, giving the building an almost churchlike atmosphere. A thirty-eight-foot-tall mural depicting smoke billowing from Mount St. Helens reached high into the cathedral ceiling. Parker started looking around the center as dignitaries, politicians, and Forest Service officials began their speeches.

Gritty, ash-gray mannequins greeted her as she walked into the exhibit hall. They depicted people involved with the mountain, from early natives to present-day geologists. Hands-on exhibits with twinkling lights, levers, and talking interactive displays gave the hall an almost arcade feel. She strolled past a seismograph depicting the volcano's current activity. The seismograph stylus jerked nervously on the roll of paper but did not rise up and down in the wide swaths that could indicate a coming eruption.

Parker took notes as she went from display to display. She stared at a large photograph of a Red Zone warning sign. She knew that no such sign had ever been posted. Beneath the sign the caption told about the "20-mile Red Zone" designed to keep people away from the danger. She wrote furiously. She entered the darkened theater just in time to catch the Forest Service film on the eruption. She sat dumbfounded as she heard the narrator tell about the victims who recklessly risked their own lives in ignoring authorities and entering the Red Zone.

When she found the director of the center, she could hardly talk. Her body tightened as she clenched her fist. She demanded changes. The director said he needed proof that the warning-sign exhibit and the film had errors. She brought him an armload of material to make her case.

Weeks later she got a call. The Forest Service had decided to change the exhibit. The director also said the agency would review its other displays for incorrect references to the victims. Then he made sure Parker knew how much it was going to cost. "To just change that photograph of the sign and the caption on the wall, they are going to charge us twelve hundred dollars." She was not impressed. To change the film was going to be another matter, however, the director told her. That would involve thousands of dollars and could not be accomplished right away.

She called Jeanette Killian with the good news of their partial victory. But this was just the beginning, Parker thought. Other agencies and media also had to be set straight.

PART 4
1989-1991

A Restless Slumber

It's not over by any means.

—Rocky Crandell

CHAPTER 25

1989–1990

FISHERMEN WERE EVERYWHERE. They were strung along the shore, shoulder to shoulder in places, casting their lines to hook one of the hungry trout literally jumping out of the water. Other people were walking along the Elk Bench trail where Peter Frenzen now led a group of journalists to get a better view of this spectacle on the banks of Coldwater Lake. The state had planted the lake with thirty thousand rainbow trout, a nonnative species favored by fishermen.

He had mixed feelings as he monitored this invasion. Frenzen was literally watching the destruction of another scientist's nine-year study. A few short years ago he probably would have been on the banks of the lake holding a picket sign decrying this outrage. But then he had been selected to become the first monument scientist, in charge of overseeing all but geological research on Mount St. Helens. His focus, by necessity, had broadened.

After a long search, Forest Service officials had discovered that the best person for the job had been right under their noses all along. The twenty-six-year-old Frenzen was selected for the breadth of his scientific knowledge, his proven ability to work with fellow scientists, and his experience in dealing with the public and the media. The new position meant realignment of some of his thinking and brought to the surface conflicting views about the mountain's future.

On the one hand were those who believed the eruption presented a once-in-a-lifetime opportunity to study the rebirth of a mountain and its ecological systems in a pristine state. Others thought the mountain and its lessons needed to be shared widely with the public. The more he observed in humility the wonder of St. Helens, Frenzen, like a proselytizing preacher, wanted to spread the word. His new position gave him the best forum. He talked about educational programs related to the mountain "providing a more informed, ecologically aware citizenry. You couldn't have a better payoff than that."

It was not easy for him. It was one thing to voice his lofty beliefs on a philosophical level and quite another to see the very real side effects on scientists and their work. St. Helens was now one of the most climbed peaks on earth, with twenty thousand people receiving permits each year in a system that had resulted in fistfights over the coveted permission slips. Litter could be found miles away from any trail. Stakes around research plots had been pulled up and experiments destroyed.

The monument's scientific advisory board had okayed the stocking of Coldwater Lake with the trout, in line with an agreement between state and federal governments. Yet try as he might, Frenzen couldn't justify the action. Coldwater Lake was one of the few pristine bodies of water created as a result of the May 18 blast. "How often do you get a lake created by a volcano that you can study from the beginning of its existence?" he asked himself. Biological changes in the lake had been studied closely for nine years. The introduction of these nonnative predators, the rainbow trout, would forever alter the natural life of the lake and halt the scientific work. The unique chance to study the birth and growth of a lake was being lost before his eyes.

Donna Parker pointed to the edge of the ridge. "Just below that is where my son and I found those tents in the ditch," she told Jeanette Killian. They were driving on roads through Weyerhaeuser land where people were killed May 18 but few bodies had been found. "Let's stop and look around," Killian said.

They walked across the gray ground and stopped at a mound of earth and ash. Parker said it looked like "someone tried to bury something here." They had no shovels, so the women began clawing at the mound with their hands.

"Look," Parker yelled. She pulled out a roll of exposed film that looked like it had been torn from a canister. They found other items that they thought indicated a campsite—cups, pans, clothing, bottles, even diapers.

In Parker's mind, this was additional evidence that, perhaps, more people had died on the mountain than officials had accounted for and made public. As far as she knew, there had never been a report of a campsite here or of any missing people from this spot. Now she visualized a family of campers fleeing the site May 18, leaving their things behind. They would have had no chance for escape this close to the exploding mountain.

Then she pictured someone, in the aftermath of the eruption, throwing the remains of the tents into the ditch and trying to cover them up. Then the bulldozer had come, Parker theorized, scraping all remaining evidence of the campsite into this mound.

"I have something I want to show you," Roald told Venus over the phone. "Can I come over to see you?" She hadn't heard from him in three years.

When Venus opened the door, she was shocked. Roald stood there, a sly grin on his face. He was carrying a one-year-old baby. "I want you to meet Baby Roald."

Roald's wife had left him and he planned to get a divorce, he said. He told Venus not only about the problems in his marriage, but also about how he had been injured at his job cleaning the insides of oil tankers. He was now on permanent disability. He heard that she was separated from her husband and in the midst of getting a divorce.

Venus held the baby boy and mussed his hair. He looked like his father, she thought, and he was adorable. The irony of the situation did not escape her. One of the reasons she and Roald had broken up was that Roald feared a deeper commitment, while she wanted more from the relationship, eventually including children.

Then Venus had met a man who said he wanted not only her, but children as well. However, what she found after marriage was that what the man really wanted was vastly different from her dreams.

Roald's wife also was not the person he thought he had married, he told her. He said that Venus looked just as attractive as she had the last time they met. They talked comfortably together. It was like they were picking up a conversation from the day before. If any woman had ever been his soulmate, Roald thought, it was Venus. They sat up talking late into the night while the baby slept.

In the days ahead they called and saw each other often. Gradually their mutual support for each other was being swept into something much more. One night he leaned over and kissed her and for a moment the years dissolved. It was as if nothing had happened to them since that balmy day at the Toutle River campground almost ten years before when they had been so carefree and in love. Soon they were spending so much time traveling between each other's houses to be together that they decided Roald and his baby boy should move in with Venus.

Flurries of shallow earthquakes swarmed across the mountain in the middle of August. Instruments detected no change in gas emissions. Swanson found no swelling of the crater floor or the dome. But the quakes did not end.

There had been no eruption of the mountain in three years. Two scientists from Arizona State had detected what they considered to be a consistent pattern in past eruptions. Based on their study, the mountain should have erupted again in June 1987. The fact that it was now 1989 and the last eruption had occurred in October 1986 led the Arizona scientists to conclude there was no explosive magma left. They concluded that the mountain was again dormant. Their findings were to be published in the scientific journal *Nature*.

Swanson gave a reporter a tour of the mountain and said he agreed with the Arizona scientists. "The mountain is almost out of gas," he said. Inside the crater, the lava dome had filled the floor to its edges with new rock. The dome was now more than a half-mile wide and 1,150 feet high. Steam wafted from hundreds of fumaroles. The echo of constantly falling rocks from the crater walls and the dome filled the natural amphitheater from all sides, like movie stereo.

"In another fifty years the rock from the walls will have halfway buried the dome if there are no further eruptions," Swanson said. "This sequence of eruptions I think is over." Deformation had ceased. Instruments no longer detected sulfur dioxide in gas samples. Tiltmeters had stopped tilting.

How could he explain the recent earthquakes? He couldn't. One theory held that they were the result of pockets of trapped gas whose sudden release made the mountain shake, much like stomach gas can cause a belly to quiver. Not long after the scientist and the reporter returned from the crater, the earthquakes began anew. Ash sprayed from the volcano onto newly fallen snow. The mountain kept shuddering.

She woke up from sleeping in the back of her Ford van, grabbed her overnight bag, and headed for the hotel entrance. The hotel was one of the state's most exclusive, in one of its wealthiest cities, the Seattle suburb of Bellevue. Donna Parker figured there could be no safer place to spend the night than the parking lot of this posh establishment. The bellboy greeted her as she walked in. Other staff members smiled and waved as she headed to a bathroom to wash up and brush her teeth. She had been here for several nights now, driving into Seattle during the day to pore over legal papers related to Mount St. Helens in the King County Courthouse.

During a layoff, after her job was shipped overseas along with thousands of others, she had taken advantage of a government assistance program and gone to school to become a private investigator. After becoming qualified, she found she wasn't interested in pursuing domestic cases and she knew she was ill-suited by nature to work for the government in criminal investigations. But her education was proving valuable in her dogged pursuit of what she considered to be some sort of cover-up on the St. Helens deaths.

When the courthouse closed that evening, she started her drive home to Oregon. By now she was in the habit of stopping along the way to search local phone books for the names of people who might be relatives of volcano victims. Most of these relatives had been excluded from the Mount St. Helens lawsuit so she had to search them out on her own. She wanted to keep relatives informed about the work of her group, SHAFT. And she wanted to let them know about her plans for a memorial gathering on the tenth anniversary of the eruption.

More important, she kept hoping that someone might offer new information to help her understand what had happened on the day Billy and the others died. If the person she was calling didn't answer or if she didn't have time to complete her calls, she would rip out pages from the phone book or simply take it with her. Her house was littered with phone books from across the Northwest.

The huge mat of floating logs, some as large as eight feet in diameter, towered over the rubber raft that carried Peter Frenzen and a British Broadcasting Corporation film crew. The logs, gray and weathered after ten years' exposure, had been blown into Spirit Lake by the May 18 eruption. As he helped guide the crew, Frenzen was wary of the massive blanket of ever-shifting logs. The logs, caught in the wind, could float faster than he could swim and could easily crush a man between them.

The crew wanted underwater footage after they filmed the log mat. They had invited Frenzen along on the dive. As he donned a wet suit, mask, and air tanks, he wondered what bizarre creatures and underwater landscape the volcano might have spawned in Spirit Lake? Long ago, native people had given the lake its name after claiming to see ghosts and monsters.

Since the eruption it had gained an even more otherworldly appearance. When history's largest avalanche rumbled through the lake, an 850-foot-high wall of water flew up the adjacent valley walls to come sloshing back with thousands of trees, dirt, rocks, and even animal carcasses. Hot volcanic rock rained down on the lake, raising the water temperature and killing off what life had been there.

High concentrations of manganese, iron, phosphate, sulfur, ammonia, and chloride mixed with nutrients from plants and soil, simmering like a witch's brew. Slime, molds, and oxygen-eating bacteria resulted, along with fungi that exploded into contorted rings of yellow, green, and red. When the oxygen was depleted, anaerobic organisms took over. Gradually, though, oxygen returned to the lake in snowmelt, rain, and new groundwater. But it was a seesaw battle as ash and rock on surrounding hillsides continued to erode into the lake, resulting in a highly soluble metal content.

As Frenzen followed the other divers, he let his imagination run wild as he thought about Loch Ness-like monsters lying in wait. Under the water he saw nothing but a thick, brown soup. An irrational fear grew and his body tensed for an unseen attack. At the bottom he could just barely make out the film crew. As he sat in the muck, eventually he decided he would be seeing nothing that day, monsters or anything else.

When he quit trying to look in the distance and focused instead on the area just in front of his mask, he grew instantly alert. He could see creatures stranger than any his imagination might have conjured up, only millions of times smaller. A world of diminutive animals danced, darted, and fed on each other in a water column a few inches in front of his mask. He focused in on the zooplankton that fluttered through the water, looking like white butterflies with stunted wings. He sat in wonder, transfixed.

—✳—

Stepankowsky stood, looking up at the hundred-ton bulldozer spreading a final layer of rock on top of an eighteen-story-high earth and concrete dam on the Toutle River. It was the Army Corps of Engineers' answer to what it called "the world's largest sedimentation problem." In true Corps fashion the agency had decided that a monumental problem deserved a monumental public works solution.

The dam cost over $80 million and was designed to store 253 million cubic yards of debris, or "enough to cover a football field 23 miles deep," Stepankowsky calculated. This new man-made monster dam downstream from Spirit Lake would prevent eroding silt from entering the Cowlitz and other rivers, according to the Corps. The alternative to the dam, the Corps told Congress and the fifty thousand people living along the Toutle and Cowlitz Rivers, was years of continued flooding, even with dredging.

The sixteen miles of avalanche debris and ash left by the May 18 eruption was like a "giant sandbox," a USGS hydrologist told the reporter. To date only 5 to 10 percent had eroded away, with serious flooding being the result. The rest of the ash and debris would eventually erode away, he said, and would have brought decades of flooding if the dam had not been built. Construction of the dam with its two hundred highly paid workers had been an economic godsend to the local area. A dedication ceremony was planned for May 18, the tenth anniversary of the eruption.

Scymanky tinkered with the engine of the aged Mustang and wondered if he could make a living for his family restoring old cars like this. He had almost completed three years of school to become a building inspector, but he wasn't sure that was how he wanted to spend the rest of his working life. The workmen's compensation assistance had run out and he would need to do something soon. Helen had found a job with an insurance company but finances were still tight.

Any lingering hopes of a monetary settlement from the state for negligence had been dashed. The Washington State Court of Appeals ruled in 1988 that the state wasn't to blame for an act of God, which is how the state characterized the eruption. A year later the court decided it would not reconsider its ruling. Texas attorney Ron Franklin refused to pursue the case further, contrary to what Scymanky thought had been agreed to when he accepted the paltry settlement from Weyerhaeuser.

The Scymankys had one big hope to ease their financial plight. For years people had told Scymanky he should write a book. Now in their spare time he and Helen were doing just that. They had even picked a title for it. *Ash Sunday*.

It came as a shock. There behind glass, on display at the Forest Service's Silver Lake visitor center, was one of Billy's lost cameras. Parker had no doubt about it. The camera even had the license number of her brother's truck scrawled on it, which is how he marked his photography equipment.

The director was not at the center. Later she called him.

"I know you have the camera," she said.

"What camera?"

"My brother's camera you have on exhibit at the center."

"I didn't know we had your brother's camera."

"Where are the rest of his cameras and guns?"

"I don't know."

"What do you mean you don't know?"

"We don't have them."

"You didn't know you had the camera. How do you know what you don't have?"

The director was silent.

"Well, you must have a storeroom where you've got all this stored."

"No, we don't"

"Then where did it all come from?"

"I don't know." He said he had no way of tracing back the items on exhibit.

On January 6, 1990, after months of earthquake swarms, the mountain shook with its biggest explosion in six years. The burst knocked out seismographs and tiltmeters as it sent ash up to ninety miles away to dust the Washington cities of Yakima and Toppenish. The study by the two Arizona scientists, claiming the volcano had run its course for now, had just been published. But the new burst involved no magmatic gases and did not qualify officially as an eruption, according to the USGS. Swanson and other scientists continued to believe that the volcano was winding down into a deep slumber.

It was storming and the Portland bridge had been shut down as high winds buffeted waves up and over its sides. At each end of the bridge a cluster of police, firemen, and other workers stood transfixed by the huge waves, ready to spring into action if the bridge started to fail.

Rogers and a friend donned official-looking hard hats as they approached one end of the bridge. Rogers had pilfered the hard hats when the local public transportation agency was handing them out to their workers at a promotional event. The throng at the bridge entrance parted to make way for the two men who looked very much like inspectors.

Rogers, still with a bad leg, and his friend walked to the middle of the empty bridge and watched the storm on the river as waves drenched them. Several times he looked up toward the mountain that had become such a part of his life. But today clouds shrouded the peak.

The little logging community of Vader with its small and aged wood-sided houses spread before Stepankowsky and photographer Roger Werth. The reporter had been here many times before to see Ralph and Jeanette Killian. He found them to be honest, straightforward people with a strong work ethic.

Stepankowsky and Werth drove down a long, unpaved driveway to the house. On one side of the house Ralph Killian had built what looked like a carport. But it was actually a shelter protecting mounds of scavenged items from the May 18 blast—vehicle parts, wooden boards, eating utensils, twisted metal, shredded clothing, camping gear—a monument to the Killians' ten-year search for their son.

The retired sixty-nine-year-old logger had the lined, weathered face that could be expected of a man who had spent over thirty years in the woods working for Weyerhaeuser. His wife, Jeanette, at fifty-nine, still had her good looks and greeted the two men with the genteel manner and slight drawl of the Southern belle she once had been.

The four of them drove out to Fawn Lake, nine miles northwest of the peak. The huge, bleached root wads of fallen trees towered over them as they hiked into the site believed to be where their son and daughter-in-law had perished. The couple posed for Werth near the small monument with a plaque they had erected to their son, John, and his wife, Christy.

Even now, ten years later, the couple had trouble accepting their son's death. The first two weeks after the eruption, Ralph and volunteer searchers had desperately cut up fallen trees blocking roads and used bulldozers to make their way to search sites. His twenty-nine-year-old son was a seasoned outdoorsman and a Vietnam vet. Like his father, Killian's son was a logger who came from generations of loggers. If anyone could survive, it would be his son. When the other searchers dropped out, Ralph Killian kept on. "You don't just disappear from the face of the earth," he told Stepankowsky.

Two months after the eruption, the body of Christy, only twenty years old, was found, her hand and forearm jutting out of her grave of mud. There was no sign of John. Eventually Weyerhaeuser gave Killian two years off with pay to search for his son.

John had been more than a son to Ralph. "John was my hunting and fishing partner," Killian said. "I used to carry him on my back fishing when he was little. Just him and I would camp out in those high lakes. It was like that all his life."

The couple was at pains to emphasize their son was not a lawbreaker, a troublemaker, or a thrill seeker. "When he was little, he never needed a spanking. He was just a good boy."

When their son left on May 17 to go camping with his wife at Fawn Lake, he had little to fear from the volcano. The area was on Weyerhaeuser land that had been excluded by the state from the restricted area. Just the previous week his work for Weyerhaeuser took him several miles closer to the volcano than Fawn Lake. His father was scheduled to go to the area on May 19 to supervise the yarding of timber that cutters had felled. Weyerhaeuser assured both father and son that the area was safe. The lake was shielded from the volcano by a large ridge.

Ralph and Jeanette Killian were at first shocked and then enraged when Governor Dixy Lee Ray and President Jimmy Carter accused the dead of skipping around roadblocks and illegally entering the restricted zone. At first Ralph refused to enter the lawsuit against Weyerhaeuser and the State of Washington for negligence. Then Jeanette became convinced it would be the best way to clear their son's name.

Even after the trial, Ralph was not convinced of Weyerhaeuser's culpability. "I never did blame Weyerhaeuser for this," he told Stepankowsky. "I worked for them all those years and they were always good to me and I raised my family." The issue had strained the couple's relationship. "If Weyerhaeuser had not told him it was safe, he would not have been up there," Jeanette said.

The Killians never stopped looking for their son. For years they rode a roller coaster of hope and despair. The couple thought there was a possibility their son had been hit on the head by a rock or a tree or had fallen and had suffered amnesia before walking away to safety. Well-meaning people would call and say they had seen John. Ralph once spotted a logger riding in a company pickup who was "a dead ringer" for his son. For days he tried in vain to track down the truck. "I don't know if it was an illusion."

Now while Stepankowsky looked on and Werth took photographs, the couple started yet another search. "I've never believed he died. I don't believe it to this day," Ralph said. "He's dead, really. But it's hard to accept. It's like he walked outside and should be back."

CHAPTER 26

1990

SLIDES WERE SCATTERED all over the table—more than four thousand of them. They showed St. Helens in the early years after the 1980 eruption in all phases of development and activity, from the mountain as a whole down to individual vents and fissures, and from every conceivable angle. Robert Rogers was feeling the passage of time. He was organizing the slides, writing timelines, and racking his brain for every memory of his experiences on the mountain. The tenth anniversary was approaching, and Rogers wanted to get everything in order for the slide presentations he knew that people would be asking him to give.

Across the blast zone, bright yellow and red flags fluttered and instrument whirligigs spun and hesitated with the wind. Iron stakes marking research plots bristled from the pumice. Antennas in all shapes and sizes sprouted from the gray ground and reached to the sky like wire sculptures, paeans to the volcano god. But the monument scientist who coordinated and oversaw this array of experiments was nowhere in sight. Peter Frenzen was confined to his office in Amboy, more than twenty miles away, with piles of computer printouts, field notebooks, appointment calendars, Forest Service manuals, and a reporter.

Peter Frenzen was explaining to the reporter, Rob Carson, how he had come to realize from the mountain that the whole planet was intricately balanced, with intertwined processes stretching back over the eons. Mount St. Helens represented processes at work continuously on the earth as a whole—except that here on the mountain, scientists had the rare opportunity to see these changes greatly speeded up. A new mountain rose on the old one, which was visibly being washed out to sea. Lava rock broke down to soil. New plants and animals adapted, attracting other plants and animals. It was all part of a never-ending cycle where old became new and new became old, as it had since the birth of the planet.

Yet as much as he liked talking and thinking about the Big Picture, Frenzen tried politely to answer the reporter's questions and then get on with all his other work. The biggest adjustment he had faced in his transition from pure science researcher to official monument scientist was dealing with administrative duties and having to divide his time among so many tasks. At the beginning he would go home and complain to his wife, Denise, about all the requests tugging at his time. "But, Peter, it's all your job," she would remind him.

The position of monument scientist represented only one of the many ways St. Helens had changed his life. He had met Denise, a botanist, at Oregon State University and she had helped in his research and shared his love for the mountain. If St. Helens had not erupted he would have gone to Yale for graduate work and would never have met this pretty young woman who was now his wife.

Eventually he agreed with her view of his work as monument scientist. All of it, from helping a school group tour the mountain to solving a management problem, was important. He was beginning to enjoy the variety and spontaneity of the job.

It was another call from Roald's wife. As the woman cursed at her, Venus tried to stay calm. Roald, now living with Venus, shared custody of his child with his estranged wife. The woman was crazy, Venus thought. She would scream at Venus over the phone, trying to start a fight over Roald or the infant boy. Venus hung up.

The past year living with Roald had not been easy, with financial and emotional strains on both of them. They clung to each other through it all in the hope they could get their relationship back to where it was ten years before. She had gotten through the most traumatic period of her divorce that was now almost final. But Roald seemed stuck.

Some time back Roald had received an unexpected call from his wife. She wanted to get back together as a family again, she said. Didn't he think that would be best for their son? He told Venus about the call. Since that time she had seen Roald grow more distant. She knew that the pressure of trying to reconcile what was right for his son with his love for her was tearing him apart.

"You need to make a decision," she told him. "You need to take care of yourself and your own problems—to clean up your own house first if you want me in your life." These were hard words for her to speak but she knew she couldn't go on with the uncertainty and the tension. Soon she wouldn't have anything left to give to Roald or herself.

A few days later Roald and his son moved to his parents' house. He was going to try to clear his head. That night as Venus lay alone in bed, an unsettling quietness came over the house.

Stepankowsky liked interviewing Don Swanson. Unlike most other scientists, who seemed to attach a qualifier to every statement, Swanson was not afraid to express his beliefs and explain his findings in layman's language, without hedging. He was the only USGS scientist left who had been assigned to the mountain continuously since 1980. At age fifty-one he was at the peak of his career.

Yes, he told Stepankowsky in answer to a question, he was the first scientist to develop a method to accurately predict volcanic eruptions. "It was a first in the field of volcanology and I am very proud of that," he said. "I am cognizant, though, of the methodology's limitations at other volcanoes. But at St. Helens it works." Stepankowsky had already reviewed Swanson's amazing achievement: the accurate prediction of all eruptions of the volcano since December 1980. The only possible exception was a June 1984 event that experts now doubted was an eruption.

After all these years Stepankowsky found Swanson still combined a schoolboy enthusiasm for his work with a sobering scientific speech. "Volcanology is the opportunity to witness very fundamental processes in the development of the earth," Swanson said. "It's important for the human race to understand nature. I believe a fundamental understanding of the natural processes is of great philosophical and practical interest. . . .

"Many people look up into the heavens and wonder about the meaning of things and people can do the same about the earth. They can ask, 'Why are there volcanoes?'"

The whole science of volcanology was still in its infancy, he said, and he thought people credited scientists with more knowledge than they actually possessed. "The whole business is as mysterious to us sometimes as it is to the public," he said.

Research at St. Helens up to now had dwelled on eruptions and the resulting deposits, he said. It was now time to answer the basic question: how do volcanoes work? To do this he touted his latest and most ambitious pet project to the reporter. Swanson and other scientists wanted to drill a mile down into the volcano in an attempt to tap its underground pool of molten lava. So far the USGS had failed in efforts to get the $1.5 million needed for the project.

Then Stepankowsky asked about Swanson's friend and fellow scientist David Johnston. "Dave, more than any of us, was worried about the

potential for an explosive eruption," Swanson said. But Swanson believed
that Johnston's famous last words—"Vancouver, this is it"—were yelled not
in fear but more with the excitement of a youngster just treated to a spree
in a candy store. "His voice," Swanson said, "was not at all fatalistic. Dave
was so close it must have been an awesome sight."

Jim and Helen Scymanky stayed up late, spray-painting the three
crosses white and waiting for the paint to dry before inscribing his dead
friends' names on them. At the airport the next day, Scymanky at first felt
his innate shyness holding him back. But as he got closer to the helicopter
pilot, he picked up his pace and smiled at the man.

"Hi, Jim," the pilot shouted.

"You must be the guy I've been waiting to see," Scymanky yelled back
at pilot Jess Hagerman. He had not seen the mountain nor the man who
had saved his life since that day ten years before when Hagerman flew to
his rescue. They shook hands and Scymanky introduced Helen. She tried
to hold back the tears. This was the man who had risked his life by
repeatedly trying to land through an ash cloud to save her husband. A film
crew for a documentary production, *The Fire Below Us*, hovered around
them.

Mist draped the mountain as they flew in the helicopter above the
gray landscape. A ridge on a steep hill patched in green caught Scymanky's
eye. "That area right there looks familiar for some reason," he said, pointing
out the side of the window with his thumb. "Yeah, that could have been
it," Hagerman said. "It was on a little ridge."

They landed on a road above the ridge. It was the same road Scymanky
and his friends tried to use to escape the holocaust. The second helicopter
landed, and cameramen and sound technicians scampered out. For several
minutes Scymanky studied the vast valley. "Look at that sun over there.
Isn't that beautiful?" He pointed to the sun streaming through the clouds,
sending rays of light upon the land. He thought about placing the crosses.
"You know, this would be a real good spot, overlooking everything."

By the time the film crew had set up, black clouds covered the sky.
Scymanky and Helen took the crosses and hiked to the little ridge. Firs up
to three feet tall were growing there along with the white-flowered pearly
everlasting. With a hammer he pounded the first cross into the ground.
"Jose Dias," the lettering on the cross read. With each blow of the hammer,
he felt the emotion well up inside of him. "Leonty Skorohodoff," the second
cross read. It was all coming back to him. He was again at that moment
ten years ago, standing near Skorohodoff and seeing the quizzical look on

the face of Evlanty Sharipoff as he looked up the hill. Then Scymanky saw
once more the hysterical Jose Dias running barefoot down the hill, yelling
"¡El volcan esta explotando! ¡El volcan esta explotando!"

Scymanky finished pounding in the last cross. He stared at the carefully
drawn lettering that read "Evlanty Sharipoff." For years Scymanky had
carried the burden of their deaths, and this was the first relief he had
really felt. At that moment the clouds opened up and showered him and
the crosses in shafts of light. The film crew grew animated as the
cameramen maneuvered to get the best angle for the shot. But the director
didn't know if he would use it. It was all just too perfect.

Swanson was finishing his mapping of the dikes, or rock intrusions,
that radiated from the center of an extinct volcano he had discovered in
an area between Mount St. Helens and Mount Rainier. It was arduous
work. He had hiked for miles up and down steep landscape, thick with
vegetation, looking for bedrock with dike exposures. The work was
beginning to pay off. Analyzed samples showed that this volcano first
erupted twelve million years ago. In contrast St. Helens had come alive
only about fifty thousand years before.

Swanson had become a volcano hunter. He was spending his spare
time searching out the remnants of ancient Northwest volcanoes. He bought
a guidebook to forest fire watchtowers across the mountains of Washington.
On weekends he and his wife, Barbara, climbed the towers and took in
their expansive views. From the land formations he observed, Swanson
came to believe that the region had been more active with volcanoes than
anyone realized.

Telltale signs of hundreds of extinct volcanoes were scattered across
the landscape. In some cases he discovered the ash fallout and even the
lava flows of one volcano interwoven with that of several others. Some
volcanoes were built upon the remains of long-extinct predecessors. The
present Mount St. Helens sat on a much older volcano that had been active
for more than forty thousand years.

He came to the end of the dike he was mapping. Swanson was now
miles from the core of the extinct volcano. He studied his mapping of the
dikes radiating out from the volcano's center. This volcano, the mapping
showed, had a footprint as huge as that of Mount Rainier, one of the
largest volcanic mountains in the continental United States. The processes
of nature over a great many years had worn down this volcano that once
covered hundreds of square miles. Some day Mount St. Helens would
suffer the same fate, worn away to nothing.

The call she feared had now come. It was Roald. He had decided that for the sake of his son he should return to his wife and try to make their marriage work.

"Roald, I've been through this," Venus told him. "It never works out." He knew there was some truth to what she was saying but he also knew how much pain his son had gone through.

After she hung up, the thoughts began racing through her mind. Had Roald been using her as an emotional crutch? Maybe they had been using each other? Had she been going through all this and been there for him out of gratitude? After all, he had saved her life, not just once but three times in the muck of the mudflow. At the same time, maybe Roald had been acting not out of love but out of some type of guilt over what had happened to her.

She came to the conclusion that they really did love each other. She cursed the volcano for pulling them apart in the first place, and for now leaving them with the bond that kept their lives intertwined. Then she realized the volcano had given her something else: a will to survive. If she could survive that day ten years before, she thought, and what she had to go through to heal, she could survive anything, even this.

Stepankowsky strode into managing editor Bob Gaston's office. He slammed a fist down on the editor's desk as he had many times in the past. Gaston looked up. The reporter plopped down into a chair and let out a sigh of exasperation.

"I'm having trouble writing this May 18 story," Stepankowsky said. He had been asked to write the cover story for the newspaper's special section for the tenth anniversary of the eruption. The editors wanted him to describe "with fresh words, feeling, and perspective a decade's worth of events unfolding from that morning of May 18, 1980." But it had all been said before.

Gaston let the reporter have his lament. Sometimes the cure for writer's block was nothing more than a little venting directed at the boss.

Two days later the section's editor, Dave Rorden, came into Gaston's office holding the copy for Stepankowsky' story. "Andre did a great job on the May 18 story," he said. Gaston wasn't surprised.

"Ten years after the mountain blew, people and the land of Southwest Washington are healing their wounds and turning a tragedy into triumph," Stepankowsky wrote. "Humans matched the volcano's awesome power with prodigious energy of their own to close the history books on a disaster that seemed to have no end. . . .

"Scientists are discovering how nature rebounds from calamity. And we're all learning lessons about our relationship with the earth—that 'God's Country' can quickly become a devil's pit.

"A decade—time enough for rebirth of the land and the human spirit, time enough to learn how vulnerable we may be.

"But never enough to forget that Sunday. . . . "

The Scymankys stopped at the 19-Mile Restaurant, named for Weyerhaeuser's 19-Mile Logging Camp that was destroyed there in the May 18, 1980 eruption. Signs led them to the basement of the restaurant, to a small museum filled with volcano artifacts: newspaper clippings, damaged car parts, stuffed animals, camping gear, chain saws, and everywhere the dank smell of ash.

Suddenly, Jim Scymanky bolted backward.

"What's wrong?" Helen asked.

"I'll be damned," he said. "They've got one of my boots in that display case."

"Are you sure?"

"I wouldn't forget something like that."

He pointed to a mangled cowboy boot. Then he glanced hurriedly over the rest of the museum. "Look, Helen. It's our cooler." As she looked at the melted and contorted plastic of the cooler, she began to realize just how much heat her husband had survived in the eruption. But when he discovered his frying pan with a hole blasted through it by a rock from the volcano, she found herself saying a little prayer of thanks for his survival.

The three women had lost loved ones to St. Helens: a brother, a son, a mother, a sister. They told the *Seattle Times* reporter they still believed that Dixy Lee Ray, governor at the time, had approved an undersized hazard zone in order to help Weyerhaeuser log as much land as possible. State officials had misrepresented the true hazard area, they said, lulling the victims into a false sense of security that led to their deaths.

"These people died in areas that were open to the public, and the state has never acknowledged that," said Donna Parker, speaking up for her dead brother Billy. "They should acknowledge that on the tenth anniversary."

While the state still refused to do that, Parker said she had found some success. The Forest Service on the anniversary would be premiering a new version of the St. Helens movie it showed at visitor centers. The movie would no longer characterize the dead as curiosity seekers. However, the Service balked at a planned memorial Parker designed to be made out of a large black slab with the names of the dead inscribed. It would have been the first memorial for the dead.

"It's kind of like being a Vietnam War veteran," said Maxine Bowers, whose son was killed in the eruption. "It's like they did something wrong and no one wants to touch us." She said her son, Wally, a twenty-year Weyerhaeuser employee, was afraid to be logging while the mountain was acting up. "They told the loggers, 'You don't like it, just pick up your paycheck and go on down the road,' but they couldn't do that. They needed the work."

"I'm still very angry," said Roxanne Mansfield, the former Roxanne Edwards. Her mother, Arlene Edwards, and her sister Jolene died in the blast. "I felt ever since this whole thing started that we wouldn't get satisfaction in the courts. I still can't get over it. I go to therapy every week. Without us the state of Washington wouldn't have their tourist center, but they still can't say they made a mistake in letting them up there."

She complained about state and county government insensitivity in the chaotic days after the eruption. "The coroner called me and told me my mother wasn't there because they couldn't find her body. They found my sister and my sister's dog and my mother's dog in the truck, along with their personal belongings. Because they couldn't find my mother right away, they told me she'd probably run away and started life over somewhere else. They found her body a month later and I really lost my cool."

Mansfield told about visiting a privately owned volcano museum with Parker and seeing her mother's truck, and finding some of her mother's personal papers hidden inside. "There's this person who bought my mother's truck, to put it in a sideshow," she said. "You know, 'This is the truck of Jolene and Arlene Edwards.' I went up there once and paid fifty cents to see it. I was just livid. Sometimes I feel like I have no control. I'm trying to put my life together, but something like this pops up and sets me off again.

"I know this is just one of the things that will go on all my life and I'm going to have to learn to accept it. It's the tenth anniversary now, and someday I'll have to go through a twenty-fifth anniversary. I really dread it."

Peter Frenzen turned on the television in his living room just in time to have his own image beam back at him. As he sat there listening to his well-rehearsed description of St. Helens' rebirth chopped into sometimes nonsensical sound bites, he winced. Then he grimaced when he saw the camera pan the mountain and heard the reporter make a generalized statement about recovery for the whole area that he knew only applied to a small portion of land. As the sequence unfolded, his face continued to be a mobile map of his emotions, moving from pain to shock to exasperation. He didn't know whether to laugh or cry at the reporter's overly broad reporting of such a complex subject.

This was becoming almost a nightly occurrence as the tenth anniversary drew near and more reporters called on the monument scientist to help them with their stories. He turned to Denise, who was smiling at him.

"What are you looking at?" he asked.

"I've told you before. Your face is more fun to watch than the news."

On the morning of May 18, ten years to the day since the big blast, more than three hundred scientists and observers descended on the Red Lion Inn in Kelso for an anniversary symposium. "Before Mount St. Helens [erupted in 1980] there was a feeling that volcanoes were something from the dinosaur era, interesting, but something that happened somewhere else," said Robert Christiansen, chief of volcano hazards for the USGS. But with the May 18 eruption, St. Helens led off what was becoming known as the decade of the volcano.

Since 1980, volcanoes had killed 26,000 people and left another 336,000 homeless in a ten-year period that was "the most disastrous since the early 1900s" in terms of human impacts, according to USGS scientist Thomas Casadevall. In recorded history there had been six hundred volcanic eruptions resulting in the deaths of more than 250,000 people.

Research on St. Helens had found that the volcano erupted in 1480 with a force and resulting debris six times that of the 1980 blast. Former USGS geologist Rocky Crandell said that based on the five-hundred-year history since 1480, the mountain would have another massive eruption by the middle of the twenty-first century. But there was a 15 percent chance the mountain could erupt in any particular decade prior to that. "It's not over by any means," he told the audience.

Venus Dergan opened the *Tacoma News Tribune*, and her heart sank. Actress Jill Ireland had died at age fifty-four on this day, the anniversary of the eruption. A new set of memories rushed in to mix with those of her fight for survival ten years before. Visions of Roald and herself in Hollywood flashed through her mind. She remembered how radiant Ireland had looked the day that she and Roald were interviewed on TV by the star and her actor husband, Charles Bronson. Ireland's obvious concern for Venus and Roald's relationship was touching. She wondered about Roald. He had not called for a long time.

He was exhausted. The St. Helens special section he had worked so hard on had been published that day. A big dinner was scheduled for the evening to bring together nearly four hundred people associated with the St. Helens recovery effort of the past ten years. Andre and Paula Stepankowsky were among the invitees, this time not as reporters but as regular guests.

He didn't want to go. He was tired. The anniversary of the May 18 eruption had fallen on this day, a Friday. However, Saturday would be another big day with more events to cover including the dedication of one of the largest sedimentation dams ever built and a rally to protest restrictions on logging because of an endangered owl species. But Paula told him it would be fun to see everyone who had been involved with the volcano.

The couple made their way through the throng in the ballroom of the Kelso Red Lion. Without his reporter's mantle, Stepankowsky could be shy and even withdrawn. But he responded to the warm greetings he received from some of the officials there. In some cases these were the same people who had felt the sting of his probing news articles. The Stepankowskys took their assigned seats at a round table not far from where a movie screen had been set up.

Members of a special committee had selected ten individuals who they believed had contributed most to the area's renewal since the May 18 eruption. Before each of the award presentations, a narrated slide program laid out each winner's accomplishments. Stepankowsky politely applauded as each of the honorees walked to the dais, among them the head of the state Department of Emergency Services, the supervisor of Gifford Pinchot National Forest, a planner for the Army Corps of Engineers, a USGS hydrologist, and a top political aide who had helping swing funding bills in Congress.

An audible "Oh, Jesus," heard by the audience, was the only thing Stepankowsky could think to say when a slide appeared on the screen showing him in front of St. Helens. Slides flashed before Stepankowsky's eyes, detailing his life with the volcano, accompanied by a recorded narration from publisher Ted Natt. Then the emcee spoke. The reporter's stories about the plight of the people and the region, many of them read into the *Congressional Record*, had been instrumental in securing funds and support. His doggedness in questioning government and scientific reports and information had revealed serious flaws in the area's recovery plans and kept many of the officials now in this room on their toes. Stepankowsky was the only nongovernment employee to receive an award.

He was overcome with emotion as the applause echoed in his ears while he walked up to receive his award. For one brief moment he thought back to the young reporter ten years before, overwhelmed by the enormous power of St. Helens raging beneath him as he flew over the volcano.

The monument scientist arrived early at his office to prepare for the flow of media calls he knew would come that day, exactly ten years since the May 18 eruption. For the past six months he had been overwhelmed by the interest in St. Helens and its rebirth. But he had no idea of the intensity of the coverage he would face on anniversary day.

He started by fulfilling requests for interviews with media on the eastern seaboard. Some interviewers kept him on the phone longer than anticipated. Then other media called, demanding to talk to him. Coverage was like a wave, he thought, building in intensity as morning swept across the nation's time zones, waking more reporters or talk-show hosts. He hoped that when this wave of media interest finally crashed, he wasn't at the bottom of it, buried by disgruntled journalists, misquotes, and confused audiences.

At one point he had fleeting thoughts about how much his life had changed. Not too long ago he was counting plants for a research study in a long and sometimes lonely pursuit. Now he was interacting with some of the nation's largest television networks and major newspapers and magazines, with his words going out to millions of people.

It grew to be late at night. He was exhausted and feeling testy. A reporter for the Reuters news agency called. Other journalists waiting to talk with him were backed up on the phone line. The reporter asked a bunch of routine questions that had been asked of Frenzen a hundred times that day. He wondered if anyone had bothered to read the carefully prepared press kit that had been sent to the media.

"Can I fax you some material?" Frenzen asked.

"No, I'm on deadline. I really need this right now."

"Well, if you just read this stuff, you can call me and I can give you some quotes."

"No, no. I don't have time," the reporter said.

"Well, I don't either," Frenzen said. The reporter hung up on him.

The monument scientist sat there and tried to calm himself before going on to the next reporter. Then it hit him. "Oh god, what have I done?"

he asked himself out loud. He had just offended a reporter from the world's largest news service.

Rogers had developed a business based on his ability to fix just about anything. A few years before, he had borrowed money from his sister and bought a dilapidated house in a depressed area of north Portland. He fixed up the house, borrowed money on it, and gradually bought three other run-down houses that he remodeled and rented out. Then he started a business offering repair services to small businesses. He took care of their electronic door systems, computers, lighting, burglar alarms, ventilation, and anything else. He never refused a job. "I can solve that for you," he would say. "Sure, I can do that."

Every day as he left his house he was aware of the mountain looming over him—some days visible, other days hidden in a Northwest soup of mist, fog, and clouds. As the tenth anniversary of the eruption grew closer, he thought increasingly about St. Helens and his place in the history of the mountain. He began to wonder when he would start getting invitations to show his dramatic, unmatchable slides of the volcano. No one had called— not the media, not the Forest Service, not the USGS.

Very late on the night of the anniversary, Rogers put his slides back into the closet. The calls never came.

Hundreds of chilly people sat in chairs in nice, neat rows on the spillway of the sediment retention dam. The red, white, and blue bunting on the speakers' stand billowed in the strong breeze blowing off the mountain. The Toutle Lake High School Band played Sousa marches. Red and white tents stood ready in case of rain.

Precisely at 11 a.m., as if by cue, the clouds parted and the sun shone, and it was time to start the ceremonies for dedication of the concrete and earth behemoth. This great structure, designed to save the region from flooding, now stood as a monument to man's persistence in, if not conquering nature, then at least reaching an accommodation with it. It had taken six years of political wrangling, three years of construction, 750,000 man-hours of labor, and $80 million to build the 185-foot-high dam spanning hundreds of feet across the Toutle River.

Stepankowsky took notes as the words of optimism and pride spilled forth.

"Dedication of this sediment structure hopefully signals the end to the disaster era," a county commissioner told the crowd.

"The sediment retention structure is the final piece of the solution to the Mount St. Helens flooding threats," said an Army Corps of Engineers commander.

Sid Morrison, the area's representative to Congress, said the challenge now "is to bring millions of tourists to the mountain. This structure is to keep the mountain from going to the millions."

Pat Keough, the Corps of Engineers official who directed most of its St. Helens recovery work, said much of the agency's accomplishment would never be seen. "It takes the form of floods that will not happen, homes and communities that will not be destroyed, and river traffic that will flow smoothly. It represents unprecedented achievements of coordination and problem solving under emergency conditions."

For all his holding of the Corps' feet to the fire, Stepankowsky had to agree. In ten years the Corps had built projects that usually took anywhere from twenty-five to thirty years. The reporter had been there every step of the way, keeping the Corps under an independent scrutiny they rarely endured elsewhere. Now he felt a great sense of relief. He believed that this dam, indeed, could be the solution to the flood threat spawned by the eruption of a decade ago. Maybe he could end this chapter in his life and move on to other things, knowing that he had done his job and accomplished in part the goal he had set for himself when he became a reporter—to make a difference.

In nearby Toutle, the clown judging contest had just begun for Volcano Daze, whose theme this year was "Movin' and Shakin'." The contest was to be followed by picture-taking with Bigfoot, the International Lumberjack Show, the chili cook-off, logger poetry, and a vintage car show. The big event would be the parade of volcano floats, marching bands, clowns, and horses down Spirit Lake Highway through the main part of town. In Longview, more than 250 bowlers from across the nation had gathered for the tenth annual Mount St. Helens bowling tournament. In Kelso, workers were setting up for a concert later that night by the Hager Twins and Juli Manners from the *Hee Haw* television show.

In a line that stretched for miles, the 554 shiny log trucks roared down Interstate 5. Yellow ribbons, symbol of the loggers' movement, fluttered from antennas, grills, and bumpers. From the cabs of the trucks, loggers and their families waved to cars going past.

At the same time, Stepankowsky was driving down Spirit Lake Highway, rushing to get from the dam dedication to the loggers' rally in Kelso. The loggers were gathering to protest logging restrictions designed to protect the northern spotted owl. As he drove, Stepankowsky figured out the approach he would take on the analysis piece he was assigned to write.

The two events he covered this day were symbols of transition, he decided. The 1980s marked the decade of battling an active volcano. The 1990s would bring a showdown between the timber industry and the proponents of saving the spotted owl and the ancient forest environment it represented. Thousands of jobs hung in the balance.

The volcano had actually been a boon for the region during the national recession of the early 1980s. Flood-control projects created hundreds of new jobs. And many more logs were cut in salvage operations than could ever have been cut in normal conditions. But now the projects were mostly done. As for the stepped-up logging, it was beginning to look like a mixed economic blessing. As Paula Stepankowsky, now business editor of the *Daily News*, put it, "Though beneficial to the labor force in the first three years of the decade, the long-term impact of the hasty harvest may not be as positive. Many trees salvaged were not the optimum 45 years old, and didn't achieve full value on the market.

"The blast probably cost long-term jobs as well. 'You have to weigh the three years against the next 15 to 20 years, when people might have been employed,'" Duane Wend of the International Woodworkers of America told her.

In the rush to try to make St. Helens another Grand Canyon with its tourist dollars, other negative economic impacts were being felt. Before the May 18 eruption, the area already was a tourist magnet. Many more tourist dollars were dropped into local coffers before the eruption than after. People attracted to the region's scenic beauty before the eruption tended to stay for several days of fishing, hiking, boating, or camping. Tourists in the years after the eruption usually just pulled off the freeway for a few hours to see what they could of the volcano, then continued down the interstate.

Local and state officials still held out tourism as an economic salvation. The Forest Service predicted over two million visits annually within a few years. But Stepankowsky couldn't see it. Every tourist estimate by officials in the past had fallen wildly short of the mark.

Other factors were converging to disturb the region once known as the world's timber capital. Weyerhaeuser and other companies were finding it more profitable to ship whole logs to be milled in foreign lands rather than in domestic sawmills. This practice cost jobs in Longview and in dozens of small towns where the local mill was often the biggest employer. At the same time, new environmental standards for mills brought the specter of further cuts in operations.

As Stepankowsky turned onto the freeway and raced to get ahead of the loggers' convoy, he thought how ironic it was that the region had dealt successfully with the uncertainty of volcano-related floods only to confront a new economic uncertainty. Since the eruption there had been no chance for the region to catch its breath.

Parker looked up to see the Murphy sisters, Toni and Lori, walking across the park toward her. She felt her heart thumping as she jumped up to greet them. She hadn't known they would show up. They had lost their parents, Edward and Eleanor Murphy, in the May 18 eruption. The bodies had never been found.

For years Parker had written and called the sisters, but only one of them had ever seemed responsive. Now she embraced the sisters. "Thank you for coming," she said. As they walked to where some forty other relatives of St. Helens victims were gathered, both sisters told her how much her letters and calls had meant to them over the years. The group was meeting under a wide canopy Parker had borrowed from the mortuary that had buried her brother. The owner of this private park at Kid Valley, not far from the mountain, let them use the site for free. The sweet scent of barbecuing meat wafted through the air. Photographs and clipped news stories of lost loved ones covered the tops of tables placed under the canopy.

Several people gathered around a plaque at one of the tables that listed the dead and missing. Victims' relatives had contributed over twelve hundred dollars to have the plaque made, with Parker putting up a big part of the money. "As long as the mountain shall live, so shall the memories," it read in part. The plaque showed an outline of the restricted zone, and the message was clear: almost all the victims had met their deaths outside its boundaries. The relatives hoped to get the plaque placed in one of the monument's visitor centers.

"Donna, look over there," one of the relatives said. She looked up to see a television crew from Seattle setting up their cameras not far from the gathering. The TV reporter was getting her microphone ready. "This is a private picnic," Parker told her. "The media is not allowed here."

"We just want to ask you a few questions."

"I assured these people that they would not have to talk about the mountain to the news media," Parker said as she took the woman by the arm to guide her off the property.

The reporter pulled her arm away. "Don't touch me," she screamed.

"Get off this property," Parker said.

The crew went across the street and set up their equipment. Parker hopped into her blue van and parked it to block their view. The crew then changed position. So Parker and some of the relatives hung up tarps to block the filming. Eventually the crew gave up and left.

That evening, however, the relatives allowed Michael Lienau to film the group as part of a future expanded version of his documentary *The Fire Below Us*, which had recently been released. He had convinced Parker and the others that he would record the event tastefully for the historical record.

As the relatives sat around a campfire, some of them opened up to Lienau. For many, it was the first time they had talked publicly, or even privately, about their ordeal. The interviews started with halting, hesitating voices, but soon words gushed out as emotions pent up for the past ten years found release. The Murphy sisters and Roxanne Mansfield were among those who told their stories.

Behavioral scientists had followed many of these people and other relatives of the dead and missing for most of the decade. Studies had spawned a new psychiatric term, using the initials of the mountain, for what scientists were finding: MSH disorder, a combination of anxiety, depression, and post-traumatic stress. These symptoms were found to be more intense in relatives of St. Helens victims than in victims' relatives who had been studied after other disasters and wars. The symptoms also seemed to linger longer. The St. Helens relatives experienced more divorce, unemployment, and family dysfunction than the population as a whole.

The media's intense coverage of the eruption made the deaths public events, adding to the stress, said Dr. Shirley Murphy of the University of Washington psychology nursing department, who had studied the relatives. Also, unlike in other disasters, the victims had been vilified as though they had been responsible for their own deaths, even by the president of the United States. Through her interviews Murphy found that those with family members missing and presumed dead suffered more than those whose loved ones' bodies were recovered.

As the relatives said their last good-byes to Donna Parker that night and she poured water over the coals of the campfire, she felt good. In some way, she knew she had helped.

CHAPTER 27

1990-1991

HE CAREFULLY OPENED the package, and there were the rocks and the note. The note was in the vein of others Peter Frenzen had received, also in packages containing rocks. "My wife has left me. My car has broken down. My life has gone downhill since I took these rocks from the mountain. Please take these cursed rocks back to the volcano." And Frenzen did as he was asked.

It was illegal to take rocks or anything else off the mountain, which had special status as a national monument. It was also illegal to place anything on it. One day a scientist, working at his research plot on St. Helens, found a bag of ashes labeled as being from a funeral home. Frenzen guessed that someone had hired a plane to scatter the ashes over the volcano, but instead the whole bag had just been flung out. Forest Service personnel contacted the funeral home and told its owner that it was illegal to leave anything on the mountain.

Venus turned on the TV to see the documentary *The Fire Below Us*. As she watched, she grew furious. A reenactment, using actors, of her and Roald's fight for survival was grossly inaccurate, she thought. It even showed her and Roald fishing from a big rock when the floodwaters hit them. That's not how it happened.

Then, there was the interview with Roald. Long ago they had promised each other they wouldn't do interviews without the other being present.

Toward the end, the narrator said, "Roald Reitan doesn't see much of Venus Dergan these days. The mountain has come between them. A reminder of terrors they would rather forget."

She immediately phoned Roald. What had happened to the promise they made to each other? "God, Roald, you didn't even contact me about this."

Ten years after Frenzen started research on plant succession on Mount St. Helens, another paper updating and detailing his findings was published in the *Journal of Vegetation Science*. The research was integrated with that of other scientists in his field, who shared credit with him. The conclusion was that a variety of plants could grow on the harsh, ash surfaces in the volcano's scorched and blown-down forests. St. Helens had shown scientists that even in the bleakest areas, if there were fallen trees, snags, isolated forest stands, or a mixture, life could be regenerated.

Much of the information Frenzen and other scientists had gathered was valuable for formulation of the New Forestry principles that Frenzen's former teacher Jerry Franklin was espousing. New Forestry had applications for the recovery of distressed areas whether from burning, clear-cutting, or other man-made or natural events. This approach stressed the importance of leaving a substantial legacy of plants and trees to help forests and wildlife recover after logging, rather than continuing with the industry practice of stripping forests clean in clear-cuts.

Research results were being published in time for scientists to influence the debate over the endangered northern spotted owl and old-growth forest. Franklin said that if timber companies observed New Forestry techniques instead of clear-cutting, conditions would be created over time suitable to the owl and other wildlife thought to require old-growth forest.

Several politicians latched onto the New Forestry as providing, perhaps, a compromise between the environmentalists who wanted no logging and the timber companies that wanted to keep clear-cutting. But environmentalists and the companies were united in their dislike of the New Forestry. Environmentalists did not think it would work and would only provide loggers a pathway to finish destroying the old-growth forest. Timber interests saw it adding to their costs.

The Weyerhaeuser company even went so far in their advertising as to imply that their techniques of mass, uniform plantings on stripped land were superior to the New Forestry in supporting wildlife. Newspaper advertisements said that after the May 18 eruption, "Regeneration of the forest began almost immediately. Within seven years, Weyerhaeuser crews planted more than 18 million seedlings. . . . This work to give Mother Nature a head start has paid off. Today, the trees planted are healthy and growing; some are more than 25 feet tall. All of the Weyerhaeuser forest land in the path of the volcanic eruption is stocked with healthy, vigorously growing trees, mostly Douglas fir and noble fir.

"And every animal species known to have inhabited these acres before the blast is back. . . .

"The scene is quite different on neighboring lands unaided by humans. . . . But even these lands should return to forests eventually, most likely after several decades of slow and uncertain natural regeneration."

What the Weyerhaeuser advertisements failed to mention was that several studies had shown that the method of planting used, where uniform trees ended up tightly packed on acre after acre, actually precluded any chance for return to an old-forest eco-system and eliminated many forms of wildlife. The lands where the company had cut isolated stands of old-growth timber in the name of "salvage operations" would, under Weyerhaeuser's replanting practices, never again see the forms or levels of wildlife they had once sustained.

She hiked up the stairs to the Cowlitz County Courthouse. Now that the anniversary events were past, Parker had time to return to her passion of investigating what she saw as unanswered questions involving both the St. Helens trial and the number of deaths on the mountain.

For years she had wondered what happened to those black bags of possessions she saw in the courthouse basement a decade before when she was looking for her brother's things. There were license plates on top of many of the bags that authorities could have used to locate relatives and return the items. But had they?

She told the clerk what she wanted. The woman gave her a vacant look before leaving to ask someone else. No one knew about such bags, the clerk said when she returned. Parker was not surprised. In her experience many items related to the victims, including important documents, were said to have been lost or destroyed.

Donna Parker and Helen Scymanky had approached several attorneys about taking on new suits against Weyerhaeuser and the State of Washington. But none would take the cases on contingency. The least that any of the law firms would take as retainer to initiate action was ten thousand dollars. The women couldn't come up with that kind of money.

Parker sent a four-inch-thick document she had prepared detailing her concerns to the state attorney general. The attorney general's office wrote back saying the agency had no jurisdiction to address Parker's grievances.

The reporter never really knew the former splendor of Mount St. Helens. Now he was entranced by its present-day stark beauty. Stepankowsky was a student of poetry, and he wrote his own poetic Thoreau-like song to the mountain in his reporting for the *Daily News*.

In the middle of summer, millions of cotton-covered seeds blow through the sky around Mount St. Helens in a blizzard that sweeps new life into a once barren land.

Purple lupine is so densely packed that it prevents other alpine plants from taking root. But lupine periodically dies off, leaving rich beds for competitors to take root.

Mount St. Helens emits tons of gas that makes acid rain, which breaks sterile ash into fertile soil.

Spiders spin parachute-like webs and fly into the ashen landscape. They feed on insects that blow into the area.

Nature, as these examples show, is weaving an elegant and intricate tapestry of change that is creating beauty of its own in the land lashed by Mount St. Helens ten years ago.

Territory once labeled a 'moonscape' now blooms with color. Gray still dominates. But every summer the palette of the hills and rills becomes more flush and lush with foxglove, Indian paintbrush, mountain ash, huckleberries, fireweed, pearly everlasting, and other plants.

Nearly everywhere in the volcano's fan-shaped blast zone, life has returned to nooks of survival: near a water seep, on a little mound of original soil kicked up gophers, atop elk droppings, in tree root wads.

Erosion cuts ash from a hillside. A seed takes root. Insects take up residence in the plant, and a bird finds its prey there.

These links of life are weaving a pattern of natural recovery that in 200 years will build a majestic forest like the one the volcano swept away.

The photograph that showed light streaking down through the huge split tree onto Wally Bowers' cross, wedged into the base of the dead fir, had become almost an icon for relatives of the dead and missing. It was a large photograph, on display in the Cowlitz County visitor center. The caption provided the only acknowledgment at any of the various volcano visitor centers that almost all the dead had been killed outside of the restricted zone.

Donna Parker didn't place the cross at the split tree in order to create a photo opportunity. She simply wanted to make it easy to find the cross on her yearly maintenance trips. Less than fifty feet away she had placed the cross for Tom Gadwa. The men met their deaths together while cutting timber.

At first, not many families asked to visit the crosses. Many said it would be too painful. But after the tenth anniversary gathering of the victims' relatives, more family members asked Parker to escort them to the cross that marked the place where it was believed their beloved had died.

Now she was taking the Gadwas to their son's cross in front of a stump that rose like a finger pointing into the air. She knew it had taken a great deal of strength from this couple in their seventies to come forward. For years they had declined to visit the cross. The crosses of Gadwa and Bowers stood on a wide, barren hill with a few snags and stumps sticking out of the gray ground nine miles northwest of the volcano. The group had to hike up a steep bank to reach a somewhat level spot to traverse over the hillside to Gadwa's cross. They carried twelve fir seedlings the parents wanted to plant.

On the way up, Thor Gadwa made it plain he was doing this for his wife. But when they got there, Parker could see the man was powerfully moved. The couple stared for a long time at the cross. "Thank you. Thank you," Thor Gadwa finally said. After another moment of silence he spoke once more. "Tom would have appreciated you putting his cross near that stump." Parker walked off to give the couple privacy as they planted the seedlings in a circle around the cross.

It seemed like he had read every piece of text so far at least a dozen times, but he wasn't complaining. Frenzen found it exhilarating to work with the creative people on the exhibits and videos for the new visitor center. He had seen every mockup so far, reviewed all the plans for displays, and taken the videos home to watch.

The professionals had done a wonderful job in capturing the essence of the mountain. It was give and take, he found, to produce the displays. They had to be engaging but at the same time scientifically accurate.

"We have to have sound in that part of the video," the director told him.

"Most of the eyewitnesses told us it was silent," Frenzen said.

"You can't have the thing be silent when it erupts."

"I think it would be more dramatic and eerie just to have the silence," Frenzen said.

"Look, you go to your local video shop tonight and get a dramatic action movie. Take it home and watch it on mute. Then turn the sound on and watch the same scenes again."

Frenzen did that. The director was right. They had to have sound. But he was going to make sure it was accurate sound. They ended up using the sound of trees crunching and the roar of the avalanche that witnesses had reported hearing.

He was extremely proud as he saw the exhibits and displays come together. They told a good story in a dramatic but accurate way.

"I just got my tape of the *Fire Below Us* and watched it," Donna Parker told Helen Scymanky over the phone. "I think that Mike Lienau did a great job. Have you seen it, yet?"

"Yeah, the kids and I watched it," Helen said. "He did a good job."

"Jim did, too," Parker said. "I think it's all going to help him cope better."

"Donna, Jim won't watch the tape."

For fifteen years he had been dreading the day the phone would ring and they would tell him his father was dead or dying. He was never really sure how he was going to feel. Then, when it happened, Rogers found himself rushing to the hospital to see his father in the last moment's of the old man's life. There was still so much to say, so much anger to dispose of.

His father had gone into the hospital for heart surgery. After the operation they couldn't stop the flow of blood. "He's bleeding to death on the operating table," his sister told him. Soon a nurse took Rogers in to see his father, who was now in another room. The old man was conscious and vomiting blood. All Rogers wanted to do was run. And he did.

Later they called him to say his father had survived. He was going to be released from the hospital in a few days. But his father had a form of hepatitis that was eating away at his liver and he was not expected to live much longer. His sister asked Rogers if he would install an air conditioner in the house trailer where his father and mother lived in Orchards, Washington. It was summer and the heat would be stifling.

As he measured the space for the air conditioner, Rogers thought how ironic it was that his father had remodeled houses and built things for people all his life and now had ended up in what Rogers called "this goddamn, stupid little house trailer." Rogers went to the small shed that served as a workshop. He wanted to cut some framing on his father's table saw. Rogers looked for an operating manual, but instead found something far more valuable.

Years ago, before becoming alienated from his family, Rogers and his friends had built a pole barn for his father. His father had never expressed appreciation. Now, in looking for the manual, Rogers pulled down a clipboard that hung over the workbench. It took him a few moments to realize what he was looking at as he gazed at the paper on the clipboard. It was his old drawings for the pole barn. His father had moved several times around the country, and yet he had kept the set of drawings in this cherished place, his personal workshop.

Rogers thought he understood what that meant. It meant his father loved him, he told himself. But his estrangement had been so profound, Rogers had trouble processing this new insight.

That week, his father died. The funeral was in a small church packed with his father's friends. Rogers listened as speaker after speaker told how proud his father had been of his children. Rogers had always believed his father despised him.

Rogers thought about the bridges, the tall buildings, and the mountain where he had risked his life. He couldn't decide if he had been climbing to get to something or to get away from it. And now he wanted desperately to climb something again, to rise high above the earth, to be free of it all.

Venus gave the interview alone to the newspaper reporter. It felt strange not having Roald by her side. She had given notice at her food service job. She had decided to turn her life around. It was time she invested in herself and concentrated her energies on improving her life, Venus thought. If there was a good man involved in it, so much the better. But if there wasn't, she knew she would be all right. She enrolled in school to become a travel agent.

The research scientist had forgotten his tongs, so they were forced to hold vials in their hands over the fumaroles until hot water spurted out and they could collect a sample. Peter Frenzen had never been this close to these vents on the dome. This was the type of activity he liked best in his job: helping experts in their research on St. Helens. It exposed him to cutting-edge thinking in a variety of sciences.

The scientist he was helping was studying bacteria that lived in the hot water of the vents. They were believed to be the most primitive life forms on the planet. Scientists were studying them on St. Helens for their ability to live under harsh conditions and to derive their energy from breaking down the bonds of various compositions of minerals. These scientists thought the bacteria might be used to eliminate toxic waste from groundwater.

Scientists also theorized that the bacteria could travel on a network of molten rock from a source beneath the continental plates at sea, to and from the volcanoes of the world. But the theory that fascinated Frenzen most was that these bacteria were the basic units that had led to life on earth and eventually to humans. And they had been created, many scientists believed, in the crucible of a volcano.

Frenzen held up one of the vials to the light with fingers that had grown red from trying to capture the hot water. It amazed him that in this bubbling liquid he could be holding the beginnings of life on earth.

The news shocked Swanson. A friend and colleague of his had been killed by a volcano. Harry Glicken met his death while observing dome-building eruptions on the Unzen volcano in Japan. Two other volcanologists and forty Japanese journalists had perished with him.

They had been high on a hill across a valley from the volcano. All previous pyroclastic flows from the volcano had been diverted away by the valley. But in an event eerily similar to St. Helens, the volcano dome had unexpectedly collapsed, releasing a huge flow of superheated gas and rock. A cloud of material from the flow surged through the valley and up its side to kill Glicken and the others on the hill.

When Swanson got the news, he was no longer at St. Helens. He was still employed by the USGS, but he was now doing research at the University of Washington in Seattle.

Swanson thought back to the grieving Glicken, who could not be consoled after David Johnston's death. The three volcanologists had been inextricably connected to Mount St. Helens, and their names always would be linked. Swanson remembered the good times the three of them had shared.

Then he thought of Glicken and the pioneering work the young man had done in mapping the debris field at St. Helens. But even in his grief over Glicken's death, he had to smile to himself when he thought about the eccentric side of this scientist who absentmindedly drove through stop signs while ardently expounding on his work to a terrified Swanson.

PART 5
1999-2000

Remembrance

Where were you when the mountain erupted?

—Robert Rogers

CHAPTER 28

1999—2000

VENUS DERGAN couldn't believe what she was seeing. The site looked just as it had that serene morning twenty years before. The pristine waters of the river gurgled in a gentle flow. The wind whispered through the firs. The morning sun dappled the ground. She had seen photos of the devastated area after the flood. Now she frantically looked for something that would confirm the reality of what had happened to her and Roald, but it just wasn't there. It was as if her ordeal had all been a fantastic nightmare.

"Damn, this looks like a great place to fish," said a member of the Discovery Channel crew. Venus smiled weakly. The cable television channel was producing a documentary. They had brought her and Roald back to the place they had been camped the day the volcano blew.

The producers had found the helicopter pilot who rescued Venus and hired him to fly her and Roald to the mountain. As they waited anxiously for the helicopter to land, Roald asked Venus, "How do you thank someone who risked their life to save you?" The helicopter started its descent, and Venus shielded her eyes from the sun to watch. "It was like an angel from heaven," she said, thinking back to the day of their rescue.

They backed away from the surging rotor wash as the helicopter landed. Pilot Ray Pleasant got out and walked over to them. Roald shook hands with him. Venus held out a hand and Pleasant grasped it with both of his hands. "It was incredible flying that day," she told him.

The wrath of the volcano that caught her and Roald in its grasp, and the fury it unleashed on the land, became real again. She had never seen the devastation in the Toutle River Valley from the air. For only the second time in her life, Venus flew aboard a helicopter. Again it was Ray Pleasant at the controls. With a cameraman filming, Pleasant pointed out the huge mounds of debris along the Toutle River. The closer they flew to the mountain, the more barren the land became, with gray and denuded ridges and hills in afternoon shadow. Usually when Venus traveled in a car, the closer she got to the mountain, the greater her anxiety became. But now she felt comfortable, even relaxed. She was with the former Vietnam pilot who had helped save her life.

The mountain, stark and majestic in a clear sky, loomed before them. Venus was startled when Pleasant piloted them right into the crater, but she was still not afraid. "Oh, god," she yelled out more in excitement than fear as she looked at Roald, who appeared as awestruck as she. The helicopter flew around the dome, with the crater walls on one side so close she could see the veins in the rock. Then Pleasant hovered in the gap on the north face that had been blown out in the eruption. She looked down over the miles of the valley below and could see the path of the flood that almost killed her when she was just twenty years old.

They came around the bend of the trail in the still-gray blast zone and stopped in their tracks. The group led by Peter Frenzen gathered around the remains. The wind blew through the gristle and tufts of hair still stuck to the two elk skeletons.

"There is currently more elk in the monument than there is food," he explained. Hundreds of elk had starved to death the previous winter. The Forest Service had a policy in the monument of not interfering in what was thought to be natural cycles maintaining a balance of nature. Ultimately, however, it was up to the state's wildlife department to decide if the animals should be fed. "It was a good year to be a coyote," he said, but no one laughed. "Left alone, the elk population eventually will reach equilibrium."

He watched as the group seemed to unconsciously step back. Frenzen realized they did not view the elk bones with wonder at the natural order of things, but with pity at the sight of the remains of two dead creatures. It was a feeling he too shared even with all his years of trained objectivity as a scientist. "It was pretty sad to see elk out here last winter," he told the group. "They couldn't even get up, they were so weak."

The Longview *Daily News* had been sold to a newspaper chain. In interviewing publisher Ted Natt for a story about the sale, Andre Stepankowsky found him shaken and morose. Natt paused often and sighed to control his emotions as he tried to answer questions. Many times, he said, he had thought about backing out of the deal.

The paper's economic base had steadily eroded, even while its reputation as a great small-city newspaper had grown. The volcano had brought the *Daily News* fame and glory but it had also contributed to its long, slow decline from the effects of the eruption on the local economy. Weyerhaeuser had cut back operations. Other timber companies had shut down or struggled to survive. The volcano never lived up to its promise as a lucrative tourist mecca. Ted Natt and other members of the Natt family decided to sell before the paper's economic fortunes plummeted much further.

The new owners saw potential for the future. The surging national economy coupled with a booming Portland, Oregon, only fifty miles away placed Longview on the edge of robust growth, an executive of the Howard newspaper chain said. As for the newspaper itself, "There is an expectation that the bottom line will be improved," Bill Howard told Stepankowsky. Howard would make no commitment to filling the vacant news staff positions. Many of the paper's news reporters had jumped ship during the months of uncertainty preceding the sale.

Later two of the chain's executives wanted to talk to Stepankowsky. Would he be interested in the position of city news editor? He and Paula had started a family and put down roots in the community, and he had that continuing fascination with the volcano. He told them he would take the job.

One day managing editor Bob Gaston walked up to Stepankowsky. "This is my last day at the *Daily News*," he told the shocked journalist. The chain's executives had plans to change the *News* into a morning publication, start a Sunday edition, and introduce other sweeping changes—all at once, and without a commitment to adding news staff. He had given the new owners notice of his retirement, anyway. When he complained about the short staffing just before a scheduled vacation, they told him he didn't need to come back.

Robert Rogers was busy cleaning up a building for the nonprofit recycling corporation he had helped found. The corporation had purchased the large structure in Portland from a company that built displays and exhibits. On top of a spray booth he discovered a fiberglass model of Mount St. Helens as it looked after the 1980 blast. Rogers took the model home and showed it to his wife, Janelle. The twentieth anniversary of the eruption was approaching. He thought he could use the model if he was called upon to make a presentation of his experiences.

His wife had gained some idea of what St. Helens meant to him when he insisted their newborn daughter, Molly, be given the middle name Tephra—the technical word for the solid material ejected during a volcanic eruption and transported through the air, commonly as ash. But she had no idea of the scope of his past involvement with the mountain until she now watched him bring out thousands of volcano slides and photographs, notes from forays into the volcano, assorted memorabilia, technical books and articles, and scientific paraphernalia.

As she watched him scatter his treasures across the living room, she realized this was the sort of thing that drove her crazy and attracted her to him at the same time—his ability to immerse himself in his latest passion. She was a natural resources economist and a graduate of Harvard's Kennedy School of Government. She was rational, analytical, organized, while he could be impetuous and adventuresome, impervious to the chaos in his wake. They balanced each other. She was also attracted by his intellect. She told friends that Rogers was one of the two most brilliant people she had ever met, the other one being an internationally renowned Harvard professor.

While he was the one who usually came up with the surprises in their relationship, she almost one-upped him the year before. They had gone together for years and he had agreed with her that they should marry. But he always seemed to be too busy. She decided to shanghai Rogers for a surprise wedding. She invited guests and made preparations.

Two weeks before the wedding, she became afraid he might suddenly go off on one of his expeditions at the last moment and miss the surprise ceremony. So she told him about it. He agreed to the wedding but wanted to pick the location. He selected the floating dock in Cathedral Park beneath the St. Johns Bridge, a 1933 suspension span that was one of Rogers' favorites. The bridge crossed the Willamette River in Portland. On the cold, damp December day of the wedding, fog swirled around the bridge above them. John Law, former head of the now defunct San Francisco Suicide Club, read to the gathering from an article by the engineer responsible for construction of the bridge.

Then the crowd parted to allow the bride, wearing a white, traditional wedding gown and cowboy boots, to pass as they sang "Here comes the bride." She joined Rogers, clad in a tuxedo, and the minister at the edge of the dock. He gave her a ring made out of metal scrap left over from renovation of the bridge. The captain of a giant tugboat coming down the Willamette saw the wedding. The tug tooted its whistle and used its side thrusters to pirouette three times in honor of the couple.

"Not everyone can say they had a tugboat dance at their wedding," Rogers told Janelle. On his wedding night he climbed the bridge with John Law. Janelle didn't climb bridges or volcanoes.

On a little-used road in the blowdown zone, Frenzen drove the Forest Service truck through a thicket of firs and brush. He stopped and checked his map. The study plot should be right here, but the landscape was now so overgrown that it was difficult to tell.

Frenzen was in a race against time to save what he could of the hundreds of ongoing scientific studies conducted on Mount St. Helens over the previous two decades. Funds had steadily declined for research and many scientists had been forced to abandon their studies.

He had been given a small appropriation to mothball what studies he could in cooperation with Oregon State University. His office and other rooms at his headquarters were filled with boxes stuffed with study results. The content of the boxes represented countless years of scientists' lives spent under often harsh conditions studying the effects of volcanoes on the environment. What he was fighting to save might never be duplicated.

He pushed aside a tall plant. There was the rebar that marked the edge of the plot. Frenzen took the GPS coordinates with a handheld device and noted them on a clipboard. Later he would have steel fence posts replace the rebar that elk had been known to kick out of the ground. He climbed back into the truck. He had many other locations to find today.

"Ted's missing." The call that Sunday morning was from John McClelland III, Ted Natt's cousin. "He was supposed to fly back in his helicopter from Oysterville on the coast and he is missing."

"Oh, my god," Stepankowsky said. As soon as he finished speaking with McClelland, he started making phone calls from his home. He felt conflicted. On the one hand he now had a job to do as a journalist but on the other he wanted to help the family of the former *Daily News* publisher as much as he could. As he talked to people he found himself falling back into his old habit of referring to Natt as "my boss."

The next morning Stepankowsky started coordinating coverage before he went into the office. He heard that Natt's wife, Diane, had gone to wait at the local airport. He assigned a reporter to interview her briefly. When he got to the office Stepankowsky learned that ten Civil Air Patrol planes were involved in the search along with several planes piloted by volunteers. Natt had relatively few hours of flying time in a helicopter, but he had extensive experience piloting other aircraft.

A man came out of nowhere, yelling at Stepankowsky. "Get your goons off my mother. Get your goons off my mother." It was Natt's son, David, who had remained as operations manager at the newspaper after its sale. For a moment Stepankowsky was dumbfounded. He looked at the man and tried to control his own rising anger. "What are you talking about?"

Stepankowsky asked. Then he remembered the reporter he had assigned to get a quote from Ted Natt's wife. "This is a story I've got to cover," he told David Natt. "I'm assuming we're being sensitive about this." Ted Natt had been the consummate newspaperman and Stepankowsky was determined to cover this story with the thoroughness and professionalism he knew his former boss would have demanded.

Over the next week hundreds of volunteers joined in a fruitless search. Finally it began to sink in that the straight-talking but popular newspaperman had perished. For weeks the *Daily News* carried page after page of letters in testimony to the high regard in which Natt was held by journalists, other colleagues, and citizens of the region. Six weeks after he was reported missing, Ted Natt's body and the wreckage of his helicopter were found by a hunter on a hillside near Clatskanie, Oregon, not far from Longview. He was fifty-eight.

Hundreds of mourners turned out for Natt's memorial service at St. Stephen's Episcopal Church. Speaker after speaker told of his life, his kindness, and his love for journalism. They also spoke of his remorse for selling the *Daily News*.

"Take a look at this," the *National Geographic* photographer told Frenzen as they stood on top of the lava dome behind the man's tripod and camera. Frenzen looked into the viewfinder. He didn't know it was possible to see what he was seeing. The top of the dome seemed to dissolve into almost the entire crater rim that surrounded it. "That's a *Geographic* lens," the photographer said. "That's not my lens."

So that's why the pictures in *National Geographic* look so good, he thought. Their special lenses and top-notch photographers gave a scene an aura all its own, capturing the spirit of a place that went beyond what the naked eye could see or a regular lens would show.

For days he had been camping with the photographer and his team. Frenzen had shown him his favorite spots on the mountain. He had taken the photographer ten miles away from the peak on the northeast side to the blown-down forest where, looking back, the entire valley lay with huge, stripped trees lying on their sides. Here the distance from the mountain and the scale of things attested to the power of the volcano and what it had wrought on May 18, 1980.

From Johnston Ridge he had marveled with the photographer that the entire top of St. Helens had been blasted five miles across a valley, its pieces traveling over ridges, hills, and even another mountain, its path still visible twenty years later. It was like seeing a long luge run down the valley, high on the sides and with the center scooped out by the force of the avalanche.

Standing on the pumice plain brought a completely different perspective. They looked up over the huge slope of gray gravel to the great, gaping hole in the crater and tried to imagine the immense amounts of matter that had traveled this way with the force of many atomic bombs behind it. He took the photographer to Loowit Falls, spilling hundreds of feet down the north slope of volcano, carving its path through a hard rock canyon that an unknowing observer might have thought had taken thousands of years to form but had been cut in less than a decade.

Frenzen's favorite spot was on top of the crater rim looking out to the north. That view helped put it all together, looking down at the dome, and then up through the gap, across the pumice plain, out to Spirit Lake and beyond to where the blast zone merged with the blown-down forest. The mountain had become part of him, as if he carried its spirit within him. He had grown very protective of it.

"The news media started really dumping on the people that died up there," Donna Parker told the reporter for the Discovery Channel. "And my mother would have friends come up to her in town and they would say, 'We're sure sorry to hear about your son.' And she would thank them, and they would say, 'Well, he should have known better. Everybody knew that thing was going to blow.'"

"It took us over a year to find out that was not a closed area where he was," she said.

She told the reporter everything. She told about the skimpy Red Zone, about the government refusal to confess its true boundaries, about her belief the St. Helens negligence trial had not been fair, about her suspicions that more people had died on the mountain than authorities reported. She told about her frustrations in failing to get officials or the media to believe her, in failing to find any attorney to take on a new lawsuit. She showed the boxes of material she had collected over twenty years—stacks of notes from interviews, hundreds of newspaper clippings, thousands of pages of documents from private and government sources.

She didn't tell the reporter what her sometimes lonely quest had cost her personally. She knew she had become obsessed. Over the years friends and relatives had drifted out of her life, telling her she needed to put it all behind her. She had spent tens of thousands of dollars of her own money helping the relatives of victims and financing her investigation. That was no mean feat for a single mother who had never had a job paying more than ten dollars an hour.

She couldn't help but think that the Dixy Lee Ray government and others had gotten away with what she called the "big lie"—that most people still believed the victims had brought on their own deaths, that government and vested interests carried no blame. Or she thought that most people no longer cared and the victims were being subjected to the worst injustice of all, to have the lie of their deaths embedded as history. Time had become her enemy. Many of the key players in the St. Helens saga, people like Ted Natt, Dixy Lee Ray, and Judge James McCutcheon, were now dead.

The Discovery Channel crew filmed her walking into the visitor center at Silver Lake. The Forest Service had removed language characterizing the victims as thrill seekers from its movies and displays. The agency also had allowed her to place the plaque with the names of the dead just inside the entryway to the center. Now the cameraman filmed the plaque and scanned the names as she pointed to them. "They did nothing wrong," she said. "They did nothing wrong."

CHAPTER 29

2000: Looking Back

THE ROILING, BLACKISH-GRAY TOADSTOOL column listed to one side, like a cyclone, as it rose miles into the sky. Cauliflower clouds blossomed out of the volcano while the plane flew closer to the eruption stem. As Swanson watched the film on his monitor, lightning streaked across the clouds. A wall of black filled the screen as the plane tried to penetrate the hellish mass to get to David Johnston.

The film had taken on a yellowish brown cast, like a faded photograph. For the first time in years he was reviewing the footage he had taken during the eruption. The multitude of hundred-foot rolls of film he had taken that day with an old Bell and Howell windup movie camera had been spliced together and recorded onto videotape. Now he was editing the tape down to a shorter version, to show during events leading up to the twentieth anniversary.

It had been a busy time. Swanson had been appointed to one of the USGS's most prestigious positions in 1996—scientist in charge of the Hawaiian Volcano Observatory. In the late 1990s he had launched an ambitious project, a stratigraphic analysis of the Hawaiian volcanoes.

The passage of years brought a freshness to Swanson's viewing of the film. As the view on his screen pulled back from a close-up of the cauliflower clouds, a thin ruffle of ash collapsed off the eruption column and poured down the mountain at great speed. He stopped the video and rewound the tape a short way. He played the scene again. He thought he was seeing a pyroclastic flow, a fluid cloud of hot gases and volcanic ash, formed by the densest material falling out of the eruption, that could move at hurricane speeds as it hugged the ground. He repeated the scene several times. He was now certain of it. Ever since the eruption scientists had thought the pyroclastic flows only occurred later. While this flow did not have the vigor of pyroclastic flows recorded at the zenith of the eruption, it was still of significant size.

Swanson marveled at how he and other scientists were still learning about the eruption. This tape had been viewed hundreds of times by specialists and no one, to his knowledge, had ever caught the scene of the earlier pyroclastic flow. It was not a major breakthrough, but one more bit of scientific understanding.

Stepankowsky was beginning to feel like he was in a surreal nightmare that started with rumors about the sale of the newspaper, continued through the trauma of Ted Natt's death, and now seemed never-ending as he walked into the *Daily News* office being remodeled. Workers were drilling and hammering, and construction dust floated in the air. His reporters sat in a row behind their desks in a jerry-built newsroom, reminding him of call-takers on a Jerry Lewis telethon. He resented what the Howard newspaper chain was putting the staff through. There was the remodeling of the building, the start of a Sunday edition, the change to morning publication, and other orders issued from on high with little regard to impacts on the staff or community.

He sat at his desk and absentmindedly ran his hand through thinning hair spotted with construction dust while he looked at a story on the volcano. The Forest Service had announced that a budget shortfall would cause it to close its most popular St. Helens visitor center, the one at Silver Lake. The center was to be closed just before the twentieth anniversary. The story fit right in with Stepankowsky's view of a world gone haywire around him.

"You can't believe how devastated it was. There was nothing out here," Peter Frenzen said with a sweep of his arm over the hummock mounds and valleys that were so barren that some reporters thought little had changed.

"On one trip a scientist found a small part of a leafy plant and became ecstatic as about twenty of his colleagues gathered around him," Frenzen said. "They looked at it for a while before one of them figured out it was an alfalfa sprout from the sandwich he had eaten for lunch that day." The story always drew a laugh from journalists, but Frenzen found it hard to laugh at his own humor. He had already told the story several times that day. Media interest in the volcano had been much more intense than he expected.

Frenzen walked farther up the dusty trail amid rolling hummocks, or miniature mountains, of spewed rock, volcanic clods, and pieces of old lava domes that St. Helens had purged from its north-flank innards. "When the eruption is over, it's not over—the consequences go on," Frenzen said. A gray, sandlike mantle covered the ground everywhere with bursts of white blossoming lupine, red fireweed, and green-needled pine, poking up seemingly at random, like a painter's afterthoughts on a monochrome canvas.

To Frenzen, though, the apparent random placement of plants had order and reason. After almost twenty years, he could explain to an inquiring reporter why a certain plant had taken root in a certain place. Frenzen dug his fingers into a mat of lupine, a plant that hugs the ground, producing white, dainty, four-petal flowers. "A few lupine seeds blew in here after the eruption. They grew in the nutrient-poor ash because their roots contain bacteria that helps pull nitrogen out of the air." Dead insects and dirt particles catch in the plant's tight weave, which acts as a net, he explained.

After the eruption about two million tiny spiders fell per day on the barren land, carried aloft by web fragments acting as parachutes. They did not survive. They provided food for numerous species of beetles that soon invaded the area. In turn both dead spiders and beetles supplied nutrients on which wind-blown seeds could take root.

At first the lupines were thought to be nursing plants that would help in the growth of other species. But research found they were actually jealous landowners that permitted no intrusion by other plants. It was only after the lupine had died off in a few years that other vegetation began appearing to feed off the organic material the dead plant provided.

Between the mounds thicketed with patches of red alder, fireweed, and scotch broom, the 2.3-mile Hummocks Trail now slithered like a long, grayish brown snake up to its crest, with the pumice plain and snow-covered St. Helens dominating the background. On the other side of the crest, the trail leveled out. He motioned to a grove of alders twenty-five feet high. Five years before in this same spot, they weren't even five feet tall, he said. "Give this place twenty more years and you might have trouble seeing the volcano's cone from here and other lower trials."

In twenty to fifty years, the alders, willows, cottonwoods, and other hardwoods now starting to thrive here will die out, he said. Lodgepole pine and eventually Douglas fir would begin to take over as the mountain returned slowly over the next 150 years to an appearance it had before the eruption.

A flock of geese rose from a nearby pond. Birds from many different habitats, to scientists' surprise, had colonized both the pumice plain and the blast area. Western meadow larks, horned larks, and rock wrens typically found in the Great Basin of Nevada and Utah and birds such as American pipits and gray-crowned rosy finches, usually thriving at higher altitudes in the Cascades, had joined species preferring lowland pastures, such as the red-winged blackbird and Savannah sparrow. Eventually, as the landscape matured, hermit thrushes and gray jays would dominate.

Water, where it seeped or collected in ponds, streams, and rivers after the eruption, helped provide a life-spring for the recovery of plants and animals. Many frogs, newts, and salamanders along with some other animals survived the eruption, Frenzen said, either covered over by ice or snow or buried in mud or burrowed beneath the ground. "The timing of the eruption is what really helped the frogs and some other aquatic species," he said.

The Cascades in May still have plenty of snow and many high-elevation lakes are still frozen over.

Totally new and unexpected symbiotic relationships developed between surviving species. For instance, when frogs emerged they found a hot, barren land of ash—totally alien to these creatures of the damp. Pocket gophers also survived, burrowed deep into their miles of tunnels beneath the mountain. With the loss of almost all their natural predators and an ample supply of roots from plants the eruption had sheared off above ground, the gophers thrived and built more tunnels. The frogs began sharing those tunnels to migrate from one watering hole to another. Returning elk even assisted with their weight and their hooves by unintentionally punching holes in the tunnels for the frogs' entrance.

Frenzen led the group back up toward the parking lot but stopped short of it by a few hundred feet. He could see the next group of journalists already milling around the lot. Recovery had been diverse and uneven on the mountain, he said. In some places there remained only barren desert. But within that desert might stand patches of Pacific silver fir or mountain hemlock that had been encased in snowdrifts or sheltered by lee slopes and ridges during the eruption.

"Large-scale disturbances don't leave a clean slate," he said. "You hear reports on forest fires that twenty-three thousand acres were destroyed. Well, how do you destroy an acre? Even then, there are patches spared. What was left behind is as important—is more important—because it drives regeneration. Plants and animals that survived or moved in, and the trunks and stems of blown-down trees, are all starting points for the next system."

The CBS News crew scoured Donna Parker's home and yard for the best interview settings. One of them spotted the rock. "What's that?" he asked.

"That's the rock that killed my brother," she told him. A cameraman filmed the rock from different angles and then turned the camera on Parker as a reporter stepped in to interview her. She told the story of her brother and the rock. "But why did you bring the rock down from the mountain?" the reporter asked.

She stared at the rock for a long while. "I don't know. I really don't know." It was the same answer she gave to people who asked her why, after twenty years, she was still investigating what had happened to her brother and the other victims and was still trying to bring out the truth.

Stepankowsky gave a close read to the reporter's story and asked questions to clarify some points. With less than a month until the twentieth anniversary, more than fifty politicians, businesspeople, government representatives, environmentalists, and scientists had just met in what was touted as a summit to discuss the future of Mount St. Helens.

"It's appropriate that we're having a summit on Mount St. Helens, because the mountain actually hasn't had a summit since 1980," quipped the emcee, Congressman Brian Baird, to a chorus of groans. But what was happening at Mount St. Helens National Volcanic Monument was no laughing matter, they all soon agreed. Federal dollars allocated to the monument had dwindled. The Forest Service had responded with orders to close its Silver Lake visitor center. Only a few days before the meeting, the agency relented and promised to keep the center open for a few more months. Revenues from tourists had sagged as attendance at the monument and its visitor centers plummeted.

The biggest issue dividing the group was a proposal to develop a loop highway to connect the east and west sides of the peak and the various tourist attractions and visitor centers. The route would cost about $200 million, said Don Wagner of the state Department of Transportation. "I think we're doing a disservice not to loop up the various visitor opportunities in southwest Washington and to keep the roads open as much as possible," he said.

Not everyone agreed with him. For once, Weyerhaeuser and the environmental interests could see eye to eye on something. Weyerhaeuser did not want the highway slicing through the middle of its industrial timberland. The environmentalists and some scientists bemoaned the amount of land the highway would disturb.

As he read his reporter's copy, Stepankowsky saw the same old issues emerging, but overlaid with the dashed expectations of the past twenty years. To businesspeople the mountain represented the somewhat fading dream of salvation through tourism. To timber interests it was a place that might once more become a lucrative tree farm. To scientists and environmentalists it represented lost opportunities as more of the monument was exploited for commercial and recreational gain.

It was the perfect time for a gathering Frenzen had long dreamed of. He hoped to bring together survivors of the 1980 eruption so they could meet each other privately and also tell their stories in a public presentation. For years he had watched many of these people interviewed on television, but he had met few of them. Not even many of the survivors knew each other, although they all shared in some way that horrific day twenty years before.

The event would be hosted by the Mount St. Helens Institute, which Frenzen helped found. It was dedicated to raising the public profile of the mountain through education. This might be the last chance for such a gathering. The Japanese had a proverb—that by the time the next disaster occurs, the people will have forgotten the previous one. Some of the survivors had died. It was becoming increasingly important to Frenzen and others to capture the memories of the survivors.

At first Jim Scymanky said no. He hadn't spoken before an audience since his news conference at the hospital almost two decades ago. But Frenzen argued that it would be for history and for the generations to come. The argument began to weigh on Scymanky. Maybe this would be a way to help preserve the memory of his friends and others lost on the mountain. Helen thought he should do it. Public exposure couldn't hurt the book they were trying to sell to a publisher.

Robert Rogers accepted Frenzen's invitation without hesitation. He didn't want this anniversary to end like the last big one, ten years ago, when he found himself sitting on the sidelines.

When Frenzen invited Venus and Roald to speak, they accepted. But when he also invited them to a private barbecue and campout for the survivors the night before the public presentation, Venus had to laugh. Did Frenzen think she was crazy? After her last campout on the mountain, camping anywhere near any volcano was the last thing in the world she wanted to do.

High above Weyerhaeuser's vast green timberlands, they stood on a ridge and looked into the mouth of the volcano's icy crater. Scymanky told the *Oregonian* reporter about how he and his sons now worked on vintage autos in a family business and that life was good. But on occasion he still felt an overwhelming sense of isolation and loneliness.

Helen pulled out three bundles of flowers. She began poking stems in a stump. "Evlanty," she said. She picked up another bundle and when she had finished, she called out, "Jose." After the final bouquet had been arranged, she whispered barely above the wind, "Leonty." She stepped back and stared at the flowers as she and Jim took deep breaths. Jim broke the silence as he stared down at the crater. "I don't know why I love it up here, but I do," he said in a low voice, as though he was not speaking to anyone in particular. "I always feel like they are up here."

It suddenly dawned on him that the mountain he had hated so long for what it had done to him, his family, and his friends was now a place where he could find peace.

CHAPTER 30

2000: A Matter of Time

"MOST VOLCANOES don't erupt often enough." It was Don Swanson speaking, and he was trying to highlight his concern that the lessons of the 1980 eruption might be lost. Time passes, and people forget. "By far the biggest problem we still face . . . . is sustaining public education and understanding of volcanoes," he told an audience at Washington State University in Vancouver. He feared that by the time St. Helens erupted in another massive blast, the mistakes of 1980 would be repeated, costing lives.

It was beginning to look like he might be right. Already a resort with lodge, cabins, RV park, and restaurant had been built in the blast zone just outside the national volcanic monument. And owners of a former lodge at Spirit Lake had applied for permits to build an upscale resort there. The Toutle River Valley, swept by deadly floods from the 1980 eruption, was now one of the hottest housing development areas in southwestern Washington. A young carpenter working on one of the new homes scoffed at the danger. "It would never reach here," he told reporter Rob Carson. "They might get a little ash, but that's about it. If it did go off, think of what a great view they would have."

The most immediate threat from the volcano area, Swanson believed, was from flooding. He still questioned the long-term security of the Spirit Lake drainage tunnel. He believed that dams would not be able to hold up to a major earthquake. And unstable avalanche material from 1980 was still all that held back some high mountain lakes.

The St. Helens crater was filling with rock debris from its sides and from the dome. Snow and ice over the years had packed into the crater, creating a glacier of over 80 million cubic meters, a greater volume by far than all the material swept down the Toutle River in 1980. It would not take even an eruption to set this material loose. Devastating flows could be triggered by small avalanches or even heavy rain.

As to the threat of a major new eruption, it was only a matter of time. There had not been a dome-building eruption, where magma reached the surface, since October 1986. The last time the volcano had emitted a steam and ash plume was in February 1991. There were no outward signs at the moment that the volcano was coming back to life. Swanson always emphasized these words: "at the moment." There would be another devastating eruption in the future. He liked to say that of all the potentially active volcanoes in the continental United States, St. Helens was the most likely to erupt next.

Whenever the eruption came, the science he helped develop would provide the foundation for its monitoring and study. The precise prediction method he developed for St. Helens worked for a steep-sided stratovolcano with a dome that was building episodically on a relatively flat crater floor. No volcano to match that description had erupted since St. Helens in 1980. However, much of Swanson's science and that of other St. Helens scientists was adapted for use before the eruption of Mount Pinatubo in the Philippines in 1991. Geologists rushed into action weeks before the eruption and installed monitoring devices, then called for an evacuation of 250,000 people. When the volcano erupted shortly thereafter with a force ten times that of St. Helens, it was estimated that the discoveries of Swanson and other St. Helens scientists had saved tens of thousands of lives. The science had also been used on active volcanoes in such countries as Ecuador, Mexico, Nicaragua, Indonesia, Guatemala, and Papua New Guinea.

And now this science was even more refined. New tiltmeters on St. Helens could detect changes in slope down to a fraction of a microradian—the equivalent of slipping a dime under one end of a half-mile-long carpenter's level. Global positioning systems were measuring motions of less than half an inch on St. Helens' surface .

One scientific instrument that promised exceptional benefits on St. Helens was not being used because bureaucrats and politicians deemed it too costly. The device, known as a bore-hole strain meter, could sense changes in volume of one part in a trillion of magma moving ten to twenty miles beneath the earth's surface, rather than at shallower depths where the magma had already entered the volcano. The government determined that several of the instruments would be needed at St. Helens and, at $100,000 apiece, ruled out the purchase of even one.

Yet for all the high-tech gear now in use, no one had been able to duplicate Swanson's feat. He remained the only person ever to make a long-range eruption prediction with any degree of accuracy. He did it not once, but fifteen times, sometimes to the day, relying in large part on an everyday tape measure you could buy at any hardware store.

"You want to go to New York?" Roald asked Venus on the phone. "CBS just called. They want us both to appear on the *Early Show.*"

"What?" she said. He had caught her off guard. On the show, there would be a mini-documentary about the mountain before Bryant Gumbel interviewed them. She agreed to go. She had lived in New York right after high school, and there were all sorts of attractions she would like to share with Roald.

When Frenzen picked up the newspaper, he thought he was going to read a story about the rebirth of St. Helens' lakes. But mixed in with the story in the *Seattle Post-Intelligencer* was criticism of the Forest Service's science and research programs on the mountain. The critic was not a newcomer just shooting off his mouth, but a respected scientist who had made the study of the lakes a good part of his career for the past twenty years.

Lake scientist Douglas Larson was adjunct professor of biology at Portland State University. At the time of the 1980 eruption he was a researcher for the Army Corps of Engineers. He was one of the first scientists on the scene, where he established baseline information for studying how a lake recovers from such devastation.

Through the reporter, Larson disclosed that he was one of the scientists who had discovered Legionella bacteria in the lakes, bacteria that made him extremely ill. The government "covered up" the discovery, Larson said, in order to not panic the public.

Larson had critical words for the Forest Service, which had jurisdiction over the national forest that included the mountain. He said the Forest Service did little to support research at St. Helens immediately after the eruption and had done little since. "There's been very little research done up there since 1986," he said. "They haven't had much interest in the science. What we have up there is not so much an area for scientific study as a tourist theme park."

"What ever happened to Billy's truck?" the reporter asked Donna Parker.

She thought back on the rescue of the truck nineteen years before. After they retrieved it, she had given it to Steve Gianopoulos, who promised to restore it. When he finished, he drove the truck back up to where Billy and Jean were killed on the mountain. There he drank a toast to them with the Yukon Jack he had found there, as he had vowed to do. After that Parker had kept tabs on the truck.

"The truck is now *living* in Alaska," she told the reporter.

Frenzen drove to a plot in the blast zone. It had been years since he had been here, and it took him a long time to find it. The seedlings he had planted almost twenty years ago had grown into a forest. As he touched the trees almost with reverence, he felt a sense of loss. This plot had once been at the center of his life. It had been a small part in the studies that had revolutionized forestry practices and man's knowledge of how the land heals itself after devastation. He had been fortunate enough to have been involved in the pioneering work of Jerry Franklin and other top forestry scientists.

Their research at St. Helens had turned scientific belief on its head. Before St. Helens, scientists tended to adhere to a philosophy of, "If something's dead it doesn't count anymore," as Franklin put it. The extent that downed trees, scorched snags, and other seeming devastation could play in an area's recovery was only truly recognized in studies after the mountain's 1980 eruption. What was learned at St. Helens was driven home and confirmed in research after the fierce 1988 Yellowstone National Park fire.

In 1994 this New Forestry emphasizing the role that "biological legacies" play in regeneration became a key part of President Clinton's Northwest Forest Plan. The plan dictated management of most federally owned lands within the range of the northern spotted owl, a huge hunk of the nation's forests. Live trees, scattered and in patches, had to remain on at least 15 percent of an area after harvest. On each acre, large logs and snags had to be kept. These dead and live trees were now recognized as lifeboats for species preservation.

"We want an interview," the producer from Portland's Channel 8 television told Rogers over the phone.

"Well, I've got to work tomorrow."

"Listen. We're in a production jam. We tried to get hold of you a week ago."

"I know. We kept playing phone tag."

"We need to talk to you *now*. We're doing this documentary and we want to show it before the anniversary. We've got deadlines."

"Well, I'm going to work Monday. You can come to work with me."

"What do you do?"

"I climb radio towers."

And so it was that on Monday, Rogers started climbing toward the 150-foot level on a radio tower, followed by a television reporter and a cameraman. Rogers climbed easily, the torn ligaments in his leg now healed. The reporter was doing fine, but the cameraman huffed and puffed as they worked their way upward.

At the platform the exhausted cameraman rested. Rogers had planned to be filmed with Mount St. Helens in the background, but haze hid the mountain. As he was interviewed on camera and told about the day of the volcano's fury, Rogers was not too disappointed. He would have a good story to tell about the day he made a news crew really work for their interview.

Stepankowsky sat down at his computer and words tumbled out for the lead editorial for the edition of May 18, 2000. The words had been bottled up inside him for a long time.

> *It's time to fess up.*
>
> *Twenty years after Mount St. Helens snuffed out 57 lives, the families of those victims never have received an official apology.*
>
> *Only three of the people killed in the blast are known to have been inside the restricted 'Red' and 'Blue' zones around the mountain the morning of May 18, 1980. And those three— including Mount St. Helens Lodge owner Harry Truman—had permission to be there. Yet, in the days following the eruption, officials, including President Jimmy Carter, accused those who died of being "lawbreakers," thrill seekers and fools.*
>
> *The truth is so much different.*
>
> *Vader logger John Killian died at Fawn Lake. He had worked several miles closer to the mountain the week before with his Weyerhaeuser Co. logging crew. His bride of seven months, Christy, was with him.*
>
> *Terry Crall and Karen Varner of Longview were camped in the Green River Valley, separated from Mount St. Helens by 14 miles and a whole mountain range, when they were killed by falling trees. There were no road blocks on the way for them to skip around.*
>
> *Wally Bowers of Winlock and Tom Gadwa of Montesano were presumed to have been logging in the Shultz Creek area nine miles northwest of the volcano. Hundreds of Weyerhaeuser employees had crossed the area the week before because company lands were excluded from the restricted area.*

Kelso's Jerry Moore and his wife, Shirley Ann, died at Tradedollar Lake, well outside the restricted area about 11 miles northwest of the volcano.

Sure, a few victims got closer to the volcano than was advisable. But this handful was there legally.

What's all the more maddening about this whole saga is that state officials had misled the press and the public about the true size of the restricted area. It wasn't until the fall of 1980 and the spring of 1981 that it became widely known that the restricted zone had not extended much beyond the flanks of the volcano on the west, northwest and north—the side where the mountain's flank was bulging 5 feet a day. State officials thus compounded their insensitivity and incompetence with a cover-up.

Many relatives of those who died never have put the eruption behind them, in part because of these old wounds.

Cowlitz County is hosting a dedication of a memorial grove of trees to those who died in the eruption. Amid the green of the recovering forest, state officials can do a great deal to encourage human healing if they utter two simple words: "We're sorry."

CHAPTER 31

2000: Survivors

SMALL CLUSTERS of St. Helens survivors milled around picnic tables where many of them had spread out photos, newspaper clippings, and other tokens of their ordeals. Robert Rogers brought out a projector and a screen, ready to show his volcano slides to the survivors gathered at Seaquest State Park.

Peter Frenzen scanned the group of about twenty guests for reporters. He had promised that no media would be present at this barbecue and campout—that everyone would be able to relax and share their thoughts and stories without being quoted. He was feeling good. Things had gone well at the public symposium that day on the scientific aspects of St. Helens. Tomorrow promised to be an emotion-charged day, with selected survivors telling their stories at the symposium.

Then he spotted Jim Scymanky. He hadn't been sure he would show up. Frenzen had never met him in person. Now as he greeted Jim and Helen, he felt a certain awe to be here with this man whose name was often linked with the mountain in the same breath as the almost mythical innkeeper Harry Truman and the scientist David Johnston.

Rogers placed his fiberglass model of Mount St. Helens on a picnic table. He had also brought modeling clay and tiny flags on toothpicks. He rolled a piece of clay between his hands, stuck a flag on it, and placed it on the fiberglass mountain to show where he had been when the volcano erupted.

When Helen Scymanky saw the fiberglass model, she gasped. Memories flooded back to her of the Mount St. Helens trial fifteen years before. The model looked exactly like the one used in court. She became suspicious of Rogers, whom she had never met. Did he have something to do with Weyerhaeuser, the lawyers, or other authorities?

The Scymankys introduced themselves to Rogers and his wife. Helen could not take her eyes off the map. She tried to make small talk without arousing Rogers' defenses while she led up to her question about how this man had come into possession of the map model. She soon found that she didn't need to be so tentative with Rogers, who rarely missed an opportunity nowadays to tell a good story when it had something to do with the volcano. "That's quite a map," she said, pointing down at the model. Rogers took this as his cue to launch into his story, in detail, of how he had found the model.

Rogers picked up a piece of modeling clay and a flag and turned to Jim. "Where were you when the mountain erupted?" Jim thought for a moment. Then he showed Rogers the approximate location, and Rogers planted another flag. Rogers had been seven miles from the mountain and Scymanky had been thirteen miles away. But, while Rogers was at the edges of the blast, Scymanky caught its full force.

After the Scymankys walked off, Rogers studied Jim. The former logger, now volcano hero, seemed to stand alone, even in a crowd. Jim Scymanky had what Rogers called "that thousand-yard stare" that Rogers had seen in Vietnam veterans. He had considered telling Scymanky about therapy techniques he had learned, to help Scymanky relieve the mental load he seemed to be carrying. But Rogers found Scymanky reserved and contained within himself, offering no openings into this most famous survivor's psyche. Rogers started a discussion with his wife, Janelle, about what the definition of a Mount St. Helens survivor should really be. "I'm no survivor," he said. "But Scymanky sure as hell is."

Richard Waitt, the USGS scientist who had worked over the past twenty years on a timing study of the May 18 eruption and was writing a book about it, approached Jim Scymanky. "I think I found it today," he told them. "I think I stood in the exact same spot today where your pickup had been." The pickup had been hauled off the mountain long ago. Its exact location had never before been noted.

Frenzen shook hands with Rogers. Rogers held up his daughter. "This is Molly Tephra." Frenzen smiled broadly. "That's the third Tephra I know about that resulted from the volcano," Frenzen said. "Francisco Valenzuela named his daughter Tephra." Rogers was happy to hear about his old Forest Service friend who had witnessed the eruption with him. Rogers hadn't been in touch with Valenzuela, who was not attending the anniversary events, in well over ten years.

"By the way, did you get hold of Ty Kearney?" Frenzen asked.

"What do you mean?" Rogers asked.

"Remember, I asked you to let him know about the gathering tonight."

In a panic Rogers searched the faces of the crowd but couldn't spot the Kearneys, the couple that he and Valenzuela had been with when they witnessed the eruption.

"You mean, you didn't try to call him, too," Rogers asked meekly. He felt a sudden surge of anger directed toward Frenzen. How could Frenzen have entrusted him with such a duty, when everybody knew Rogers had no sense of time, deadlines, or responsibility, not to mention his dislike of using the telephone. Then he realized that Frenzen *really* didn't know him.

The crowd spilled out from under the shelter while they waited for the meat to be barbecued. After they ate, Rogers set up his screen and slide projector. But no one paid any attention when he suggested they find seats for his show. Their buzz of conversation, punctuated with laughter, continued until a light mist began to fall and the crowd huddled under the shelter. It had grown dark. As people gathered around the campfire to get warm, Rogers saw his chance. "Hey, time for a slide show," he shouted. This time he caught their attention.

Rogers presented the slide show as his own personal view of what happened on May 18, knowing that other people would have other views. He was right. "No, you didn't see that," said Keith Stoffel, a geologist who barely escaped death when the volcano erupted beneath his light plane. "It couldn't have happened that way," he said about Rogers' remark that he saw the volcano jetting steam from vents on its sides just before the eruption. "What you saw must have been dust kicking up." Stoffel's abrupt manner made Rogers think he had finally met someone with fewer social skills than himself.

Richard Waitt stepped in at this moment. In his gentle manner he said that in his interviews to determine the timing of eruption events, he had discovered that many survivors in viewing the same thing had different descriptions. He didn't believe any one version to be necessarily truer than another. Waitt didn't say so, but in fact there were tapes of transmissions from Gerald Martin, the ham radio operator who in the last moments of his life gave a vivid moment-by-moment description of the eruption hurtling toward him, and he too had seen the vents of steam.

Waitt's comments left an opening for varying views, and soon the survivors came alive with memories they had forgotten or suppressed. Waitt grabbed his notebook and began writing. It was a euphoric moment, one that Frenzen, Rogers, and others considered magical. Frenzen had heard some of the descriptions before, but now he found himself enraptured as eyewitnesses from all vantage points—the east, west, north, and overhead in an airplane—compared notes.

At first Frenzen took an almost clinical view of his part in the proceedings—a "fly-on-the-wall" approach. He listened intently and closely watched the interaction of people. As he saw several of the survivors leave the group to regain their composure while scientists continued explaining in technical terms what had happened during the eruption, he came to realize the differences in how people viewed the event. To scientists who had seen the eruption, it was a once-in-a-lifetime "wow event." To the survivors who had lost friends and nearly died themselves, it was a moment of tragedy. But as the night wore on, he noticed that many of the survivors seemed to gain a certain satisfaction in finding the answers in scientific terms to what happened to them that horrific day. And he found himself and the other scientists growing emotional as they heard the survivors' stories and put a human face to the events of May 18.

A few miles away, at Kid Valley Park, Donna Parker sat alone in front of a campfire, leafing through the scrapbooks of newspaper clippings and photographs about St. Helens and her brother. Then she picked up the photo album that showed her and Billy when they were children together and played in the shadow of the mountain.

They drove up Spirit Lake Highway, Venus Dergan with her boyfriend Mark and his two children. She turned to Mark at the steering wheel and studied him. Venus knew he adored her, and she was very much in love with him. When she had told him what happened to her on the mountain, she held out her scarred hand and he took it and told her "that's not so bad," and she quit trying to hide the hand from him.

Now driving up to the mountain that was so much a part of her life, she desperately wanted him to understand what it meant and how it had changed her. But deep down she knew that no one else could truly understand except for Roald, and maybe that was all right.

She watched St. Helens loom closer. Her anxiety grew as it had the few times she had driven close to the mountain since the eruption. But now she was going to push through her distress. She had to. She was going to the mountain not to conquer it, that couldn't be done, nor to be at peace with it, that would never happen, but to be able to move on, knowing its legacy would always be with her.

She had grown close to the children in the backseat. Her experiences on the mountain made her a hero in the eyes of Mark's son and daughter. Whenever the subject of Mount St. Helens came up with their classmates, they told them about her.

Venus and Roald weren't scheduled to speak until midafternoon at the St. Helens public symposium. She and Mark decided to take the children beforehand to the Johnston Ridge Observatory, only five miles away from what was left of the mountain's peak. At the observatory they viewed the movie and played with the interactive displays. They watched the seismograph needle that measured the quaking in the ground beneath the volcano as it moved only slightly. They came around a corner, and there on what was called the Survivors' Wall, the children looked up in awe at a large photograph of Venus standing next to Roald, with their story beneath, enshrined for posterity.

When Don Swanson rose to speak at the symposium, movie cameras followed him. A German television network that had started filming him in Hawaii for a documentary now stalked his every move. He was still disturbed by the fact he had been grouped with the survivors for his talk rather than with the scientists of the day before. The scientist Don Swanson had found new ways to predict eruptions, and he was now scientist in charge of the Hawaiian Volcano Observatory. But it was his personal story that he would tell today.

He began with a disclaimer. "I was an eyewitness to part of the eruption," he said, "but not the most vigorous part of the eruption. And I was no way a survivor of what happened on May 18, because my life was never in danger. So I am sort of the odd man out in today's session.

"But I do have a story to tell. . . . And that is that my life was really determined a couple of days previous." He told the audience how David Johnston as a favor had taken Swanson's observation duty at Coldwater II the night before the eruption. And how Johnston had perished in the blast of May 18.

"On May 17, both of us, both Dave and I, visited the top of the mountain. I was brought up by the helicopter pilot, Lon Stickney, and it was a beautiful midmorning, 10:30. We shut down the helicopter on top of the volcano. I put in a new reflector that was being used for the geodetic monitoring by the team that I was part of at the USGS. And the weather was so nice, that once I put the target up, Lon and I walked over to the lip of the crater and peered down. . . .

"Then I retreated and Lon that afternoon brought Dave Johnston up to the summit of the volcano. Dave climbed down into the crater. He got a sample of water from the lake that was in the crater and took it back down with him to where he was killed at Coldwater II. . . . Neither Dave nor I had any sense, any premonition, any scientific reason to think that the mountain was going to be erupting soon."

Swanson showed the video copy of the 16-millimeter film he had taken with the movie camera as he flew near the eruption. The awesome black clouds roiling off Mount St. Helens in a huge column still had the power to mesmerize an audience. "It was kind of an isolated, cocoonlike environment on the airplane. Here was this mighty dark column rising many miles into the sky, and yet we just had no direct part in it. We were observers only and yet felt no endangerment, no fear, no involvement with the situation."

At the end of the film, Swanson said that while he was aboard the plane, he was worried about Johnston "because he had not returned the radio message that I had made to him. So we kept trying to fly out over Coldwater II, but couldn't because the ash was so great. So in the back of my mind, and I am sure in the backs of minds of others, was the thought that David perished."

"It was only that afternoon when I got back to Vancouver . . . during this unwinding period of the afternoon, that the impact of what had happened, the enormity of it, the fact that there were several casualties which then blossomed into many casualties later, that all hit home then."

"I was the trespasser," Rogers told the symposium audience at the Silver Lake visitor center. "I went up there because forty-six miles from Portland, volcanoes don't pop out of the ground very often. . . . I wanted a volcano, an earthquake, and a hurricane to see during my lifetime. So here was the volcano." Rogers' hands pumped, jutted, flailed, spread, and pointed as though mere words were not enough.

The lights went down and Rogers showed his slides. At times he became the professor, peppering his talk with terms like lapelli and fault scarp and phreatic, educating his audience on the workings of a volcano even as he waved his own considerable knowledge in their faces. At other times he became the comedian, bringing down the house in laughter. But for the most part he was a P. T. Barnum, beckoning the audience into his tent to share in the rousing adventures and clandestine experiences of an American maverick, thumbing his nose at authority and even the elements, challenging danger at every turn.

Rogers told his listeners that he had "two size-8 self-issuing permits" that gave him all the permission he needed to be in the Red Zone. His friends in the USGS were "my spies, my geologic consultants." On the mountain, Rogers made it clear, he had been witness to a feast of images reserved only for the gods and for the few mortals whose fearlessness made them worthy. "As the sun set, I could see color start to show in that thing, and I realized I was going to get to witness something that other people hadn't been able to see, which is color in the floor of a crater from incandescent rock. . . . You are looking down into the vent and these were like the fires of hell." And it was all true, validated by the photographs he had taken, photos that many scientists now considered to be invaluable to the understanding of Mount St. Helens.

But to the audience now, the most striking photos were those of Rogers himself at the center of the drama. There he was, standing on a stump as a broiling, black volcanic cloud bore down in the background. There he was, dangling his feet over the edge of the crater as the volcano glowed red below. There he was, standing on the side of the volcano and looking down at the steaming new lava dome.

He couldn't resist packaging his descriptions of the volcano with stories that promoted his image as a rebellious adventurer caught up in hair-raising escapades of his own making. He took his listeners back to the day of the eruption when he and Francisco Valenzuela, rushing to escape the holocaust, decided to go back to retrieve Valenzuela's car and to recover a roll of film that Rogers wanted.

"As soon as that column went up, there's deep, rich, purple lightning strikes that were striking down through the cloud in the area around the volcano and up at the top of the volcano. And we were witnessing these magnificent lightning strikes in the clouds, roaring, rumbling, raging noise, flapping, blowing wind, and decided to go back and drive underneath it to dig his car out and find that roll of film."

As they stood together before the audience, Roald dwarfed Venus. Roald had bulked up considerably in the past twenty years while she remained lithe and petite. She barely passed his shoulder in height. She looked down as Roald spoke, content to let him narrate their basic story without interruption. But when they started fielding questions—questions that called for a more emotional tone—Venus became the dominant one.

"So what was it like from your perspective? Do you remember much?" a woman asked.

"I do remember," she said. "I remember it seemed like it was taking a long time, but it all happened in a matter of seconds. From the time we walked out of the tent till the mudflow. It had to be a minute or two. We just jumped onto the car and I just remember this one big flow of mud and debris just washed us right into the middle and the next thing that I knew, we were just fighting for our lives. . . .

"I remember Roald screaming for me to hang on. And in amongst all the turmoil going on, if you had ever heard a grown man scream in terror, it's enough to put the hair up on the back of your neck."

"The back of my neck, too," Roald said, smiling.

"He was screaming like I had never heard him scream before . . . and just desperately asking me to hang on, that's all I remember and that's what I did. It gave me the will to keep trying. Even though you are fighting forces that are way beyond your control. I mean, there was no thought that we were ever going to survive this."

"And luck was on my side," Roald said, "finding her three different times. I still don't know how that happened."

"Right, he was extremely adrenalized through all this," she said in a deadpan manner, and the audience burst into laughter.

Another woman stood up to ask a question. "What went through your mind when all of this was happening and did you do some praying?"

"I thought it was the end of the world," Venus said. "And I honestly believe that it was God who decided it wasn't our time."

"Twenty years later it's hard to believe it happened," Roald said at one point. "But we have the scars and the memories."

"We were involved in a unique experience and we were one of the lucky ones to survive," Venus said. "But you can't let that fear overtake your life. You got to move on. And that's what we have tried to do."

Scymanky began his presentation, and the audience was soon caught up in the story. The soft-spoken Scymanky seemed more like a person just sitting across from you at the breakfast table, involved in conversation. The talk was charged with an emotion only underscored by its low-key nature. Here was a man who had seen the death of three of his friends, a man who had himself survived only at a terrible physical and emotional cost.

He told about that day as if he were there again, struggling to lead his fellow workers through the ash cloud that had turned day to night. He relived again his attempt to keep Evlanty Sharipoff from wandering off to his death. At times his speech would slow and he would look out into the distance and pause.

> *Evlanty, one of the Russians, said he was going to cross the valley. The valley was to our right at this time. The avalanche was straight ahead. That sheer rock wall was to our left. I mean we're trapped, and to go across the valley? "Well, where are you going to go across the valley? There is nothing." I told him, "Hey, we've got water here." But he just couldn't stay still. He had worked up in that area, I guess, several years prior to that. And he started walking, falling down, crawling. Leonty and I were right behind him, trying to pick him up, trying to grab him by his arms. "Ow," it hurt. Don't touch his arm. Couldn't fight him any longer. I don't know, maybe, not even a quarter-mile. You just kept trying to . . . couldn't talk him into coming back. Didn't want any reasoning. Yelled not to touch him.*

Scymanky paused and let out a long sigh. "And I don't know what the time frame was. So Evlanty disappeared out into the ash."

"Dias wasn't going to stay there," he told his listeners. "He was going to try to climb the avalanche." Scymanky again visualized the avalanche debris that had blocked their way, and he pointed up as if he could touch it. "So he disappeared up there, somewhere." Scymanky and Leonty Skorohodoff ended up staying together. "Thank God, at least there was two of us left."

After Scymanky's talk, a member of the audience asked him, "Do you harbor some resentments to the company for having you in there so close to the mountain?"

"No," he replied. "I mean, I'm just damn glad to be alive, you know. . . . It's incredible I'm even here. I'm just thankful every day I'm here to relate what happened twenty years ago."

Venus and Roald's interview in New York on the CBS *Early Show* went well. But the flurry of shopping and sightseeing that followed left her in a melancholy mood, bringing back memories of their visit to Hollywood twenty years before and of the way the two of them had been together then.

They got to the airport in New York only thirty minutes before their flight was scheduled to take off for Seattle and found it was delayed by a storm. Torrential rains accompanied unrelenting thunder and lightning. The airport shut down. They were surprised when they got a boarding call an hour and half later despite the storm. They sat in the plane on the tarmac for hours.

The lightning strikes hit ever closer to their jet. Venus was tired and hungry, but she prayed the plane would not take off. Between claps of thunder she told Roald, "I do not want to take off in this. I do not want to take off in this.... I don't want to take off in the middle of the night in a storm. I've got a real bad feeling about this flight. Our luck has not been good together and I do not want to take off."

"It's not my fault," Roald said.

"My god, I'm never going anywhere with you again." Another clap of thunder shook the plane. Venus was terrified and near tears.

"I'm never going anywhere with you ever again," she now screamed at him. "Every time we go somewhere there's a natural disaster happening. Every time we get together there is a natural disaster somewhere."

"It's not my fault," he mumbled.

CHAPTER 32

2000: Anniversary Day

EARLY THAT MORNING of May 18, just before the Hawaiian sun began its ascent in the sky and the birds began chirping in the lingering predawn darkness, Don Swanson drove to the Hawaiian Volcano Observatory, where he was scientist in charge. The building was empty and quiet. He went to his office, which looked out over the vast Hawaii Volcanoes National Park, a land where over the ages volcanoes grew dormant as others sprang to life, spewing streams of lava to the sea, leaving island-building deposits for lapping waves to erode in a never-ending cycle. Years ago he had made this land his playground while David Johnston had staked out his own volcanoes, as in their youth they tried to pry open the mysteries of erupting mountains.

It was precisely 8:32 on Mount St. Helens, twenty years to the minute since the eruption. He thought of Johnston, the mountain, the passage of time, and death.

Swanson had once written an article that touched knowingly on scientific curiosity, volcanologists, and personal danger.

> *Volcanologists who are curious will get themselves in trouble and sometimes die of it. It is often stated that we must weigh the potential benefits and risks before doing something that may be perceived as risky. Of course we must, but . . . in the end, it comes down to common sense, which varies among individuals and in any case is far from foolproof. Let it be no other way, and let us praise the curious as we mourn the dead.*

But at times Swanson found he could not keep death at arm's length no matter how hard he tried to objectify it. In a few years he was due to retire. His work on Mount St. Helens had been the crowning achievement of his life. During that time it had been as if Johnston were there with him. In reaching decisions, Swanson often found himself asking, "What would Dave do?" And now as Swanson watched lava from the Kilauea volcano stream red to the sea, he realized his friend's presence would never leave him.

Venus and Roald slept during most of the flight back to Seattle. They were silent on the drive from the airport.

Venus drove to the home of Roald's parents, where Roald and his son were staying, and pulled into the driveway. They just sat in her car for what seemed a long while. A lot of people had wanted a Cinderella ending for their story, Venus thought. She and Roald would get married and live happily ever after. But it was not to be.

What had happened on the mountain exactly twenty years ago helped forge a bond. It wasn't destined for them to live their lives together, but maybe what they had was even better and stronger. She was seriously involved with Mark and she had explained to him that their life as a couple would be intertwined with Roald's in some ways, and he had accepted it. Mark had asked her to marry him and she had said yes.

Now as she and Roald sat in the car, Venus finally turned to him. "We should at least wish each other a happy anniversary."

"Happy anniversary," he said.

"Happy anniversary," she replied as she pulled him to her for a long, lingering hug.

Donna Parker drove past the Cowlitz County volcano visitor center at Hoffstadt Bluffs. The county's new memorial to the dead was to be dedicated there on this twentieth-anniversary day. The memorial included the names of every person who had perished as a result of the eruption.

She traveled up the mountain to the Johnston Ridge Observatory. She watched the movie in the auditorium and later met Ty and Marianna Kearney outside. Media swarmed over the area.

"I want to get a picture of all three of us standing with the mountain in the background," Ty said. As Kearney set up the photo, Parker glanced to her side and saw Mike Donahue, a Portland television news anchor. She was afraid Donahue would recognize her and ask for an interview. "I do not want to talk to Mike," she told Marianna. "This is not a celebration, a happy anniversary. This is my memorial day. I'm leaving."

Parker drove a short distance from the observatory building and parked. She grabbed a small American flag and hiked down a path until she came to one of the crosses she had placed to honor victims whose bodies had never been found. The cross was for Gerald Martin, the former Navy radio operator whose words heralded the eruption in the moments before the exploding mountain took his life.

She bent in front of the cross and planted the flag. Memorial Day was less than two weeks away. She decided that military veterans killed on the mountain whose bodies had never been found should be honored just like the war dead. A wind blew down off the mountain and the flag flapped as gray dust swirled in the air.

Parker would not be going to the dedication ceremony at Hoffstadt Bluffs. She considered the event and the memorial to be nothing more than a tourist attraction devised by the county. The memorial was sponsored by the Cowlitz County government, and she couldn't get out of her mind the shabby way the county treated relatives of the dead when they first inquired about their loved ones. She was not about to support anything put together by county officials. "Piss on them," she thought. "Piss on them."

Like ghosts resurrected from eternity's slumber, voices of the crusty old innkeeper Harry Truman and the cocksure young scientist David Johnston were heard again as the radio show turned to taped file interviews. Jim Scymanky, sitting for an interview at the Portland radio station, listened intently to the recording featuring the two men whose names were most often linked with his in remembrances of the mountain.

"Well, the damn thing, it's a perfect cone," Truman said in the file tape, speaking of the lovely mountain that rose above his Spirit Lake lodge. "That would be a horrible thing to see that son of a bitch blowed all to hell, wouldn't it?"

About leaving the mountain because of the danger, he chuckled and said, "No, I'm not going to leave. Damn right I'm not going to leave. I am going to stay here. If I left, it'd kill me. If I left this place and lost my home, I'd die in a week. I couldn't live. I couldn't stand it. So I'm like that old captain, and by god, I'm going down with the ship."

Johnston's prediction of the coming eruption was still haunting in the tape made a few days before his death. He was asked when St. Helens might explode. "Well, if it occurs, it would probably be no more than a few months on the outside. But it could be as soon as a few hours."

The voices of Johnston and Truman seemed so real to Scymanky, as if the men were in the studio joining in the conversation. Scymanky told the radio host about his recovery at Emanuel Hospital and the deaths of his three comrades. The host wanted to know if doctors had figured out why Scymanky lived while the others died.

"They had a lot of doctors asking me different questions . . . about what I might have done differently," he said. "And I just told them, we all drank fresh water, we were all together all the time and they knew everything else, the burns and everything was equal, but they couldn't understand why I survived, and I certainly don't."

Near the end of the show, the host asked Scymanky for any closing thoughts.

"Well, the next time you're out in our great Northwest outdoors," he replied, "whether you're fishing on a lake or skiing down a slope, maybe you're fly-fishing in a stream, just take a good look at the trunks of some of those great big fir trees and let your eyes kind of wander up the branches to the tops, and take a good look at those snow-laden peaks, because it may be your last."

On this warm and mostly windless day, not unlike the day the mountain exploded, four hundred people gathered in a huge tent overlooking the river and the volcano beyond. They were at the county's Hoffstadt Bluffs visitor center to dedicate the memorial to the dead of St. Helens.

On a helipad next to the tent, television helicopters mingled with rental choppers all ready to fly tourists, for a hefty fee, to get a bird's-eye view of the volcano. The nearby Alaskan dogsled ride featuring huskies pulling a wheeled sled had been suspended for the ceremonies, along with the park's llama ride. Satellite news trucks with their generators humming sat close by in the parking lot. Technicians set up television cameras. Reporters scurried through the gathering crowd, trying to ferret out interviews with relatives of those killed. It was soon apparent that few relatives were there and even they, for the most part, refused to be quoted.

A Boy Scout color guard led the Pledge of Allegiance. A sheriff's deputy sang the "Star-Spangled Banner." The county chaplain asked God to be with the victims' friends and families. Politicians from across the state sat in the audience along with USGS representatives and the uniformed brass of the National Guard, Army Corps of Engineers, and the Forest Service.

Nowhere to be seen was the man the victims' relatives most wanted to hear from, the man who had added to their grief after the eruption by basically calling their kin curiosity seekers who somehow brought on their own deaths. Former President Jimmy Carter had elected to stay away. Dixy Lee Ray had also faulted the victims, but she was no longer alive. Organizers had asked Carter to attend, through two avenues—the official one through the Carter Center and another more informal invitation through the group Habitat for Humanity. A request for a letter from Carter to be read at the ceremony also got nowhere. It was obvious that the former president wanted nothing to do with the mountain or the dedication ceremony.

The highest state official at the ceremony was Transportation Secretary Sid Morrison, a former member of the U.S. House of Representatives. Far from acknowledging any failures in the state's planning before the eruption,

Morrison put a positive spin on the disaster. "We remember those lives lost," he said. "But we also remember the challenge that was gained. So I hope that those people that we lost in this event . . . were pleased that we did the right thing with the challenge that was presented on that day. The cooperation between all levels of government was absolutely unbelievable. It seems when the pressure is on, something dramatic is in front of us, we respond to that."

The keynote speaker was Michael Lienau, who nearly lost his life with a film crew during the second eruption of the mountain and went on to make *The Fire Below Us* that touched on the stories of the victims and their families. "I've had an opportunity to hear those stories and to sit with them and cry with them about the difficulties that they've faced," Lienau said. "For a lot of them the shadow of Mount St. Helens is still heavy upon them. A lot of mistakes were made." He spoke of forgiveness and the need for healing. "I think this memorial grove and this plaque is a great place to start."

He showed a short clip from his film. In the well-lit tent, the image appeared faded on the screen as Jim Scymanky pounded a cross into the ground in remembrance of his fallen friends. With just the cross as a backdrop, names of the dead began appearing on the screen. The fifty-seven names scrolled on for several minutes. The film affected the audience more than any words could. In the silence of the tent, only sobbing was heard.

The chairman of the Cowlitz tribe gave the audience an insight into the mountain's history and tradition. "You call her Mount St. Helens," Ray Wilson said. "She has only been known by that name for a very, very short time on the historical time line. For millenniums . . . she was known by her real name, Lawelata . . . 'person from whom great smoke comes.'"

Five young girls in Indian costume danced to taped tribal chants and music. The shells jingling on their dresses made an almost hypnotic sound. Wilson walked to the tent entrance and began beating a round, thin drum he held in one hand. Two other tribal members with drums, followed by the dancers and the color guard, lined up behind him as he left the tent chanting "*aie yahhh, aie yahhh, aie yahhh* . . . " Soon the hundreds of people in the audience followed in a line that extended for a hundred yards as they moved slowly toward the memorial grove of trees. Nearby, a helicopter roared to life and shut down almost immediately. Robert Rogers had persuaded a TV reporter and his pilot to give him a ride over the volcano as he told his story, but a producer nixed the idea at the last moment. In the distance, a few wispy clouds wreathed the crown of the mountain.

At the grove surrounding the plaque that recorded the names of the dead, the tribal chairman blessed the memorial. Below the grove, the Toutle River that had taken so many lives on this day twenty years before murmured peacefully. A Cowlitz County commissioner gave a short speech.

Another commissioner read the names of the dead. Then he asked if any relatives of the victims wished to speak.

"I just want people to remember," said Sheryl Bales, as she broke down in front of the plaque that bore the name of her sister, Karen Varner. "The deaths were always a footnote," said Paul Nickell, who lost his mother and stepfather to the mountain. "This is a long time in coming." Nearby, a long-haired man who looked like an aging hippie stood listening, his potbelly covered by a faded T-shirt with the words, "Mt. St. Helens/ A Saint No More / 1980."

When the dedication ended, reporters swarmed in with questions for the few relatives on hand. Television reporters and camera crews were at the core of one group surrounding Nickell. Andre Stepankowsky decided to listen to his comments.

"Donna Parker is the one who really deserves a lot of the credit for this," Stepankowsky heard Nickell say. "She has been the one who has kept the survivors alive. I wonder where she is?"

"I didn't see Donna today," Stepankowsky found himself saying to Nickell. Then Stepankowsky turned to look at the dispersing crowd. The determined woman who had become the champion and friend of so many grieving relatives was nowhere to be seen.

A few days after the anniversary, Donna Parker drove up the steep hill to Hoffstadt Bluffs. As she walked across the broad expanse of grass, she noticed the Russians in their traditional colorful garb leaving the memorial site. When she got to it, she was alone. The site with its grove of trees, bench, and bronze plaque with names of the dead seemed peaceful to her. She was thankful as she looked out over the dried mudflows along the Toutle River that she could not see the mountain from here. Parker took a few photographs of the plaque on its metal post. Then she ran her fingers over her brother's engraved name, WILLIAM PARKER, and wept.

The flash of green caught Peter Frenzen's eye. In the gray and barren crater, a tiny fir seedling sprouted from the boulder-strewn ground. The boulders shaded the young tree, providing protection for the seedling to take root as the ashen earth nurtured its growth. He walked over for a closer look, again marveling at the miracle of nature's healing. This was the first tree seedling he had seen in the crater. Life had returned even here, to the heart of the volcano's devastation.

Epilogue

THE FREQUENCY OF EARTHQUAKES grew until at 6,000 a day the temblors were almost continuous as the mountain seemed to shudder itself awake after slumbering eighteen years. Crevasses opened in the huge glacier in the crater. On October 1, 2004, the volcano exploded with a burst of steam and ash that rose 16,000 feet. Rocks two feet in diameter sprayed like shrapnel within the crater. A helicopter landing spot on the dome was now a cavity five stories deep. Once again Mount St. Helens was a threat.

The blast demolished scientific instruments vital to foreseeing future activity. Scientists told throngs of media from as far away as Norway and Japan that this explosion had been a relatively minor throat-clearing of old ash and steam. They predicted more activity but no one could say exactly what or when. Within hours of the eruption, swarms of earthquakes returned.

Instruments flown over the crater detected highly elevated amounts of carbon dioxide, signaling that explosive gas was making its way to the surface. The volcano now rattled in a cadence of two beats per second. The U.S. Geological Survey issued a Level-Three alert, its most urgent volcano advisory indicating the possibility of an eruption threatening people and property. More than 2,500 visitors and journalists were evacuated from the Johnston Ridge Observatory, five miles from the crater. Weyerhaeuser suspended logging operations within twelve miles. Aviation authorities created a no-fly zone around the mountain so that emergency crews would not be hindered if medical aid or rescues became necessary.

Still, thousands more people were drawn to the mountain. Daily, they packed to capacity the Coldwater Ridge Visitor's Center, just three miles from the Johnston Ridge Observatory. They clogged roads to the mountain. They spilled over onto outlying ridges where they pitched their tents or parked motor homes and sat in lawn chairs with binoculars and telescopes hoping to witness an eruption. In the first two weeks of renewed activity, a "volcano-cam" trained at the mountain received more than 30 million hits, causing the host monument website to crash repeatedly.

Meanwhile, rising magma pushed up a new dome in the crater, just south of an older dome, at the rate of over thirty feet per day. Scientists nicknamed the dome, "the Loaf" because of its appearance. It glowed pink. Temperature readings made in flights over the new dome showed temperatures in excess of 500 degrees Fahrenheit. No one had expected that much heat. Explosive gas oozed from the dome.

Soon, 1,200-degree lava began to pour out of the new dome at the rate of one truckload per second, day after day. In barely a month, the dome covered more than seventy acres and had risen nearly 800 feet. In a quick foray into the crater, scientists hammered off a chunk of the new dome that even with special protective gloves was almost too hot to handle.

Analysis of the rock confirmed the scientists' worst fears. It closely resembled samples taken in 1980. Scientists determined that the sample had come from miles beneath the mountain instead of from a shallow reservoir of old magma, as some scientists had speculated. The new analysis revealed that the magma was much more volatile and virtually unlimited in amount. "It is the kind that could drive the large explosive eruptions that St. Helens is well known for in the past," said Willie Scott, scientist in charge of the Cascades Volcano Observatory, responsible for monitoring the mountain.

In two years, if the dome continued to grow at its present rate, it would be high enough to be seen from Portland, sixty miles away. In a decade, it would fill the crater and Mount St Helens would return to something of its pre-1980 form and elevation—a cone-shaped peak.

However, betting among scientists was that the mountain would not reach that point. The USGS now faced a much more dangerous volcano than previously believed. The thick, paste-like magma could plug the throat of the volcano, resulting in a massive explosion. As the new dome grew, so would the danger. Unstable slopes could collapse, triggering avalanches of gas and highly heated rock. The avalanches could melt the glacier in the crater resulting in mudflows that would surpass even the massive flows from the 1980 eruption. Even more ominous, the increasing weight of the dome would force pressure onto the magma below, increasing the chances of an explosive eruption. In what appeared to be an understatement, Scott said, "This is a rather unpredictable volcano. It's very important we monitor the situation closely."

Signs now indicated to scientists they could face an eruptive phase lasting years and possibly decades.

Coincidentally, Don Swanson stepped down as scientist in charge of the Hawaiian Volcano Observatory on the very day that Mount St. Helens came alive again. There was speculation, which Swanson did not discourage, that ultimately he would end up back at Mount St. Helens.

Swanson had spent a good part of his career since 1980 warning about the dangers of volcanoes. He believes volcanoes are often the most dangerous and unpredictable when awakening from a long slumber. Nearly every active volcano in the world has unique characteristics, often requiring scientists to customize their predictive tools. Even the same volcano can display differing characteristics from one eruptive period to another. Scientists said the patterns of activity since Mount St. Helens rumbled back to life in 2004 were unprecedented. Predicting the future of the mountain is difficult. But Swanson can say with confidence: "Another catastrophic eruption of Mount St. Helens on the scale of May 1980, or larger, is just a question of when, not if."

50,000–40,000 BC

Mount St. Helens is formed as the youngest peak of the Cascade Mountains, a range containing fifteen major volcanoes extending from Northern California to British Columbia. The Cascades are part of the "ring of fire" volcanic ranges that circle the Pacific Ocean.

2500 BC–AD 1800

The mountain erupts during at least twenty separate intervals. As various tribes settle in the area, Mount St. Helens becomes a central part of their religion and myths.

1800–1857

Mount St. Helens is active in a series of eruptions that do little to mar the mountain's beauty. William Clark of the Lewis and Clark expedition calls the mountain "the most noble looking object in nature." Thomas J. Dryer, editor of the Portland *Oregonian,* makes the first recorded climb of the mountain, in 1853.

1857–1980

The mountain goes dormant. Mining and logging begin. Mount St. Helens becomes a mecca for recreation and an icon of Northwest beauty. USGS scientists Dwight Crandell and Donal Mullineaux determine that St. Helens has been the most active and explosive volcano in the continental United States during the previous 4,500 years. In 1978 they forecast an eruption before the end of the century.

May 18, 1980

Mount St. Helens erupts in a massive explosion that kills fifty-seven people and devastates 230 square miles of land.

1980–1986

The volcano records twenty-one more eruptions, all much less damaging than the May 18 blast. Scientists on St. Helens revolutionize the field of volcanology, developing prediction techniques that save hundreds of lives around the world.

1986–2004

Mount St. Helens again goes dormant.

Fall 2004

The mountain awakens once more with eruptions of steam, ash, and lava.

TIME AND AGAIN, in doing my research, when I mentioned St. Helens, I found a connection and usually a story to go with it. And that was part of my problem initially. Everybody had a story. I spent months researching and talking to potential subjects. I was looking for people who not only had strong personal ties to the mountain, but who collectively could tell the complete story of Mount St. Helens. Many times, scientists Don Swanson and Peter Frenzen emphasized that their research was a collaborative effort often involving teams of associates. Likewise, Andre Stepankowsky pointed out he was just one member of the staff of reporters and photographers who won the Pulitzer Prize for the Longview *Daily News*. Donna Parker noted some relatives of victims had suffered far worse than she.

And they were all right. There were many stories about people whose lives were changed greatly by the eruption. But I was able to concentrate only on a limited number of people.

When possible, I corroborated information from my subjects in a variety ways, including interviews, letters, diaries, photographs, and media accounts. Unless this additional research provided new information, these sources are not cited.

In conducting interviews, I discovered what social scientists and law enforcement officials have known for years—that people witnessing the same event from the same spot can have varying views on what happened. When I encountered differing recollections, which occurred on surprisingly few occasions, I tried to reconcile them. When that was not possible, I relied on the memories of the people whose stories I tell in this book. After all, *Echoes of Fury* is, for the most part, a chronicle of the mountain as seen through their eyes.

We did the best we could in placing and dating the scenes. Allied material helped a great deal in this task, but we sometimes had to rely solely on a subject's memory, often more vivid right after the eruption than in subsequent years. Though every attempt was made to put scenes in chronological order in the book, some minor adjustments became necessary for clarity and readability.

In researching printed matter, I was struck by how many differing facts and figures I encountered for the same event. Facts and figures used in this book were cross-checked by me and reviewed by experts in various fields. Almost all the people I have written about extensively reviewed for accuracy sections of the manuscript that described their experiences. That said, I take full responsibility for any errors.

In the prologue and early chapters of the book, minor portions of my previous writing on Mount St. Helens have been modified, rewritten, and incorporated here. Original sources are cited in the notes that follow.

PROLOGUE: 1980

12. **No one noticed**   "Mount St. Helens Holocaust, A Diary of Destruction," produced by the Vancouver *Columbian* (Barron Publications, 1980), p. 4. The earthquake registered on a seismograph at the University of Washington in Seattle but it wasn't recognized until later as being the start of the volcano's awakening.

12. **One of the world's**   Dick Thompson, *Volcano Cowboys, The Rocky Evolution of a Dangerous Science* (St. Martins Press, New York, 2000), p. 71-72; Portland, Channel KGW News, April 9, 1980.

12. **A team of Dartmouth**   Thompson, *Volcano Cowboys*, p. 64, 65.

13. **the mountain erupted with a force**   Steven R. Brantley and Bobbie Myers, "Mount St. Helens—From the 1980 Eruption to 2000," USGS Fact Sheet no. 036-00, March 2000. The bomb dropped on Hiroshima had a force of 20,000 tons of energy. The lateral blast of the volcano released twenty-four megatons of energy, (seven by blast, the rest through release of heat). However, this is only a rough comparison. Different forms of energy don't translate equally (see Decker, "The Eruptions of Mount St. Helens" cited elsewhere in these notes). Figures for the amount of energy the eruption released and comparisons are numerous and divergent, depending on the types of energy compared, methods of release, and durations. For example, the *National Geographic* ("Mount St. Helens: nature on fast forward," May 2000) says the eruption was "500 times the force of the Hiroshima blast" but this would have to include energy released by the volcano initially and then only part of the thermal heat released with the blast. Other publications include the total energy released over its nine hours of eruption to come up with much higher figures.

13. **"Vancouver, Vancouver**   Johnston's famous last words never reached the USGS headquarters in Vancouver. But they were heard and taped by a ham radio operator.

PART 1
CHAPTER 1: DEATH AND THE MOUNTAIN

16. **Don Swanson turned**   Interview with Don Swanson.

19. **a reporter approached**   I was the reporter and many of my personal observations appear in this chapter.

CHAPTER 2: SUNDAY, MAY 18, 1980

21. **That morning**   Interview, Don Swanson.

22. **After the Sunday briefing**   Dick Thompson, *Volcano Cowboys, The Rocky Evolution of a Dangerous Science* (St. Martins Press, New York, 2000), p. 109.

22. **Even with the chainsaw**   Interview, Jim Scymanky.

22. **The man working**   Alan K. Ota, John Snell, and Leslie L. Zaitz, "A Terrible Beauty," special section published by the Portland *Oregonian*, p. V24, October 27, 1980.

23. **For weeks they**   Interview, Scymanky.

23. **Robert Rogers was on**   Interview, Robert Rogers.

26. **The man, Francisco Valenzuela**   Interview, Francisco Valenzuela.

26. **The men located**   Todd W. Powell, "Blast from the Past," *Alaska Airlines* magazine, May 2000, p. 18.

27. **Martin believed** "A Terrible Beauty," Portland *Oregonian*, V20.
28. **The two reporters** Interview, Andre Stepankowsky.
29. **It took Peter** Interview, Peter Frenzen.
30. **The young couple** Interview Roald Reitan and Venus Dergan.
31. **Later that morning** "A Terrible Beauty," Portland *Oregonian*, p. V26.
32. **The news flash** Interview, Donna Parker.
32. **Now Parker stood** Jerry Bowen segment on Mount St. Helens, CBS *Early Show*, May 18, 2000.

CHAPTER 3: MAY 18: HOLOCAUST

33. **The sound was a hiss** "A Terrible Beauty," Portland *Oregonian*, p. V24.
33. **Scymanky convinced them** Interview, Scymanky.
34. **Don Swanson and the pilot** Thompson, *Volcano Cowboys*, p. 109.
35. **At 12:17 p.m.** USGS paper 1250, p. 809; "A Terrible Beauty," Portland *Oregonian*, p. V28.
35. **When Robert Rogers camera** Rogers, interview.
36. **Its boiling black cloud** "A Terrible Beauty," Portland *Oregonian*, p. V21.
36. **Valenzuela was not so lucky** Rogers, interview.
37. **Rogers wanted to reload his camera** Powell, "Blast from the Past," *Alaska* magazine, p. 18.
37. **Andre Stepankowsky had lost his view** interview, Stepankowsky; Kathleen Reinholdt, "Charter pilot: 'I felt like I was with the Valkyries'," Longview *Daily News*, May 18, 1990.
38. **imagined the ash cloud** Andre Stepankowsky, "Plane followed eruption's path." Longview *Daily News*, May 19, 1980.
38. **"What the hell** Interview, Reitan.
39. **they ran for the car** "A Terrible Beauty," Portland *Oregonian*, p. V24.
40. **Roald landed on a log** Interview, Reitan.
40. **Venus looked up** Interview, Dergan.
40. **The hot brown muck** Carol Perkins and Darrell Glover, "Campers Flee Wall of Mud," May 19, *Seattle Post-Intelligencer*, May 19, 1980.
41. **Robert Rogers spotted** Interview, Rogers.
42. **The Governor's helicopter** Interview, Stepankowsky.
43. **The loggers sat** "A Terrible Beauty," Portland *Oregonian*, p. V25-26.
43. **Scymanky tried to keep** Interview, Scymanky.
44. **They found a spring spurting** Julia McDonald, "Painful memories," Longview *Daily News*, May 18, 1990.
45. **When Don Swanson returned** Interview, Swanson.
46. **Glicken told Swanson** Dick Thompson, *Volcano Cowboys*, p. 111-112.

CHAPTER 4: MAY 18: DELIVERANCE

48. **"Hold on, hold on,** Dergan, Interview.
48. **Roald Reitan was surprised** Reitan, Interview.
50. **Roald had found the bridge** Dave Rorden, "Campers deluged, survive river's rage," Longview *Daily News*, May 19, 1980.
51. **Scymanky looked down** "A Terrible Beauty," Portland *Oregonian*, p. V27.
51. **For a long time** Interview, Scymanky.
52. **The pilot, Captain Jess Hagerman** Kathy McCarthy, *Associated Press*, "Rescuers: Copter pilots were drafted heroes" in *Year of the Mountain*, Yakima *Herald-Republic*, May 17, 1981, p. 12 E.
52. **watched the volcano** Interview, Rogers.
53. **The small helicopter** Interview, Reitan.
53. **Venus Dergan lay on a mat** Interview, Dergan.
54. **The hospital emergency area** Interviews, Reitan and Dergan.

57. **They quickly took control**   Interview, Life Flight nurse Lin Coon.
59. **A nurse undressed Scymanky**   Interview, burn nurse Jane Carol.
59. **A nurse who introduced herself**   Interview, burn nurse Melissa Metcaff (who requested that her real name not be used).
59. **The National Guard pilot**   "Rescuers," *Year of the Mountain, Yakima Herald-Republic*, p. 14E.
59. **Nurses found a wallet,** "A Terrible Beauty," Portland *Oregonian*, p. V40.
60. **Don Swanson finally**   Interview, Don Swanson.
60. **At his apartment,**   Interview, Peter Frenzen.
60. **He learned that**   For an accessible and concise account containing many of these same details, see "Volcano vomits ash and death," *Bellevue Journal-American*, May 19, 1990.
60. **Donna Parker did not stay**   Interview, Parker.
61. **He couldn't believe**   Interview, Stepankowsky.
61. **Water was rising rapidly**   Jay McIntosh, Andre Stepankowsky, "Lexington folks returning home," Longview *Daily News*, May 19, 1980.
61. **"I own these apartments,"** ibid.
61. **where a logjam**   "The Volcanic Eruptions of 1980 at Mount St. Helens: The First Hundred Days," USGS paper 1249, p. 69.
61. **Helen Scymanky** made a big dinner   Interview, Helen Scymanky.

CHAPTER 5: May 19-23

63. **The scientists and technicians**   "A Terrible Beauty," Portland *Oregonian*, p. V14-15.
63. **From a helicopter**   Dick Thompson, *Volcano Cowboys*, p. 113-116.
64. **Everybody was talking**   Interview, Donna Parker.
64. **The burn center was**   Interview, Metcaff.
64. **Helen Scymanky and Jim's**   Interview, Helen Scymanky.
65. **Stepankowsky couldn't sleep**   Interview, Stepankowsky.
65. **The house had been knocked**   Andre Stepankowsky, "Neighbors pitch in at Green Acres," Longview *Daily News*, May 20, 1980.
66. **Robert Rogers called**   Interview, Rogers.
66. **When Roald Reitan woke**   Interview, Reitan.
66. **Venus Dergan, too**   Interview, Dergan.
67. **As they flew**   speech by Don Swanson, "Eyewitness Accounts of the Eruption," Mount St. Helens Institute symposium, Silver Lake Visitors Center, May 14, 2000.
67. **The helicopter swept over Spirit Lake**   Dick Thompson, *Volcano Cowboys*, p. 115.
68. **With the sun setting**   Interview, Don Swanson.
68. **When he returned**   Interview, Andre Stepankowsky.
68. **Orders had been issued**   Andre Stepankowsky, "Agencies scrambling—some waiting for worst," Longview *Daily News*, May 20, 1980.
69. **Jim Scymanky often**   Interview, Scymanky.
69. **Donna Parker was now fairly certain**   Interview, Parker.
70. **On the Tuesday**   Dick Thompson, *Volcano Cowboys*, p. 139.
70. **The next day Don Swanson**   "Mountain sending new danger signs; Carter views devastation," *Seattle Times*, May 22, 1980.
70. **The Red Cross office**   Peter Lewis, "Complete tally of dead must wait for the living," *Seattle Times*, May 21, 1980; "Four mystery people still counted on peak missing list," Longview *Daily News*, March 17, 1981.
70. **The nurse in the burn center**   Interview, Metcaff.
71. **When Dwight**   Interview, Rogers.
72. **At first Venus**   Interview, Dergan.
72. **Children with dirty**   Interview, Stepankowsky.
73. **"It's better to be a safe chicken**   Andre Stepankowsky, "'Chickens' camp out," Longview *Daily News*, May 22, 1980.

73. **The young married couple**   Interview, Swanson.
73. **blue Datsun he spotted**   Thompson, *Volcano Cowboys*, p. 114.
74. **Scymanky was riding backward**   Interview, Scymanky.
74. **By Thursday 118 people**   "Missing 118," Longview *Daily News*," May 22, 1980.
74. **Some families**   "Terrible Beauty," Portland *Oregonian*, p. V33.
74. **That morning**   Interview, Donna Parker.
75. **"I think we will get millions**   Donna duBeth, "Visitor centers planned," Longview *Daily News*, May 23, 1980.
75. **Washington governor Dixy**   "Terrible Beauty," Portland *Oregonian*, p. V34.
75. **Donna Parker couldn't,**   Interview, Parker.
75. **Every day Peter Frenzen**   Interview, Frenzen.
76. **Stepankowsky found the family**   Interview, Stepankowsky.
76. **His dedication to his patients**   Andre Stepankowsky, "Family rescues treasures," Longview *Daily News*, May 24, 1980.
77. **The inside of her brother's**   Interview, Parker.
77. **He looked up**   Speech by Jim Scymanky, "Eyewitness Accounts of the Eruption," Mount St. Helens Institute symposium, Silver Lake visitors center, May 14, 2004.

CHAPTER 6: MAY 24-JUNE 1

78. **Earthquakes shook**   Interview, Swanson.
78. **The volcano, he learned**   For a good history of the mountain with information presented here, see *Mount St. Helens, the Eruption and Recovery of a Volcano*, by Rob Carson (Sasquatch Books, Seattle, 1990) and "Terrible Beauty," Portland *Oregonian*, p. V2.
82. **To Swanson and other scientists**   Interview, Swanson.
82. **The next day**   Jay McIntosh, "Peak quiet after ash falls here", "Wet ash causes power outages," and "Ashland," Longview *Daily News*," May 26, 1980.
82. **It was a terrible day**   Interview, Stepankowsky.
83. **Everhart Funeral**   Interview, Parker
84. **Scientists determined**   USGS Fact Sheet 036-00, March 2000.
84. **Some of his colleagues argued**   Dick Thompson, *Volcano Cowboys*, p. 145-146.
84. **Whenever St. Helens**   Interview, Swanson.
84. **The crane lowered**   Interview, Scymanky.
85. **Day after day**   Interview, Dergan.
85. **Often the three women**   Interview, Helen Scymanky.
86. **He was on his way**   Interview, Stepankowsky; Andre Stepankowsky, "Ash traps: Panty hose precaution could save air filters, engines," Longview *Daily News*, May 26, 1980.
86. **"How do you deal**   Andre Stepankowsky "Ash cleanup has officials baffled," Longview *Daily News*, May 26, 1980.
86. **The afternoon *Daily News***   Marion Villa, "Daily News photo makes *Time*'s cover," Longview *Daily News*, May 27, 1980.
86. **The next day**   Interview, Stepankowsky.
86. **He was now having**   Interview, Reitan.
87. **When Leonty**   Interview, Metcaff.
87. **Helen Scymanky had also**   Interview, Helen Scymanky.
88. **Peter Frenzen had not**   Interview, Frenzen.
88. **Authorities had stopped**   "Air hunt for survivors ended," *Seattle Times*, May 29, 1980.
88. **Terry Crall and Karen Varner**   Andre Stepankowsky, "Eruption turns weekend fishing trip into life and death ordeal," Longview *Daily News*, May 29.
89. **The Army Corps of Engineers**   "Corps ups dredging estimate," Longview *Daily News*, May 29, 1980; "Corps wants to dredge to Toutle," Longview *Daily News*, May 28, 1980.
89. **Donna Parker was trying**   Interview, Parker.

CHAPTER 7: JUNE 2-19

91. **The vastness of it all**   Interview, Stepankowsky.
91. **When the doctor**   Interview, Scymanky.
91. **Venus Dergan pulled**   Interview, Dergan.
92. **They were celebrating**   Interview, Rogers.
94. **pick up her brother's**   Interview, Parker.
94. **He and the photographer**   Interview, Stepankowsky.
95. **Lee owned the Kid Valley store**   Andre Stepankowsky, "Valiant valley: Merchant awaits his fortune in shadow of the mountain," Longview *Daily News*, June 3, 1980.
96. **Everything medically**   Interview, Metcaff.
96. **The burn victim**   "Terrible Beauty," Portland *Oregonian*, p. V40, and Interview, Scymanky.
96. **He was all over**   Interview, Stepankowsky.
96. **The lever man**   Andre Stepankowsky, "Dredge: Floating 'vacuum' cleaning river," Longview *Daily News*, June 7, 1980.
97. **After the funeral**   Interview, Parker.
97. **Stepankowsky in his**   Interview, Stepankowsky.
98. **Patrick Keough, a planner**   Andre Stepankowsky, "Dredge Cowlitz, people demand," Longview *Daily News*, June 4, 1980.
99. **who predicted mere inch**   Andre Stepankowsky, "Inch of rain could cause Cowlitz to flood," Longview *Daily News*, June 4, 1980.
99. **"If two usually uncontrollable**   Andre Stepankowsky, "Flood fight: Corps engineers want safe Cowlitz by fall," Longview *Daily News*, June 4, 1980.
99. **It seemed like Scymanky**   Interview, Scymanky.
100. **Except for minor ash explosions**   "USGS chief foresees rumblings on peak," Longview *Daily News*, June 7, 1980.
100. **using tiltmeters and geodimeters**   Thompson, *Volcano Cowboys*, p. 58-59.
101. **The long history of volcano eruptions**   B.A. Bolt, W.L. Horn, G.A. Macdonald, R.F. Scott, Geological Hazards, (Springer-Verlag, New York, Heidelberg, Berlin, 1977) p. 128-129.
101. **Swanson now had**   Interview, Swanson.
102. **Swanson and other USGS geologists**   Donna duBeth, "Friday 13 eruption would set science back," Longview *Daily News*, July 12, 1980.
102. **Billowing clouds**   "Quakes ease as volcano steams," June 14, 1980; "Scientists spot dome through clouds," June 17, 1980, Longview *Daily News*; Carson, *Mount St. Helens*, p. 79.
102. **The helicopter hovered**   Interview, Rogers.
103. **When Donna Parker returned,**   Interview, Parker.
104. **With his dust**   Interview, Frenzen.
104. **Authorities had**   Interview, Rogers.
105. **"Take the thing**   Interview, Parker.

CHAPTER 8: JUNE 20-30

106. **"It's time for you**   Interview, Scymanky.
106. **The twelve of them sat**   Interview, Rogers.
107. **The woman's husband**   Todd W. Powell, "Blast from the Past," *Alaska Airlines* magazine, May 2000, p. 77-78.
107. **Stepankowsky wanted**   Interview, Stepankowsky.
108. **The helicopter carrying Swanson**   Interview, Swanson.
109. **"Venus if you don't do**   Interview, Vergan.
109. **"Gray do you still have**   Interview, Rogers.

110. **They came mainly**   Interview, Parker.
111. **Swanson's trips**   Thompson, *Volcano Cowboys*, p. 145-146.
112. **Swanson found the structure**   Interview, Swanson.
112. **The nurse left**   Interview, Scymanky.
113. **Hordes of townspeople**   Interview, Stepankowsky; Andre Stepankowsky, "Cowlitz dredges to be trucked up river," Longview *Daily News*, June 23, 1980.
113. **In her nightmare**   Interview, Parker.
114. **Roald Reitan could get**   Interview, Reitan.
114. **He had his best man**   Interview, Stepankowsky.
114. **Swanson and his crew**, Interview, Swanson.
115. **Rogers lashed**   Interview, Rogers.

CHAPTER 9: JULY

117. **Do you need any**   Interview, Dergan.
117. **A grey haired woman**   Interview, Swanson.
118. **Three ancients made**   Howel Williams; Alexander McBirney, *Volcanology*, (Freeman, Cooper & Co., San Francisco, 1979) p. 13-18.
119. **"The progress made since,**   ibid. p 13.
120. **Swanson and others hoped**   Thompson, *Volcano Cowboys*, p. 142, and Carson, *Mount St. Helens*, p. 76-80.
120. **"Have you seen how little**   Interview, Stepankowsky.
120. **The Corps was trying to squeeze**   Andre Stepankowsky, "Going through channels: Corps cuts red tape to clear river quickly," Longview *Daily News*, June 24, 1980.
121. **But now none other**   Andre Stepankowsky, "Engineers can't rule out floods," Longview *Daily News*, June 30, 1980. (An inch of rain did cause some flooding in early September. See Andre Stepankowsky's, "Inch of rain swells Toutle, causing minor floods and damage," Longview *Daily News*, September 2, 1980.)
121. **The changes in Scymanky**   Interviews, Jim and Helen Scymanky.
122. **"I want you to help me buy**   Interview, Donna Parker.
123. **The dormant Mount Hood**   Interview, Swanson.
123. **What concerned Swanson**   "Scientists watch dome for lava flow," Longview *Daily News*, June 18, 1980.
123. **The picture in the newspaper**   Interview, Jim Scymanky. The grisly photograph taken by Mike Carin appeared in several newspapers on July 10 and 11, 1980. (For related news story, see Donna duBeth's "Treed body identified as Sharipoff, father of five," Longview *Daily News*, July 10, 1980.)
123. **"Roald, right before**   Interview, Venus Dergan.
123. **From the first day**   Interview, Gary Eisler.
124. **Then he started thinking**   Interview, Scymanky.
124. **The room grew**   Interview, Eisler.
124. **"Is anyone missing**   Interview, Reitan.
125. **Donna Parker called**   Interview, Parker.
127. **He was sure of it now**   Interview, Swanson; Rick Seifert, "Scientists get clear look at new crater," Longview *Daily News*, July 24, 1980.
127. **Scymanky was apprehensive**   Interview, Scymanky; USGS scientist Richard Waitt; and pilot Jess Hagerrman.
128. **The mountain had taken its toll**   "Search yields another volcano victim," Longview *Daily News*," July 18, 1980. (This is the official number of dead at the time. The figure later was lowered to 57.)
128. **"The mountain just blew up**   Interview, Stepankowsky.
129. **"My wife told me**   Andre Stepankowky, "Nice timing: Peak's pop perks up tourists," Longview *Daily News*, July 23, 1980.
129. **Drake thought**   Interview, Stepankowsky.

131. **"A reporter and photographer** "Reporters cited in Red Zone," Longview *Daily News*, July 23, 1980.

131. **The volcano, in the series** Lew Pumphrey, "Eruption replaces dome with pit," Longview *Daily News*, July 23, 1980.

132. **In the dark** Interview, Rogers.

133. **Swanson had reestablished** Interview, Swanson; D.A. Swanson et al., "Predicting Eruptions at Mount St. Helens," *Science*, September 30, 1983.

CHAPTER 10: August

134. **On the morning of August 6,** Donna duBeth, "Harmonic tremors signal latest eruption," Longview *Daily News*, August 8, 1980.

134. **To celebrate** Thompson, *Volcano Cowboys*, p. 135-137.

134. **But Don Swanson was not** Interview, Swanson.

134. **The world was now** Carson, *Mount St. Helens*, p. 29-71; Thompson, *Volcano Cowboys*, p. 107-132; "Terrible Beauty," p. V23-24; Jelle Zeilinga de Boer, Donald Theodore Sanders, *Volcanoes in Human History* (Princeton University Press, Princeton and Oxford, 2002), p. 236-239; USGS Fact Sheet No. 036-00.

136. **The region and beyond** "The Acoustic Effects of the Eruption of Mount St. Helens May 18, 1980," Clara M. Fairfield, *Mount St. Helens: One Year Later*, S.A.C Keller, Editor, (Eastern Washington University Press, 1982), p. 83-85.

137. **For nine hours the volcano** Robert Decker and Barbara Decker, "The Eruptions of Mount St. Helens," *Scientific American*, May 1981, p. 68.

137. **Ash blanketed everything** Washington State Department of Game estimates; Carson, *Mount St. Helens*, p. 83. (Author's note: I have used the final official figure here for the number of persons killed.)

137. **The flight over** Andre Stepankowsky, "Landowners take hard line on dredging," Longview *Daily News*, June 10; Stepankowsky, "Flood fear: River landowners leery about dike options," Longview *Daily News*.

138. **"I can get you real money** Interview, Rogers.

138. **Jim Scymanky** had lost strength Interview, Scymanky.

139. **What impressed Peter** Frenzen, Interview.

140. **A wall of water** Andre Stepankowsky and Rick Siefert, "Studies predict massive erosion, winter floods," Longview *Daily News*, October 24, 1980.

141. **"We do a show** Interview, Reitan.

142. **Whenever he could spare** Interview, Swanson.

142. **That summer Johnston's spirit** Thompson, *Volcano Cowboys*, p. 129-132.

CHAPTER 11: September

143. **Donna, it's George**, Interview, Parker.

144. **Rogers left** Rogers, Interview.

145. **The eighty-seven** Andre Stepankowsky, "Flood fear: Some riverside folks shrug, others worry about river," Longview *Daily News*, September 25, 1980

146. **Her friends were coming** Interview, Dergan

146. **Peter Frenzen hadn't** Interview, Frenzen

147. **For days now** Interview, Swanson

147. **But he didn't believe in using** Carson, *Mount St. Helens*, p. 79.

148. **Rogers was hit hard** Interview, Rogers.

149. **Roald drove to Venus's** Interview, Reitan.

149. **Venus and her mother** Interview, Dergan.

149. **Swanson had almost no** Interview, Swanson.

149. **They drove to the house** Interview, Reitan; interview, Dergan.

150. **Late that night** Interview, Helen Scymanky.

150. **This was his first helicopter trip** Interview, Frenzen.

151. **But the Utah team made** Ibid; Carson, *Mount St. Helens*, p. 84.

152. **Venus still couldn't believe it**   Interview, Dergan; interview, Reitan.
153. **As he lay there**   Interview, Rogers.
154. **They were working**   Interview Frenzen.
154. **Stepankowsky was listening**   Interview, Stepankowsky.
155. **they found a tattered pillow**   ibid.; Andre Stepankowsky, "More bodies: Three volcano victims remains recovered," Longview *Daily News*, October 1, 1980.

CHAPTER 12: October

157. **Donna Parker wasn't daunted**   Interview, Parker.
157. **Billy was not**   "Terrible Beauty," Portland *Oregonian*, p. V2.
157. **Despite what state officials**   ibid., p. V12-13.
157. **Officials of both**   ibid.
158. **Several members of**   ibid.
159. **All records of the meetings**   ibid.
159. **Loggers made numerous protests**   ibid, p. V10-11.
160. **Coordination of rescue efforts**   ibid, p. V28-31.
160. **The Washington Department**   ibid, p. V36-37.
161. **He pulled the "Terrible Beauty"**   Interview, Stepankowsky.
162. **Something in the *Oregonian***   Interview, Parker; "A Terrible Beauty," p. V22.
163. **Not all readers**   Interview, Parker; "Weyerhaeuser is biggest by far," Longview *Daily News*, September 25,1980.
163. **When Peter Frenzen returned**   Interview, Frenzen.
163. **On Wednesday, October 8**   "Peak shoots ash, steam to 16,000 feet," Longview *Daily News*, October 9, 1980.
164. **The gritty taste**   Interview, Frenzen.
164. **It was the morning of Friday,**   "3rd blast rattles mountain; scientists see no quiet soon," Longview *Daily News*, October 18, 1980.
165. **The final, deep earthquake**   Interview, Swanson.
166. **He was out partying**   Interview, Reitan.
166. **Andre Stepankowsky and Rick Seifert**   Interview, Stepankowsky; "Andre Stepankowsky and Rick Seifert, "Studies predict massive erosion, winter floods," Longview *Daily News*, October 28, 1980.
167. **That night they tracked down**   Rick Seifert and Andre Stepankowsky, "Army Corps chief: 'I can't see giving up'," Longview *Daily News*, October 24, 1980.
168. **When the hearing was held**   Linda Wilson and Rick Seifert, "Concerned: Flood forecasts glazed with ifs, wary crowd finds," Longview *Daily News*, October 30; Rick Seifert and Andre Stepankowsky, "Discrepancies: Federal agencies change, omit flood prediction, erosion figures," Longview *Daily News*, October 31, 1980.
169. **The snow would**   Interview, Frenzen.
170. **This would be her**   Interview, Parker.

CHAPTER 13: November and December

172. **His photographs**   Interview, Rogers.
173. **At his desk**   Interview, Swanson.
173. **St. Helens had been one big joke**   Carson, *Mount St. Helens*, p. 32-33.
173. **Swanson and his fellow scientists**   Interview, Swanson; Thompson, *Volcano Cowboys*, p. 30-31.
174. **Now that Donna**   Interview, Parker.
175. **The first heavy storm**   Andre Stepankowsky, " 'Amazing things happened' in Toutle valley's rain," Longview *Daily News*, November 17, 1980.
175. **It was very cold**   Interview, Swanson; W.W. Chadwick, Jr., D.A. Swanson, E.Y. Iwatsubo, C.C. Heliker, T.A. Leighley, "Deformation Monitoring at Mount St. Helens in 1981 and 1982," *Science*, September 30, 1983; U.S. Geological Survey Professional Paper 1250, "The 1980 Eruptions of Mount St. Helens."

177. **The fog shrouded**   Interview, Scymanky.

177. **With winter near**   Andre Stepankowsky, "Snow poses danger," Longview *Daily News*, November 24, 1980.

178. **The man was elfish**   Interview, Parker.

178. **Snow filled the crater**   Thompson, *Volcano Cowboys*, p. 148-149.

178. **The man pointed**   Andre Stepankowsky, "Cowlitz: River surveyors unsure of dredge progress," Longview *Daily News*, December 12, 1980.

179. **On December 16**   Interview, Swanson; "Eruption Summary—Mount St. Helens Eruptions Since May 18, 1980," compiled by Lyn Topinka, USGS/CVO, June 17, 1997.

180. **On Christmas day**   Interview, Stepankowsky.

180. **The National Weather Service**   *Associated Press*, "Big floods forecast for nine rivers," Longview *Daily News*, December 26, 1980.

PART 2: 1981: Aftershock
CHAPTER 14: January and February 1981

200. **The heavy wind**   Interview, Swanson.

201. **Jim Scymanky leaned back**   Interview, Scymanky.

201. **Robert Rogers found a copy**   Interview, Rogers; "Robert Landsburg's brave final shots," *National Geographic*, January 1981, p. 27-28.

202. **The Corps of Engineers**   "Corps says dredging prevented flooding," Longview *Daily News*, December 30, 1980.

202. **"We haven't seen**   Rick Seifert, "Crisis chief: Threat here still serious," Longview *Daily News*, January 3, 1981.

202. **Stepankowsky was getting mixed**   Rick Seifert, "Weather Service forecasters: Worst flood danger over," Longview *Daily News*, January 9, 1981; Rick Seifert, "USGS: Flooding chances lessen," Longview *Daily News*, January 14, 1981.

202. **Then the Corps announced**   Andre Stepankowsky, "Lower Cowlitz to lose dredges by mid-April," Longview *Daily News*, January 30, 1981.

202. **Robert Rogers was not**   Interview, Parker.

203. **Stepankowsky had long**   Interview, Stepankowsky.

203. **Environmentalists had proposed**   Andre Stepankowsky, "Peak monument plan gets praise, criticism," Longview *Daily News*, January 16, 1981.

204. **Stepankowsky wrote his**   Interview, Stepankowsky; Andre Stepankowsky, "Survivors: Tower Road tough, despite peak, river," Longview *Daily News*, January 17, 1981.

205. **"I think we're going**   Interview, Swanson.

206. **Don Swanson and thirty-five**   Interview Swanson; Andre Stepankowsky, "Scientists say peak film distorts Johnston's life," Longview *Daily News*, January 30, 1981.

206. **On February 5**   Andre Stepankowsky, "USGS predicts non-violent eruption within hours," Longview *Daily News*, February 5, 1981.

206. **Some residents near**   *Associated Press*, "Ariel resident says lasagna sets off peak," Longview *Daily News*, February 5, 1981.

207. **Swanson was right**   Interview, Swanson; Rick Seifert, "Dome keeps growing," Longview *Daily News*, February 7, 1981.

207. **Back in 1923**   "John M. McClelland: Newspaper, community leader dead at 95," Longview *Daily News*, February 6, 1981; Suzanne Martinson, "Active and ageless," Longview *Daily News*, February 7, 1981.

208. **He was in search**   Interview, Rogers.

209. **Roald Rietan heard**   Interview, Reitan.

210. **Scymanky found a lawyer**   Tom Paulu, "Eruption victim files suit against county," Longview *Daily News*, February 19, 1981.

210. **On the night of February 13**   *Associated Press*, "Volcano fault shifts may have triggered Friday's quake," Longview *Daily News*, February 16, 1981.

210. **Rains that weekend**   Andre Stepankowsky, "Swollen Toutle threatens home, bridges, and freeway"; *Associated Press* and Longview *Daily News*, "Wind, trees cut power to 6,200," Longview *Daily News*, February 19, 1981.

210. **Residents were vocal**   "Swollen Toutle cuts into banks; Cowlitz not expected to flood," Longview *Daily News*, February 16, 1981.

211. **For months she**   Interview, Parker.

CHAPTER 15: MARCH AND APRIL

212. **They stood there**   Interview, Reitan and Dergan; photograph.

212. **It was late**   Interview, Stepankowsky; Andre Stepankowsky, "River projects," March 25; "Corps options"; "Debris dams," March 26, Longview *Daily News*.

214. **They were lost**   Interview, Parker.

216. **At the roadblock**   Interview, Rogers.

217. **It was a clear day**   Interview Parker.

218. **Don Swanson stared**   Interview Swanson.

219. **This was to be**   Interview, Rogers.

220. **It would be his first**   Interview, Frenzen; Carson, *Mount St. Helens*, p.83-107; "Life is returning to the devastated blast zone," Longview *Daily News*, May 18, 1981; *Associated Press*, "REBIRTH: Life starts over in ash land," *Year of the Mountain*, p. 44B; Les Blumenthal, *Associated Press*, "Blast zone comes alive, amazing experts," Longview *Daily News*, May 18, 1981.

221. **As he exercised**   Interview, Scymanky.

221. **"Do you have a pass**   Interview, Parker.

222. **At 8:21 on Friday**   Andre Stepankowsky and Rick Seifert, "More eruptions likely after morning steam, ash burst," Longview *Daily News*, May 10, 1981.

223. **That night**   Interview, Swanson.

223. **Newspapers called**   Rick Seifert, "Eruption forecast hits mark," Longview *Daily News*, April 13, 1981.

224. **What was ironic**   Dick Thompson, *Volcano Cowboys*, p. 145-146.

224. **Swanson didn't particularly**   Interview, Swanson.

224. **It was shortly after noon**   Interview, Andre Stepankowsky; "The Daily News and Werth win Pulitzer Prizes," Longview *Daily News*, April 13, 1981; "Depth of Peak coverage impressed Pulitzer jurors," Longview *Daily News*, April 14, 1981.

CHAPTER 16: MAY

227. **"They were big,**   Laurie Smith, "Logged off: Weyerhaeuser, accused over controversial Kid Valley cut, calls it mistake"; "Weyco asked to stop cutting—and it didn't,"; "County would forgive Weyco illegal clearcut," Longview *Daily News*, May 4, 1981.

228. **Again Donna Parker**   Interview, Parker.

228. **Ash pottery**   Marion Villa and Andre Stepankowsky, "Ash enterprise: Toutle Valley merchants spruce up for onslaught," Longview *Daily News*, May 2, 1981; Andre Stepankowsky, "Officials already fretting about summer tourist traffic," Longview *Daily News*, May 5, 1981.

230. **They were going out**   Interview, Dergan.

230. **The man on the phone**   Interview, Parker.

231. **Relatives of eight**   Interview, Stepankowsky; "Sieber, Mullineaux, Ray argue on TV," Longview *Daily News*, May 13, 1981.

231. **Stepankowsky headed to**   Andre Stepankowsky, "Victims kin sue the state," Longview *Daily News*," May 13, 1981.

232. **At the office, Stepankowsky started**   Andre Stepankowsky, "Lawsuit: Volcano widow not motivated by money," Longview *Daily News*, May 13, 1981.

232. **As the first anniversary**   Interview, Dergan.

232. **The national and international**   "Bleak volcano flight weekend predicted," Longview *Daily News*, May 16, 1981.

232. **Stepankowsky noticed a headline**   Steve Twedt, "Volcano celebrations to erupt this weekend," Longview *Daily News*, May 13, 1981.

233. **"Life must go on**   Dick Pollock, "Views of the News: For those who grieve," Longview *Daily News*, May 13, 1981.

233. **"What are you going to**   Interview, Donna Parker.

235. **Swanson and the scientists**   Interview, Swanson.

236. **There was one event**   *Associated Press*, "Degree: Geologist killed in eruption honored," Longview *Daily News*, May 18. 1981.

236. **"Let's compare him**   Andre Stepankowsky, "Johnston just doing his job, his kin say," Longview *Daily News*, May 18, 1981.

237. **Washington Governor**   "Spellman declares Daily News Day," Longview *Daily News*, May 16, 1981.

237. **A panel of public officials**   Rick Seifert, "Emergency officials fret over media's roll in hazards reporting," Longview *Daily News*, May 15, 1981.

237. **Of the seven geographical areas**   Rick Seifert," Delayed stress: Volcano's blast still lingers in our minds," Longview *Daily News*, May 18, 1981.

238. **Helen helped him**   Interview, Helen Scymanky.

238. **An advance story**   Les Blumenthal, *Associated Press*, "Blast zone comes alive, amazing experts," Longview *Daily News*, May 18, 1981.

238. **The three of them**   Interview, Swanson.

239. **For the first time**   Rogers, Interview.

243. **Shortly after Rogers**   "Red Zone law violator given year in Skamania jail," Longview *Daily News*, June 16, 1981.

CHAPTER 17: JUNE AND JULY

245. **He could hear the building up**   Interview, Swanson.

245. **When Frenzen reported**   Interview, Frenzen

246. **As he entered**   Interview, Stepankowsky; Andre Stepankowsky, "Otto Sieber's lawyer called him crazy," Longview *Daily News*, June 8, 1981.

247. **a week before the state's**   Andre Stepankowsky, "Suit by volcano victims' relatives links politics, safety," Longview *Daily News*, June 8, 1981.

248. **Donna Parker couldn't get**   Interview, Parker.

249. **"We think we have**   Interview, Scymanky.

250. **With a blast from St. Helens**   D.A. Swanson et al., "Predicting Eruptions at Mount St. Helens," *Science*, September 30, 1983.

250. **The race was on**   "Timber cut 'regrettable,'" Longview *Daily News*, June 17, 1981.

251. **They sat together**   Interview, Dergan.

252. **It was morning**   Interview, Parker.

252. **The critical letters and phone calls**   Interview, Stepankowsky.

253. **All that changed**   Lila Fujimoto, "Paper honoed for 'fire' that brought Pulitzer Prize," Longview *Daily News*, July 14, 1981.

CHAPTER 18: AUGUST

254. **The two geologists**   Interview, Swanson; Thompson, *Volcano Cowboys*, p.150-154.

255. **most detailed and important work**   *Journal of Volcanology and Geothermal Research*, no. 66, 1995, xxii.

256. **They're cutting off**   Interview, Jim and Helen Scymanky.

256. **Oregon State University**   Andre Stepankowsky, "Legionnaire bacteria spurs OSU warning," Longview *Daily News*, August 12, 1981; Andre Stepankowsky, "Scant Legionnaire details fan flap," Longview *Daily News* August 13, 1981.

257. **As he worked**   Interview, Frenzen.
258. **When they saw the sign**   Interview, Parker.
259. **"See, the alder's coming back**   " 'Zapped to zero': Blast zone study may teach clearcut lessons," Longview *Daily News*, August 25, 1981.
259. **The USGS scientists**   Swanson, et al., "Predicting Eruptions at Mount St. Helens, June 1980 Through December 1982," *Science*, September 1983, Vol. 221, No. 4618.
260. **Venus Dergan could tell**   Interview, Dergan.
261. **Lately it seemed**   "Souvenir shops contend with bust, volcano T-shirt glut," Longview *Daily News*, August 31, 1981.
261. **the Army Corps of Engineers now estimated**   Andre Stepankowsky, "Corps to curtail dredging, concentrate on Toutle dams," Longview *Daily News*, August 19,1981.
261. **The Reagan administration came up**   Andre Stepankowsky, "Reagan official wants peak land use plan reconsidered," Longview *Daily News*, August 31, 1981.
262. **The plan had been a compromise**   Andre Stepankowsky, "Special interests face off over volcano plan," Longview *Daily News*, October 16, 1981.
262. **Peter Frenzen listened**   Frenzen, interview.

CHAPTER 19: SEPTEMBER AND OCTOBER

263. **Robert Rogers was so**   Interview, Rogers.
263. **The USGS issued an advisory**   "Mountain set to have an eruption," Longview *Daily News*, August 26 1981.
263. **Scymanky threw the soccer ball,**   Interview, Scymanky.
264. **When the volcano erupted**   Interview, Swanson; "Eruption so quiet it fools scientists," Longview *Daily News*, September 7, 1981.
264. **the face of a fictionalized**   Marion Villa, "St. Helens: Moviegoers line up to relive eruption," Longview *Daily News*, September 4, 1981.
265. **Rogers and a friend**   Interview, Rogers.
266. **Stepankowsky no longer**   Interview, Stepankowsky.
266. **Stepankowsky had learned**   Andre Stepankowsky, "Huge 'laboratory' lures scientists," Longview *Daily News*, October 12, 1981.
267. **One theory behind**   Andre Stepankowsky, "Out of gas: Volcano is running low on its eruptive 'fuel.' " Longview *Daily News*, Oct. 12, 1982.
267. **Some of the agency's elite**   Kathy McCarthy, *Associated Press*, "Geologists caution that second big blast possible," Longview *Daily News*, May 18, 1981.
267. **It was possible a large**   ibid.
267. **The reporter and photographer**   Andre Stepankowsky, "Links of life," Longview *Daily News*, Oct. 12, 1981.
268. **Twice she drove**   Interview, Parker.
269. **From the ridge**   Interview, Swanson; Carson, *Mount St. Helens*, pgs. 124-125.
270. **Hydrologists provided**   Carson, *Mount St. Helens*, p. 123.
270. **More than a hundred**   Marion Villa, "Volcano search volunteers look in vain for more bodies," Longview *Daily News*, September, 1981.
270. **On October 20, 1981**   "It's official: Mount St. Helens nation's first Volcanic Area," Longview *Daily News*, October 20, 1981; Andre Stepankowsky, "Final volcano plan protects 109,000 acres," Longview *Daily News*, October 15, 1981.
272. **When Stepankowsky sat**   "Non-explosive eruption likely within two weeks," Longview *Daily News*, October 24, 1981.
273. **some scientists were calling**   Andre Stepankowsky, "Scientist: Allow public in peak blast zone," Longview *Daily News*, October 22, 1981.
273. **the mountain erupted**   "Volcano's lava dome growing, glowing," Longview *Daily News*, October 31, 1981.

PART 3: 1984-1986: THE TRIAL
CHAPTER 20: 1984-1985

276. **The trail narrowed**   Interview, Dergan.
276. **Suddenly, the plume**   Andre Stepankowsky, "Molten rock flows from peak dome," Longview *Daily News*, June 18, 1984.
277. **The headline in**   Timothy Egan, "Ray: State agency was to blame for Red Zone," *Seattle Post-Intelligencer*, June 27, 1984; Timothy Egan, "Ray changes story on Red Zone limits," *Seattle Post-Intelligencer,* June 27, 1984.
278. **In a telephone interview**   Andre Stepankosky, "Ray's testimony may aid case of volcano victims," Longview *Daily News*, June 27, 1984.
278. **While the former governor scrambled**   Egan, "Ray: State agency was to blame...."
278. **As he made his way**   Interview, Frenzen.
279. **As the rock-chewing monster**   Interview, Stepankowsky; Andre Stepankowsky, "Chomp: In the darkness, a rock-chewing monster feeds," October 15, 1984.
279. **All the letters were**   Interview, Parker; various letters written by Parker.
280. **The attorneys**   Interview, Jim and Helen Scymanky.
280. **Don Swanson drove past**   Interview, Swanson; Carson, *Mount St. Helens*, p. 121-125.
281. **A new government study**   Andre Stepankowsky, "Volcano Watch: Report shows why we should be glad Spirit Lake threat is gone," Longview *Daily News*, Oct. 28, 1985.
282. **Venus Dergan walked up**   Interview, Dergan.
282. **The Toutle River raged**   Interview, Stepankowsky; Andre Stepankowsky, "Volcano's eruption spared their home but eroded their marriage," Longview *Daily News*, May 18, 1985.
283. **The call from Roald**   Interview, Reitan; interview, Dergan.
283. **The scientist stood**   Photograph accompanying "Life amid the Ruins," *Seattle Post-Intelligencer*, May 12, 1985.
284. **that determination was debatable**   Interview Swanson; Andre Stepankowsky, "Molten rock flows from peak dome," Longview *Daily News*, June 18, 1984.
284. **At the Monticello Hotel**   Andre Stepankowsky, "Red carpet rolls out for tunnel's premiere," Longview *Daily News*, April 28, 1985.
284. **Donna Parker nailed**   Interview, Parker.
285. **It had been years since Rogers**   Interview, Rogers.
285. **"Since the mountain blew**   Andre Stepankowsky, "Follies and fiascos brought us some fun, frustrations"; Andre Stepankowsky, "Volcano Watch: One-year limit challenged," Longview *Daily News*, May 27, 1985.
286. **The bugling of elk**   Interview, Frenzen; Karen Adams, "Landscape 'zapped to zero' shows signs of recovery," May 13, 1985.
287. **Five years after**   Andre Stepankowsky, "May 18, 1980: A day Cowlitz County will not forget," Longview *Daily News*, May 13, 1985.

CHAPTER 21: 1985: VOLCANO WATCH

288. **Swanson wasn't too sure**   Interview, Swanson.
288. **After hearing**   Interview, Parker.
289. **When Stepankowsky looked**   Bob Gaston, "News Notes: County's economic hard times shrink newspaper's size, staff," Longview *Daily News*, May 20, 1985.
289. **That night shallow earthquakes**   Interview, Swanson; "Data shows new eruption on the way," Longview *Daily News*, May 24, 1985.
290. **It was just after 4 a.m.**   Interview, Swanson.
291. **There was a lot to report**   Andre Stepankowsky, "Volcano Watch: Actor who portrayed ill-fated geologist in movie dies in stabbing," Longview *Daily News*, June 10, 1985.

291. **The lawyers sat**   Interview, Scymanky.
292. **At first**   Interview, Stepankowsky.
293. **It was a major blow**   Interview, Scymanky.
293. **The plaintiffs' legal brief**   Superior Court of Washington for King County, "Barbara Karr, et al., Plaintiffs vs. State of Washington and Weyerhaeuser Company, Defendants, No. 81-2-07109-1, Plaintiffs' response to defendants' motion for summary judgment," September 18, 1985.
294. **McCutheon ruled**   Julie Emery, "State out of Mt. St. Helens Suit," *Seattle Times*, September 23, 1985.
294. **Five years after he started his study**   Peter M. Frenzen and Jerry F. Franklin, "Establishment of Conifers from Seed on Tephra Deposited by the 1980 Eruptions of Mount St. Helens, Washington," *American Midland Naturalist*, 114(I), 84, 1985.
294. **At the end of his leave of absence**   Andre Stepankowsky, "Pre-eruption warnings a key in wrongful death trial," Longview *Daily News*, October 11, 1985.
296. **The jet swooped down**   Interview, Helen and Jim Scymanky.
297. **"What was Weyerhaeuser's response,**   Andre Stepankowsky, "Attorney for volcano victims says Weyco ignored dangers," Longview *Daily News*, October 17, 1985; Don Duncan, "Volcano Suit Pits Little Guy Against Giant," *Seattle Times*, October 18, 1985.
298. **Clark maintained that evidence**   Andre Stepankowsky, "Weyco: Peak dangers not foreseeable," Longview *Daily News*, October 18, 1985.

CHAPTER 22: 1985: IN THE COURTROOM.

299. **When Judge James**   *Associated Press*, "Author testifies as jurors begin volcano 'education'," Longview *Daily News*, October 21, 1985.
299. **By the end of the day, Donna**   Interview, Parker.
299. **The logic of some of Judge McCutheon's**   Interviews, Parker, Jim and Helen Scymanky.
299. **With testimony of the man**   *Associated Press*, "Volcanologist says no warning was guaranteed," Longview *Daily News*, October 23, 1985.
300. **Don Swanson was taken aback**   Ibid; interview, Swanson.
300. **The Weyerhaeuser attorney, Clark**   *Associated Press*, "Scientist says he feared volcano job," Longview *Daily News*, October 24, 1985; Interview, Swanson.
301. **The most damning evidence**   *Associated Press*, "Geologist says state didn't heed his advice on volcano hazard zones," Longview *Daily News*, October 26, 1985.
302. **The trial, entering its third week**   *Associated Press*, "Volcano trial focus shifts to Weyerhaeuser," Longview *Daily News*, October 28, 1985.
302. **Donald Vernon, a Weyerhaeuser employee**   *Associated Press*, "Blast Notice was promised logger says," Longview *Daily News*, October 29.
302. **Two days before the massive eruption**   *Associated Press*, "Weyco feared loggers would want extra pay," Longview *Daily News*, October 31, 1985.
302. **The story poured forth**   Interview, Scymanky; *Associated Press*, "Logger claims Weyco assured him working near the volcano was safe," Longview *Daily News*, November 4, 1985.
303. **This was Franklin's last day**   *Associated Press*, "Boss: Weyco loggers not told of zone expansion," Longview *Daily News*, November 5, 1985.
303. **Mark Clark wasted no time**   *Associated Press*, "Scientist foresaw 'deadly' eruption," Longview *Daily News*, November 7, 1985. (Author's note: This is an incorrect headline. The scientist actually "failed to foresee" the eruption.)
304. **Clark scrambled to bring**   *Associated Press*, "Weyco VP: Blast wasn't anticipated," Longview *Daily News*, November 8, 1985.
304. **"There was no concept**   Ibid.

305. **It had now been over a month** Associated Press, "Volcano trial: Weyco says it relied on reports from USGS," Longview *Daily News*, November 14, 1985.

305. **As Andre Stepankowsky worked** Associated Press, "Eruption in Columbia buries city: 20,000 are feared dead," November 14, 1985; AP, "Rescuers dig for volcano victims," AP; "Eruption survivors tell of terror that smashed through their city," November 15, 1985, all Longview *Daily News*.

306. **News of the disaster** Interview, Swanson; Thompson, *Volcano Cowboys*, p. 180-190.

307. **A 1982 incident** "Eruption Triggered Avalanche, Flood, and Lahar at Mount St. Helens—Effects of Winter Snowpack," *Science*, September 30, 1983; Thompson, *Volcano Cowboys*, p. 155-156.

307. **Stepankowsky sat transfixed** Andre Stepankowsky, "Jury begins deliberations in Weyerhaeuser volcano trial," Longview *Daily News*, November 15, 1985; interview, Stepankowsky; Don Duncan, "An Act of God? Jury will Decide was Weyerhaeuser liable for Death, Injury in Mt. St. Helens Eruption," *Seattle Times*, November 15, 1985.

308. **When Scymanky heard** Interview Scymanky; and jury instructions, Barbara Karr et al., Plaintiffs.

308. **"I'm glad its over** Associated Press, "Killian says volcano trial set the record straight," Longview *Daily News*, December 2, 1985.

CHAPTER 23: 1985-1986

309. **Eighteen days after** Larry Lange, "Act of God' factor splits the jury in volcano suit," *Seattle Post-Intelligencer*, December 3, 1985.

310. **For days the Longview area** Interview, Stepankowsky; Tom Paulu, "Russian roots influence Andre Stepankowsky's music," Longview *Daily News*, November 21, 1985.

310. **Venus had known the man** Interview, Dergan.

311. **Judge James McCutcheon refused** Associated Press, "Judge clears way for new Weyco volcano trial," Longview *Daily News*, January 24, 1986.

311. **The blue and grey Hughes 500** "Searchers find no trace of copter missing near peak," Longview *Daily News*, February 12, 1986; Jeannie Kever, "Photographer had feeling about last flight," Longview *Daily News*, February 13, 1986.

312. **It had rained hard** Interview, Stepankowsky; Andre Stepankowsky, "Virtuoso pianist plays in spirit of masters," Longview *Daily News*, February 24, 1986.

312. **As soon as the concert ended** Interview, Stepankowsky; Andre Stepankowsky, "New storm tonight may add to flooding woes," Longview *Daily News*, February 24.

313. **This time the Federal** Dell Burner, "Clogged Cowlitz added to flood problems," Longview *Daily News*, April 15, 1986.

314. **The attorney** for the families Interview, Parker; interview, Jim and Helen Scymanky.

314. **At 259 days** "New lava may be forming on peak," Longview *Daily News*, May 12, 1986; Interview, Swanson.

314. **Still, St. Helens'** "Fewer scientists watching quiet mountain," Longview *Daily News*, March 3, 1986.

314. **Stepankowsky's wife told him** Bob Gaston, "News Notes: A phone call, an offer—and Daily News' fate hung in balance," Longview *Daily News*, April 21, 1986; Paula LaBeck Stepankowsky, "Natts buy newspaper, buck trend to chains," Longview *Daily News*, March 3, 1986.

315. **Swanson rushed** Interview, Swanson; "Crater Swelling, but scientists don't know if another eruption is brewing," April 12, 1986; "Volcano tossing around big blocks of lava dome," April 22, 1986; "Volcano puffs ash over town of Morton";

"USGS scientists might issue eruption alert," May 7, 1986; "Dome sprouts new lobe," April 14, 1986; photo caption, "Avalanche on dome," April 15, 1986, all Longview *Daily News*.

CHAPTER 24: 1986

316. **"The volcano gets so hot**   Andre Stepankowsky, "Students erupt with volcano theories six years after blast," Longview *Daily News*, May 17, 1986.
317. **As her son Jim**   Interview, Donna Parker.
317. **"Did they settle yet?"**   Interview, Donna Parker.
318. **Ron Franklin's office**   Interview, Jim and Helen Scymanky; interview, Parker.
318. **It provided between**   Order Approving Reasonableness of Settlement with Defendant Weyerhaeuser Company, Kevin Bowers, et al., Plaintiffs vs. Weyerhaeuser Company, February 6, 1987 (hearing September 22, 1986).
318. **Robert Rogers watched as one**   "Blast from the Past," *Alaska* magazine, May 2000; Interview, Rogers.
319. **Parker and Jeanette Killian continued**   Interview, Parker.
319. **Parker searched public records**   Karr et al. vs. Weyerhaeuser Company, Otto Sieber deposition, p. 31-32.
319. **Parker then discovered**   Office of the Attorney General of the State of Washington memorandum in support of brief for motion of dismissal of the state as a defendant in Karr et al. vs. the State of Washington and Weyerhaeuser Company, p. 56-59, l985, in King County Superior Court.
319. **In a June 8, 1981, story**   Andre Stepankowsky, "Otto Sieber's lawyer called him crazy," Longview *Daily News*, June 8, 1981.
320. **They also dug up**   Mike Layton, "Defiant Ray Boosts 5 to Court Bench," *Seattle Post-Intelligencer*, July 17, 1980.
320. **It was all okay now,**   Dergan, interview.
320. **Frenzen had known**   Interview, Frenzen.
321. **Parker was driving**   Interview, Parker.
321. **The years since the 1980 eruption**   Andre Stepankokwsky, "Even with funding, '87 Cowlitz dredging is out,"; Michael Yantis, "Flood meeting takes a back seat to baseball," Longview *Daily News*, October 28, 1986.
322. **He often camped out**   Interview, Frenzen.
322. **The eruption added**   Interview, Swanson.
322. **She hadn't been invited**   Interview, Donna Parker; Suzanne Hapala, "Resetting sights on St. Helens new visitors center stirs a mountain of memories," *Seattle Times*, March 29, 1987.

PART 4, 1989-1991: A Restless Slumber
CHAPTER 25: 1989-1990

326. **Fishermen were everywhere**   Interview, Frenzen.
327. **Donna Parker pointed to the edge**   Interview, Parker.
328. **I have something**   Interview Dergan; interview Reitan.
328. **Flurries of shallow earthquakes**   Carson, *Mount St. Helens*, p. 137-138.
329. **How could he explain the series of recent earthquakes**   Andre Stepankowsky, "Volcano shakes with series of earthquakes," Longview *Daily News*, August 29, 1989.
329. **She woke up from sleeping**   Interview, Parker.
330. **The huge mat**   Interview, Frenzen.
330. **High concentrations of manganese**   Carson, *Mount St. Helens*, p. 108-109.
331. **Stepankowsky stood**   Andre Stepankowsky, "$75 million project nears end; dam to be flood-threat barrier," Longview *Daily News*, September 13, 1989. By the time the dam was finished months later, the cost had risen over $5 million.

332. **Scymanky tinkered**   Interview, Scymanky.
332. **It came as a shock**   Interview, Parker.
333. **On January 6, 1990**   Carson, *Mount St. Helens*, p. 138.
333. **It was storming**   Interview, Rogers.
333. **The little logging community**   Interview, Stepankowsky; Andre Stepankowsky, "The search goes on," Longview *Daily News*, May 18, 1980.

CHAPTER 26: 1990

336. **Slides were scattered**   Interview, Rogers.
336. **Across the blast zone**   Interview, Frenzen; Carson, *Mount St. Helens*, p. 84-85.
337. **It was another call**   Interview Dergan; interview, Reitan.
338. **Stepankowsky liked interviewing**   Interview, Stepankowsky; interview, Swanson; Andre Stepankowsky, "Scientific breakthrough: St. Helens work leads to first forecasting," Longview *Daily News*, May 18, 1980.
339. **Jim and Helen stayed up**   Interview, Jim and Helen Scymanky; documentary film, "The Fire Below Us: Remembering Mount St. Helens," Global Net Productions, 1990.
340. **Swanson was finishing**   Interview, Don Swanson.
341. **The call she feared**   Interview, Dergan; Interview, Reitan,
341. **Stepankowsky strode**   Interview, Stepankowsky; Bob Gaston, "News Notes: Journalists, special section recall fury of 1980," Longview *Daily News*, 1990.
341. **"Ten years after the mountain blew**   Andre Stepankowsky, "A hailstorm of fire and brimstone," Longview *Daily News*, May 18, 1990.
342. **The Scymankys stopped**   Interview, Jim and Helen Scymanky.
342. **The three women**   Sally Macdonald, "Searing memories: Peace long in coming to victim's families," *Seattle Times*, May 18, 1990.
343. **Peter Frenzen turned on the television**   Interview, Frenzen.
344. **Venus Dergan opened the Tacoma *News Tribune***   Interview, Dergan.
345. **He was exhausted**   Interview, Stepankowsky; "Nine who played key roles in eruption recovery honored," Longview *Daily News*, May 19, 1990.
346. **The monument scientist**   Interview, Frenzen.
347. **Rogers had developed**   Interview, Rogers.
347. **Hundreds of chilly people**   Interview, Stepankowsky; Andre Stepankowsky, "Uncertain era ends with new sediment dam," Longview *Daily News*, May 19, 1990.
348. **In nearby Toutle**   Paula LaBeck Stepankowsky, "Volcano anniversary events fill weekend," Longview *Daily News*, May 17, 1990; Brenda Blevins, "Whirlwind of events could have Toutle festival goers in a Daze," *Longview Daily News*, May 12, 1990.
348. **In a line that stretched**   Paula LaBeck Stepankowsky, "2,500 strong: Protesters fight for NW timber jobs," Longview *Daily News*, May 21, 1990; interview Stepankowsky; Andre Stepankowsky, "Spotted owl, old growth, jobs: No quick solution," Longview *Daily News*, May 21, 1990; Paula LaBeck Stepankowsky, "Eruption, recession ganged up on area's economy," Longview *Daily News*, May 18, 1990.
350. **Parker looked up**   Interview Parker; "Relatives dedicate plaque to victims," Longview *Daily News*, May 19, 1990.
351. **Behavioral scientists had followed**   Michael Yantis,"St. Helens gave its name to psychological disorder," Longview *Daily News*, May 18, 1990.

CHAPTER 27: 1990-1991

352. **He carefully opened**   Interview, Frenzen.
352. **Venus turned on the TV**   Interview, Dergan; film, "The Fire Below Us."
352. **Ten years after**   Charles B. Halpern, Peter M. Frenzen, Joseph E. Means, Jerry F. Franklin, "Plant succession in areas of scorched and blown-down forest after the

1980 eruption of Mount St. Helens, Washington," *Journal of Vegetation Science*, I, 1990, p. 181-194.

353. **Much of the information**   Andre Stepankowsky, "Mount St. Helens still has lessons to teach us," Longview *Daily News*, May 19, 1990; Andre Stepankowsky, "Job, owl proposals fly in 2 directions," Longview *Daily News*, May 9, 1990.

353. **The Weyerhaeuser Company**   Advertisment: "The Choice At Mount St. Helens Was This: Either Work With Mother Nature Or Pay The Price To Father Time," Longview *Daily News*, May 18, 1990.

354. **What the Weyerhaeuser advertisments**   Stepankowsky, "Spotted Owl, old growth, jobs."

354. **She hiked up**   Interview, Parker.

354. **The reporter**   Andre Stepankowsky, "New life: Nature weaves tapestry of change in area once labeled a 'moonscape.' " Longview *Daily News*, May 18, 1990.

355. **The photograph**   Interview, Parker.

356. **It seemed like he had read**   Interview, Frenzen.

357. **"I just got my tape**   Interview, Parker.

357. **For fifteen years**   Interview, Rogers.

358. **Venus gave the interview alone**   Interview, Dergan.

358. **The research scientist**   Interview, Frenzen.

359. **The news shocked**   Interview, Swanson; Thompson, *Volcano Cowboys*, p. 249-250.

PART 5, 1999-2000: Remembrance
CHAPTER 28: 1999-2000

362. **Venus Dergan couldn't believe**   Interview Dergan; interview Reitan; film, "Mount St. Helens Fury," Fisher Entertainment for Discovery Communications, 2000.

363. **They came around the bend**   Interview, Frenzen; Carson, "Blast zone a study in creation, renewal," Tacoma *News Tribune*, May 15, 2000.

363. **The Longview *Daily News* had been sold**   Interview, Stepankowsky; Andre Stepankowsky, "Howard family assumes ownership of Daily News, Longview *Daily News*, June 8, 1999; interview Bob Gaston.

364. **Robert Rogers was busy cleaning up**   Interview, Rogers.

365. **On a little-used road**   Interview, Frenzen; Rob Carson, "A battle for the soul of St. Helens," Tacoma *News Tribune*, May 16, 2000.

366. **"Ted's missing,"**   Interview, Stepankowksy; Christy McKerney and Andre Stepankowsky, "Ex-Daily News publisher missing," Longview *Daily News*, August 9, 1999; Allen Brettman, "Hunters find Natt's body near Knappa," Longview *Daily News*, September 19, 1999; Don Jenkins, "Natt remembered as outspoken, generous," Longview *Daily News*, September 25, 1999.

367. **"Take a look at this,"**   Interview, Frenzen.

368. **"The news media**   Interview, Parker; film, "Mount St. Helens Fury."

CHAPTER 29: 2000: Looking Back

370. **The roiling**   Interview, Swanson; film and speech by Swanson, "Eyewitness Accounts of the Eruption," Mount St. Helens Institute symposium, Silver Lake visitors center, May 14, 2000.

371. **Stepankowsky was beginning**   Interview, Stepankowsky.

371. **"You can't believe**   Chris Solomon, "Lessons in life from the zone of destruction," *Seattle Times*, May 15, 2000; Tom Paulson, "From the ashes: Mountain's recovery teaches rich lessons," *Seattle Post-Intelligencer*, May 10, 2000; Carson, *Mount St. Helens*, p. 82-111.

373. **The CBS News crew**   Interview, Parker.

373. **Stepankowsky gave a close read**   Cheryll A. Borgaard, "Officials debate volcano's future," Longview *Daily News*, April 21, 2004.
374. **It was the perfect time**   Interviews: Frenzen, Rogers, Reitan, Dergan, Scymanky.
375. **High above**   Courtenay Thompson, "Lives, landscape rearranged," Portland *Oregonian*, May 14, 2000; interview Scymanky.

CHAPTER 30: 2000: A MATTER OF TIME

376. **"Most volcanoes don't**   Swanson speech, presented at "Forum: Mount St. Helens, 1980," Washington State University at Vancouver, May 15, 2000; Rob Carson, "The day the impossible happened," Tacoma *News Tribune*, May 14, 2000; Kathie Durbin, "Mount St. Helens rebirth: Cabins sprout in blast zone," Vancouver *Columbian*, May 15, 2000; Rob Carson, "Private park owner feels locked out," Tacoma *News Tribune*, May 16, 2000.
377. **much of Swanson's science**   Thompson, *Volcano Cowboys*, p. 209-295.
377. **"You want to go**   Interview Reitan; interview, Dergan.
378. **When Frenzen picked up**   Tom Paulson, "Spirit Lake came back to life," *Seattle-Post Intelligencer*, May 10, 2000.
378. **"What ever happened**   Interview, Parker; Rowe Findley, "Mount St. Helens: nature on fast forward," *National Geographic*, May 2000, p. 116.
379. **Frenzen drove to a plot**   Interview, Frenzen; Carol Kaesuk Yoon, "As Mount St. Helens Recovers, Old Wisdom Crumbles," *New York Times*, May 16, 2000; Tom Paulson, "Mountain's recovery teaches rich lesson," *Seattle Post-Intelligencer*, May 10, 2000.
379. **"We want an interview**   Interview, Rogers.
380. **Stepankowsky sat down**   Andre Stepankowsky, "Apology to most volcano victims is long overdue," Longview *Daily News*, May 18, 2000; interview, Stepankowsky.

CHAPTER 31:   2000: SURVIVORS

382. **Small clusters of**   Interview, Frenzen; interview, Helen and Jim Scymanky; interview, Rogers.
385. **A few miles away**   Interview, Parker.
385. **They drove up**   Interview, Dergan.
386. **When Don Swanson rose**   Swanson speech, "Eyewitness Accounts of the Eruption," Mount St. Helens Institute symposium, Silver Lake visitors center, May 14, 2000.
387. **"I was the tresspassser,"**   Roger's speech, ibid.
388. **As they stood**   Reitan and Dergan speech, ibid.
389. **Scymanky began his**   Scymanky speech, ibid.
390. **Venus and Roald's interview**   Interview, Reitan; interview, Dergan.

CHAPTER 32: 2000: ANNIVERSARY DAY

391. **Early that morning**   Interview, Swanson; Yoshiaki Ida, Barry Voight, "Introduction: Models of Magmatic Processes and Volcanic Eruptions," *Journal of Volcanology and Geothermal Research* 66 (1995).
392. **Venus and Roald slept**   Interview, Reitan; Interview, Dergan.
392. **Donna Parker drove**   Interview, Parker.
393. **Like ghosts resurrected**   Interview, Scymanky; Sheila Hamilton morning show, KPAM Radio, Portland, May 18, 2000.
394. **On this warm**   Erin Middlewood, "Memorial rite recalls day etched in minds," Portland *Oregonian*, May 19, 2000; film, "Mount St. Helens 20th Anniversary Ceremony," Global Net Productions, May 18, 2003.

394. **Former President Jimmy Carter**   Interview, Mark Plotkin, Cowlitz County tourism director; interview, Corrie Harmon, press aide to President Carter.

396. **A few days after**   Interview, Parker.

396. **The flash of green**   Interview, Frenzen.

EPILOGUE

397. **The frequency of the earthquakes**   Sandi Doughton, "St. Helens eruption chances put at 70%," *Seattle Times*, October 1, 2004; Mike Lewis, Chris McGann and Lewis Kamb, 'St. Helens blows off steam," *Seattle Post-Intelligencer*, October 2, 2004; Hal Berton, "Scientists wait and watch," *Seattle Times*, October 4, 2004; David Ammons, *Associated Press*, "Volcano draws crowds hoping to see it perform," *Seattle Times*, October 4, 2004; Christopher Schwarzen, "St. Helens stirring passions," *Seattle Times*, October 6; Sandi Doughton, "'Loaf' atop St. Helens now higher than dome," *Seattle Times*, October 9, 2004.

398. **Analysis of the rock**   Sandi Doughton, "St. Helens' new dome growing rapidly," *Seattle Times*, November 20, 2004.

398. **Coincidentally, Don Swanson**   Interview, Swanson.

Index

Acknowledgments ——————————————————————

I WANT TO THANK the people who allowed me to chronicle their lives in relation to the mountain: Venus Dergan, Peter Frenzen, Donna Parker, Roald Reitan, Robert Rogers, Jim Scymanky, Andre Stepankowsky, and Don Swanson. Some individuals submitted to nearly forty hours of taped interviewing over as many as a dozen separate sessions. They put up with my constant flow of follow-up questions, and it was with their help and their spirit that this book came into being.

I would like to thank geologist Geoff Clayton, who introduced me to Don Swanson; Willie Scott, scientist in charge of the USGS's Cascades Volcano Observatory, for reviewing the sections on volcanology; and Don West, a longtime friend and renowned geologist, who reviewed the manuscript as both a scientist and a reader.

Ellen Wheat, an author and editor, provided both critical direction and enthusiasm in the early drafts of the book. Iny Day Truppi, a true friend and gifted writing teacher, offered invaluable suggestions. John and Debbie Francis, close friends of mine and connoisseurs of good books, provided essential insights. My thanks also to literary agent Elizabeth Wales and to the readers at her agency for their comments on an early draft.

The administration and staff of the Longview Public Library were a great help during my dozens of visits that often started when they opened in the morning and didn't end until lights were turned off at night. I also thank the people at the University of Washington Library, the Bellevue branch of the King County Library system, and the Washington State Library, all of whom provided access to special collections on Mount St. Helens.

Michael Lienau, owner of Global Net Productions, kindly searched his files to come up with segments of video I needed. KPAM Radio in Portland, Channel 47 in Vancouver, and TVW in Olympia helped in securing video or audio recordings.

I am immeasurably grateful to my editor, Don Graydon, for his patience, understanding, and skill. His "Wow, what a book!" at the end of his editing comments meant more to me than any critical acclaim I may receive. I'm fortunate in having Kent Sturgis and Epicenter Press as my publisher. Kent's unflagging support and enthusiasm kept me going through good times and bad.

My wife, Sheila, has provided ongoing support and belief in my efforts, and this has meant a great deal to me. Often she took on my parental duties as well as her own to give me time to work. I thank my children Mary, Jonathan, Sarah, and Anna and my grandchildren Stephen and Catie for putting up with the periods when this project kept me from them.

When I started work on *Echoes of Fury*, I looked up Gary Eisler, a friend I hadn't seen in years. I discovered that Gary had terminal cancer. He offered the use of his Portland home as a base of operations for my research. At the time of the Mount St. Helens eruption, he was my top assistant at Emanuel Hospital in Portland, where I was public relations director. Gary was a gifted writer. Throughout the research and most writing of this book, he was my sounding board.

I told Gary I planned to dedicate the book, in part, to him. After that we would joke morbidly as only good friends can, betting on which would come first, his death or completion of the book. He was a fighter. Doctors said he had only a year to live, but he held on for five more. Toward the end, he would often tell me, "I don't feel so good today. You better hurry up and finish that book." Unfortunately, death was the victor. Gary died in the year before I finished *Echoes of Fury*. But somehow I think he would understand. He would want only the best I could do in a book dedicated to him.